MATHEMATICS RESEARCH DEVELOPMENTS

HANDBOOK OF GENETIC ALGORITHMS

NEW RESEARCH

MATHEMATICS RESEARCH DEVELOPMENTS

Additional books in this series can be found on Nova's website under the Series tab.

Additional E-books in this series can be found on Nova's website under the E-books tab.

MATHEMATICS RESEARCH DEVELOPMENTS

HANDBOOK OF GENETIC ALGORITHMS

NEW RESEARCH

ADALBERTO RAMIREZ MUÑOZ
AND
IGNACIO GARZA RODRIGUEZ
EDITORS

Nova Science Publishers, Inc.
New York

Copyright © 2012 by Nova Science Publishers, Inc.

All rights reserved. No part of this book may be reproduced, stored in a retrieval system or transmitted in any form or by any means: electronic, electrostatic, magnetic, tape, mechanical photocopying, recording or otherwise without the written permission of the Publisher.

For permission to use material from this book please contact us:
Telephone 631-231-7269; Fax 631-231-8175
Web Site: http://www.novapublishers.com

NOTICE TO THE READER

The Publisher has taken reasonable care in the preparation of this book, but makes no expressed or implied warranty of any kind and assumes no responsibility for any errors or omissions. No liability is assumed for incidental or consequential damages in connection with or arising out of information contained in this book. The Publisher shall not be liable for any special, consequential, or exemplary damages resulting, in whole or in part, from the readers' use of, or reliance upon, this material. Any parts of this book based on government reports are so indicated and copyright is claimed for those parts to the extent applicable to compilations of such works.

Independent verification should be sought for any data, advice or recommendations contained in this book. In addition, no responsibility is assumed by the publisher for any injury and/or damage to persons or property arising from any methods, products, instructions, ideas or otherwise contained in this publication.

This publication is designed to provide accurate and authoritative information with regard to the subject matter covered herein. It is sold with the clear understanding that the Publisher is not engaged in rendering legal or any other professional services. If legal or any other expert assistance is required, the services of a competent person should be sought. FROM A DECLARATION OF PARTICIPANTS JOINTLY ADOPTED BY A COMMITTEE OF THE AMERICAN BAR ASSOCIATION AND A COMMITTEE OF PUBLISHERS.

Additional color graphics may be available in the e-book version of this book.

Library of Congress Cataloging-in-Publication Data

Handbook of genetic algorithms : new research / [edited by] Adalberto Ramirez Muqoz and Ignacio Garza Rodriguez.
    p. cm.
  Includes index.
  ISBN 978-1-62081-158-0 (hardcover)
 1. Genetic algorithms. 2. Mathematical optimization. I. Ramirez Muqoz, Adalberto. II. Garza Rodriguez, Ignacio.
  QA402.5.H363 2011
  519.6'25--dc23
                          2012005450

Published by Nova Science Publishers, Inc. † New York

Contents

Preface		vii
Chapter 1	Changing Range Genetic Algorithm: A New Optimization Approach with Improved Performance *Adil Amirjanov and Konstantin Sobolev*	1
Chapter 2	Model-Free Deconvolution of Transient Signals Using Genetic Algorithms *László Bengi, Balázs Kovács, Máté Bezdek and Ernő Keszei*	41
Chapter 3	Application of Genetic Algorithm Optimisation Technique in Beam Steering of Circular Array Antenna *A. D. Adebola, R. A. Abd-Alhameed and M. M. Abusitta*	61
Chapter 4	Examination of Two Different Mathematical Techniques for Determining the Important Factors Affecting Indoor Air Quality *Akhil Kadiyala and Ashok Kumar*	93
Chapter 5	Optimization with Genetic Algorithms of a Hybrid Distillation/Melt Crystallization Process *Cristofer Bravo-Bravo, Juan Gabriel Segovia-Hernández, Salvador Hernández, Fernando Israel Gómez-Castro, Claudia Gutiérrez-Antonio and Abel Briones-Ramírez*	113
Chapter 6	Hybrid Genetic Algorithms: Modeling and Application to the Scheduling Problems *Jorge J. Magalhães Mendes*	135
Chapter 7	GA Optimization for a Space-Constrained Machine Room Using Multi-layer Sound Absorbers and One-layer Acoustical Hoods *Min-Chie Chiu*	159
Chapter 8	Genetic Algorithm with Optimal Genotypic Feedback: A New Evolutionary Optimization Approach *S. R. Upreti, D. Joshi, F. Ein-Mozaffari and A. Lohi*	189
Chapter 9	Geometrical Optimization in Gas Turbines by Applying Computational Fluid Dynamics and Genetic Algorithms *A. Gallegos-Muñoz, V. Ayala-Ramírez and J. A. Alfaro-Ayala*	213

Chapter 10	Study of the Influence of Forest Canopies on the Accuracy of GPS Measurements by Using Genetic Algorithms *P. J. García Nieto, C. Ordóñez Galán, J. Martínez Torres, M. Araújo and E. Giráldez*	**229**
Chapter 11	Roundness Evaluation by Genetic Algorithms *Michele Lanzetta and Andrea Rossi*	**241**
Chapter 12	Influence of Building Codes in the Behavior of a Modified Elitist Genetic Algorithm Applied to the Optimization of Steel Structures *J. J. del Coz Díaz, P. J. García Nieto, M. B. Prendes Gero and A. Bello García*	**253**
Chapter 13	Genetic Algorithms for Single Machine Scheduling Problems: A Trade-Off between Intensification and Diversification *Veronique Sels and Mario Vanhoucke*	**265**
Chapter 14	Optimal Sizing of Analog Integrated Circuits by Applying Genetic Algorithms *S. Polanco-Martagón, G. Reyes-Salgado, Luis G. De la Fraga, E. Tlelo-Cuautle, I. Guerra-Gómez, G. Flores-Becerra and M. Fakhfakh*	**295**
Chapter 15	Optimal Sizing of Analog Integrated Circuits by Applying Genetic Algorithms *Luis Gerardo de la Fraga and Esteban Tlelo-Cuautle*	**317**
Chapter 16	Application of Particle Swarm Optimization to Packing Problem *Eisuke Kita and Young-Bin Shin*	**333**
Chapter 17	Application of Advanced Grammatical Evolution to Function Identification Problem *Eisuke Kita and Hideyuki Sugiura*	**347**
Index		**359**

PREFACE

Genetic algorithms (GA) have become popular tools for search, optimization, machine learning, and solving design problems. These algorithms use simulated evolution to search for solutions to complex problems. A GA is a population-based computational method in which the population, using randomized processes of selection, crossover, and mutation, evolves towards better solutions. In this book, the authors present current research including the application of genetic algorithm optimization techniques in beam steering of circular array antenna; hybrid genetic algorithms; changing range genetic algorithms; study of the influence of forest canopies on the accuracy of GPS measurements using genetic algorithms; roundness evaluation by genetic algorithm; and optimal sizing of analog integrated circuits by applying genetic algorithms.

Chapter 1 - An attractive approach to improving a search strategy in genetic algorithms (GA) is reducing the search space towards the feasible region where the global optimum is located. This method dynamically adjusts the size of a search space and directs GA to the global optimum while significantly reducing computational cost and improving the precision of optimal solution.

By using the methods of statistical mechanics, the advantages of this new approach were proved analytically. These methods were applied to describe the effect of the size of a search space adjustment on the macroscopic statistical properties of population, such as the average fitness and the variance fitness of population. This study focuses on the interaction of the various genetic algorithm operators and how these interactions give rise to optimal parameters values. The equations of motion are derived for the one-max problem that expressed the macroscopic statistical properties of population after reproductive genetic operators and adjusting a search space size in terms of those prior to the operation.

Predictions of the theory are compared with experiments and shown to accurately predict the average fitness and the variance fitness of the final population. In addition, the developed changing range genetic algorithm (CRGA) was implemented to estimate the fractal properties of different packing mechanisms: a packing-limited growth mechanism and Apollonian packing with the random distribution of initially prepacked spheres. The fractal dimensions corresponding with the packing degree and porosity were calculated for a large range of spherical particles (in the order of millions). The results provide an experimental proof of analytically found value for a lower bound of the fractal dimension.

Chapter 2 - Model-free deconvolution of transient signals usually suffers from a wavy small-frequency component as well as an increase of high frequency noise. The more these features are damped, the more the deconvolved signal becomes distorted. If the authors can include as much information concerning the genuine deconvolved signal as possible, the quality of the deconvolved transient can be greatly improved.

The authors have developed a deconvolution procedure which does not make use of a pre-supposed (often arbitrary) functional form of the transient signal. Instead, it is based on the inversion of the distortion effects of convolution, which means temporal compression, amplitude enhancement, increasing of the steepness of rise and decay of the measured convolved signal, and cutting initial data to zero to reproduce an eventual sudden jump. As the authors results have proved that the choice of a fairly good initial population is crucial for a successful deconvolution, the authors use additional genetic algorithms to generate an initial population with a relatively high fitness.

The initial population is generated from the measured convolved signal in two subsequent stages, each consisting of a genetic algorithm and the check of the potential to improve the fitness of the population during further breeding, as a candidate for the deconvolved data set. This generation can be done "automatically" so that the user does not need to experiment much to get a good initial population and satisfactory final results.

The final genetic algorithm performing the main iteration to get the estimate of the deconvolved data is constructed in such a way that it does not enhance either the low-frequency wavy behaviour, or the high frequency noise. It is based on a smooth mutation in a range containing several data, and a dynamic adjustment of the mutation.

The method is described in connection with the deconvolution of ultrafast laser kinetic (or femtosecond chemistry) data. The advantage of this method to find an underlying molecular mechanism via statistical inference is also discussed.

Chapter 3 - Beam steering entails changing the direction of the mainlobe of the radiation pattern of an antenna in order to achieve spectrum efficiency enhancement and mitigation of multipath propagation. Beams formed by antenna arrays are categorised as either switched array beams or adaptive array beams. While switched array beams are fixed, finite and predetermined, adaptive array beams possess infinite number of patterns and can adapt in real time to RF signal environment hence adaptive beamforming arrays are always preferred. Adaptive arrays have the capability of directing or steering in real time the main beam in a desired direction or towards signal of interest (SOI) and suppressing interference or multipath signals.

A reactive loading and time modulated switching techniques are applied to steer the beam of a circular uniformly spaced six element antenna array having a source element at its centre. Genetic algorithm (GA) process is used to calculate the optimal values of reactances loading the parasitic elements for which the gain can be optimized in a desired direction. For time switching, GA is also used to determine the optimal on and off times of the parasitic elements for which the difference in currents induced optimizes the gain and steers the beam in a desired direction. These methods are demonstrated by vertically polarised antenna configuration operated at frequency of 2.45GHz. Simulations results showed that near optimal solutions for gain optimization, sidelobe level reduction, VSWR 3 over a 100MHz bandwidth and beam steering is achievable by Genetic Algorithm as optimisation techniques.

Chapter 4 - This chapter provides a comprehensive discussion on the performance of regression and regression tree analyses in determining the important factors affecting in-vehicle air quality. Indoor contaminants of particulate matter, carbon dioxide, carbon monoxide, sulfur dioxide, nitric oxide, and nitrogen dioxide were monitored inside two public transport buses. One bus operated on 20% biodiesel and the other operated on ultra low sulfur diesel in the city of Toledo, Ohio. The independent variables considered in this study include meteorological variables (ambient temperature (temp.), ambient relative humidity (RH), wind speed, wind direction, precipitation, visibility, ambient PM2.5), indoor comfort parameters (indoor temp., indoor RH), and real-time on-road variables (passenger count, bus status (bus position/door position - Idle/Open, Idle/Close, Run/Close), number of cars and buses/trucks ahead). Regression analysis was performed using MINITAB® software and regression tree analysis was performed using CART® software. CART proved to have the advantage of providing relative importance of all the independent variables considered. The important factors affecting the monitored vehicular contaminants were found to be different for each month and season on performing the regression and regression tree analyses.

Chapter 5 - Large scale purification of reaction products in a chemical plant is typically accomplished through distillation. Innovative hybrid processes offer significant cost savings, particularly for azeotropic or close-boiling mixtures, and allow the cost-efficient synthesis of new products. Hybrid separation processes are characterized by the combination of two or more different unit operations, which contribute to the separation task by different physical separation principles such that separation boundaries or inefficiencies of a single unit operation can be overcome. Despite of the inherent advantages of hybrid separation processes, they are not systematically exploited in industrial applications. A major reason relies on the complexity of the design and optimization of these highly integrated processes. In this work the authors study the design and optimization of the hybrid distillation/melt crystallization process, using conventional and thermally coupled distillation sequences. The design and optimization was carried out using a multiobjective genetic algorithm with restrictions coupled with the process simulator Aspen Plus™, for the evaluation of the objective function. The results show that this hybrid configuration with thermally coupled arrangements is a feasible option in terms of energy savings and capital investment.

Chapter 6 - The key idea of the Hybrid Genetic Algorithms (HGA) is to use traditional genetic algorithms (GAs) to explore in several regions of the search space and simultaneously incorporates a good mechanism to intensify the search around some selected regions.

Genetic algorithms are search algorithms based on the mechanics of natural selection and natural genetics. They combine survival of the fittest among string structures with a structured yet randomized information exchange to form a search algorithm with some of the innovative flair of human search.

One fundamental advantaged of GAs from traditional methods is described by Goldberg: in many optimization methods, the authors move gingerly from a single solution in the decision space to the next using some transition rule to determine the next solution. This solution-to-solution method is dangerous because it is a perfect prescription for locating false peaks in multimodal search spaces. By contrast, GAs work from a rich database of solutions simultaneously (a population of chromosomes), climbing many peaks in parallel; thus the probability of finding a false peak is reduced over methods that go solution to solution.

Problems which appear to be particularly appropriate for solution by genetic algorithms include timetabling and scheduling problems.

The scheduling problems consist of determining the starting and finishing times of the activities. These activities are linked by precedence relations and their processing requires one or more resources. The resources are renewable, that is, the availability of each resource is renewed at each period of the planning horizon.

Usually, the objective of the well-known job shop scheduling problem (JSSP), resource constrained project scheduling problem (RCPSP) and the resource-constrained project scheduling problem with multiple modes (MRCPSP) is minimizing the *makespan*.

This paper presents some approaches for the JSSP, RCPSP and MRCPSP. These approaches are compared with some other powerful heuristics and show that the authors results are competitive.

Chapter 7 - Noise control is essential in an enclosed machine room where there is a high level of reverberant and direct sound. Additionally, the noise level can be reduced using a sound absorber and an acoustical enclosure. The traditional method for designing sound absorbers and acoustical enclosures is time-consuming. Therefore, to efficiently control noise levels, shape optimization of multi-layer sound absorbers and one-layer close-fitting acoustical hoods is being considered.

In this paper, a Genetic Algorithm (*GA*) in conjunction with a theoretical model of multi-layer sound absorbers and acoustic hoods is applied in the following numerical optimizations. Before noise abatement is carried out, the reliability of the *GA* method will be checked by optimizing the sound absorption coefficient of three kinds of multi-layer sound absorbers at a pure tone (500Hz). Moreover, the noise abatement of a piece of equipment within a machine room using three kinds of multi-layer sound absorbers and one close-fitting acoustical hood in conjunction with the *GA* method has been exemplified and fully explored. The results reveal that both the acoustical panel and the acoustical enclosure can be precisely designed.

Consequently, this paper provides a quick and effective method for reducing noise levels by optimally designing a shaped multi-layer sound absorber and a one-layer close-fitting acoustical hood using the *GA* method.

Chapter 8 - Genetic algorithms or GAs are evolutionary optimization methods, which find applications in solving a variety of challenging optimization problems in chemical engineering. Different from traditional optimization methods, GAswork with populations of encoded optimization parametersetsto yield superior solutions, especially of problems that pose difficulties to conventional gradient search methods. However, GAssometime manifest slow or premature convergence, and reduced accuracies with the progression of genetic operations.

This paper addresses this problem by disseminating the genotypic information of the optimal parameter set in a GA population. After each iteration of the GA, randomly sized building blocks are extracted from the encoded parameter set that is found optimal. The blocks are then fed back or inserted into randomly selected members of the population for processing in the next iteration. Incorporated with the standard as well as an advanced GA, the new approach of optimal genotypic feedback (OGF) is successfully tested on 34 benchmark optimization functions.

The results demonstrate a significant improvement in the quality of results and computation times.When finally applied to the optimization problems of mixing characterization and minimum variance tuning of proportional controllers, OGF yields results, which are on a par with those obtained from a hybrid GA employing gradient search. Given its simplicity of implementation and quality of results, OGF is found to be a valuable enhancement to GA; better than the complex and computationally demanding add-on of gradient search.

Chapter 9 - A methodology that uses Genetic Algorithms (GA) with a computerized vision tool and Computational Fluid Dynamics (CFD) to optimize the geometric parameters of gas turbine components, is presented. The proposed methodology applies the results of the CFD simulation where the temperature and velocity contours are use to create the population of individuals. Each population generated is composed by the geometrical parameters that represent a feasible geometry. The set of parameters corresponding to the individual genotype were decoded and a script for the CFD software (Fluent ®) that describes the geometrical shape was created. The optimization process considers an initial set of individuals and a CFD simulation is created to obtain the temperature and velocity contours according to the desirable properties of the thermal behavior and subsequent populations were generated by applying selection, crossover and mutation genetic operators to the best individuals. The composition of the new population was created with 2 elite individuals, 6 individuals obtained from the application of the genetic operators and 2 new random individuals. A morphometric analysis computes several geometric properties of the temperature and velocity profiles to provide information to decide the best individuals. For the case of the transition piece of the gas turbine, the temperature and velocity profiles required at the outlet must be uniform because a non-uniformity in the temperature profile affects the useful life of the blades and nozzles of the first stage. However, a diminution in the average value of the turbine inlet temperature (TIT) produces a reduction of the thermal efficiency and power of the gas turbine. Then, by applying a genetic algorithm and CFD simulation to optimize the geometry of the transition piece it is possible to analyze the geometrical parameters that are not intuitive for a human designer.

Chapter 10 - The present paper analyzes the influence of the forest canopy on the precision of the measurements performed by global positioning systems (GPS) receivers. The accuracy of a large set of observations is analyzed in the present research. These observations were taken with a GPS receiver at intervals of one second during a total time of an hour in twelve different points placed in forest areas characterized by a set of forest stand variables (tree density, volume of wood, Hart-Becking index, etc.). The influence on the accuracy of the measurements of other variables related to the GPS signal, such as the Position Dilution of Precision (PDOP), the signal-to-noise ratio and the number of satellites, was also studied. The analysis of the influence of the different variables on the accuracy of the measurements was performed by using genetic algorithms. The results obtained show that the variables with the highest influence on the accuracy of the GPS measurements are those related to the forest canopy, that is, the forest stand variables. The influence of these variables is almost equally important without significant statistical differences. As was expected, those observations recorded in areas covered by an important forest canopy have larger errors than those obtained in areas with less canopy cover. Finally, conclusions of this study are exposed.

Chapter 11 - Roundness is one of the most common features in machining, and various criteria may be used for roundness errors evaluation. The minimum zone tolerance (MZT) method produces more accurate solutions than data fitting methods like least squares interpolation. The problem modeling and the application of Genetic Algorithms (GA) for the roundness evaluation is reviewed here. Guidelines for the GA parameters selection are also provided based on computation experiments.

Chapter 12 - In this work an elitist genetic algorithm (GA) developed by the authors and implemented in an advanced analysis program of three-dimensional steel structures (named ESCAL3D) is evaluated in order to compare the optimization results over a typical portal-frame structure. The minimum weight that satisfies the ultimate limit states of different applicable building codes (MV-103 Spanish code, Eurocode-3 and AISC-LRFD) was checked in order to obtain the influence of several parameters, such as the population's size, the number of generations, the function's evaluation, etc. Finally, the cost and weight improvements obtained using this GA for the different building codes as well as the computational effort are discussed, giving place to the conclusions exposed in this study.

Chapter 13 - In order to increase the efficiency of genetic algorithms, these algorithms are often hybridized with other heuristic procedures, such as local search algorithms or other meta-heuristics. The main purpose of this hybridization is to intensify the search process of the genetic algorithm in order to accelerate the search for high quality solutions. However, diversity is also a crucial component of the genetic algorithm that will guarantee a uniform sample of the search space. Therefore, hybrid algorithms require a careful trade-off between the diversification and intensification strategy. In this chapter, the authors discuss several techniques that take this important balance into account. To be more precise, the authors will show that the definition of a clever, often restricted, neighborhood increases the effectiveness of the embedded local search algorithm or metaheuristic. In addition, the authors will discuss how the extension from a single population to multiple populations and the use of a distance measure to define these populations can be an important stimulator to add diversity to the search process. These techniques are illustrated by means of a commonly known machine scheduling problem.

Chapter 14 - Analog signal processing applications such as filter design and oscillators require the use of different kinds of amplifiers. One kind of those amplifiers are classified to work in mixed-mode, and they can be designed by interconnection of basic analog cells. For instance, the voltage follower (VF), is quite useful in analog design, not only because it allows implementing signal conditioning circuits, but also because it can be evolved to design different kinds of mixed-mode amplifiers, namely: current conveyors, operational transresistance amplifiers, current feedback operational amplifiers, etc. Those mixed-mode integrated circuits (ICs) can be biased and sized automatically. Besides, on the one hand, IC sizing is a hard and tedious work due to the large number of parameters, constraints and performances that the designer has to handle. On the other hand, the main challenges of modern analog IC designs are oriented to solve the problem of determining the correct biases and sizes under different IC technologies. Additionally, there is a pressing need for analog circuit design automation, to meet the time to market constraints. Henceforth, this chapter shows the usefulness of the multi-objective non-dominated sorting genetic algorithm (NSGA-II) to contribute to solve the sizing problem of analog ICs. The NSGA-II is tested and linked to a circuit simulator (SPICE), to compute the optimal sizes of the analog ICs through considering several objective functions, such as: gain, bandwidth and power consumption. Additionally, a discussion on lines for future research are briefly described.

Chapter 15 - In this article two different genetic algorithms, one traditional population based with binary representation, and other steady state with real representation, are applied to solve the problem of maximize the Lyapunov exponent in a chaotic oscillator. The studied oscillator is one based on saturated nonlinear function series. The authors compute the positive Lyapunov exponent oscillators with 2 to 6 scrolls. The authors show that both genetic algorithms are suitable to maximize the positive Lyapunov exponent. As a result, the phase diagrams show that for a low value of the positive Lyapunov exponent the attractors are well defined, while for its maximum value the attractors are not well appreciated, but the higher value increases the unpredictability grade of the chaotic system. Both algorithms report almost the same results but using the steady state genetic algorithm, a reduction of eight times in execution time is obtained.

Chapter 16 - Packing problem is a class of optimization problems which involve attempting to pack the items together inside a container, as densely as possible. This research focuses on the application of particle swarm optimization (PSO) for solving two-dimensional packing problems at the arbitrary polygon-shaped packing region. Total number of items and the position vector of the item center are taken as the design variables. Then, total number of the items is maximized when all objects are included inside a two-dimensional domain without their overlapping. The problem is solved by two algorithms; standard and improved PSOs. In the standard PSO, the particle position vector is updated by the best particle position in all particles (global best particle position) and the best position in previous positions of each particle (personal best position). The improved PSO utilizes, in addition to them, the second best particle position in all particles (second global best particle position) in the stochastic way. In the numerical example, the algorithms are applied for three problems. The results show that the improved PSO can pack more items than the standard PSO and success rate is also improved.

Chapter 17 - The aim of the function identification problems is to find the unknown function representation for the given data set. Grammatical Evolution is one of the evolutionary computations which can find the function representation by using the one-dimensional chromosome and the translation rule described in Backus Naur Form (BNF). This paper describes the application of an advanced Grammatical Evolution (GE) to function identification problem. The advanced Grammatical Evolution uses two-dimesional chromosome, instead of the one-dimensional chromosome employed in the original GE. The continuous and discontinuous functions are taken as the numerical examples. The results show that Grammatical Evolution with one-dimensional chromosome can find the continuous function faster than the Genetic Programming, and that the advanced Grammatical Evolution with two-dimensional chromosomes is more effective than that with one-dimensional chromosome for finding the discontinuous function.

In: Handbook of Genetic Algorithms: New Research
Editors: A. Ramirez Muñoz and I. Garza Rodriguez
ISBN: 978-1-62081-158-0
© 2012 Nova Science Publishers, Inc.

Chapter 1

CHANGING RANGE GENETIC ALGORITHM: A NEW OPTIMIZATION APPROACH WITH IMPROVED PERFORMANCE

Adil Amirjanov[1] *and Konstantin Sobolev*[2]
[1]Department of Computer Science, Near East University, Nicosia, N. Cyprus
[2]Department of Civil Engineering and Mechanics, College of Engineering and Applied Science, University of Wisconsin-Milwaukee, Milwaukee, Wisconsin, US

ABSTRACT

An attractive approach to improving a search strategy in genetic algorithms (GA) is reducing the search space towards the feasible region where the global optimum is located. This method dynamically adjusts the size of a search space and directs GA to the global optimum while significantly reducing computational cost and improving the precision of optimal solution.

By using the methods of statistical mechanics, the advantages of this new approach were proved analytically. These methods were applied to describe the effect of the size of a search space adjustment on the macroscopic statistical properties of population, such as the average fitness and the variance fitness of population. This study focuses on the interaction of the various genetic algorithm operators and how these interactions give rise to optimal parameters values. The equations of motion are derived for the one-max problem that expressed the macroscopic statistical properties of population after reproductive genetic operators and adjusting a search space size in terms of those prior to the operation.

Predictions of the theory are compared with experiments and shown to accurately predict the average fitness and the variance fitness of the final population. In addition, the developed changing range genetic algorithm (CRGA) was implemented to estimate the fractal properties of different packing mechanisms: a packing-limited growth mechanism and Apollonian packing with the random distribution of initially prepacked spheres. The fractal dimensions corresponding with the packing degree and porosity were calculated for a large range of spherical particles (in the order of millions). The results provide an experimental proof of analytically found value for a lower bound of the fractal dimension.

Keywords: Genetic algorithms; Optimization; Statistical mechanics techniques; Fractal properties; Sphere packings; Apollonian packing

1. INTRODUCTION

Genetic algorithms (GAs) have become popular tools for search, optimization, machine learning, and solving design problems. These algorithms use simulated evolution to search for solutions to complex problems. A GA is a population-based computational method in which the population, using randomized processes of selection, crossover, and mutation, evolves towards better solutions [1]. To efficiently solve different complex problems, a GA employs a different set of operators, which include the common operators of GA like selection, mutation, and crossover operators, and the operators specific for a particular GA. There are several surveys available [1-5] that discuss in detail the implementation of a different set of operators and their advantages and disadvantages.

One way to improve a search strategy is to reduce the search space towards the feasible region where the global optimum is located. These approaches dynamically adjust a search space size and direct GA to the global optimum. These approaches are based on the idea that a parameter-space size adjustment improves the accuracy of the discrete sampling in the solution space and significantly reduces the computational time to reach the global optimum [6-10]. Amirjanov [11] analyzed these approaches and employed statistical mechanics techniques to make a mathematical model of an adjustment of a search space size. The statistical mechanics approach models an ensemble of populations to find the average of some population statistics. The statistical properties of the ensemble will not fluctuate, even though the members of the populations will. A similar approach was used by Prügel-Bennett and Shapiro [12] to derive a set of deterministic dynamical equations for describing the average behavior of simple GA. The distinguishing feature of this approach is the use of macroscopic statistical properties of population, like the average fitness μ, and the variance fitness σ^2 to model the dynamics, which predict the change of the fitness distribution of population from one generation to the next. According to this approach, to model a changing range genetic algorithm (CRGA), the following equations can be derived to take the population of a CRGA from a generation t to a generation $t+1$:

$$\begin{pmatrix}\mu(t)\\ \sigma(t)\end{pmatrix} \xrightarrow{selection} \begin{pmatrix}\mu_s(t)\\ \sigma_s(t)\end{pmatrix} \xrightarrow{mutation} \begin{pmatrix}\mu_m(t)\\ \sigma_m(t)\end{pmatrix} \xrightarrow{crossover} \begin{pmatrix}\mu_c(t)\\ \sigma_c(t)\end{pmatrix} \xrightarrow{size\ adjustment} \begin{pmatrix}\mu_a(t)\\ \sigma_a(t)\end{pmatrix} = \begin{pmatrix}\mu(t+1)\\ \sigma(t+1)\end{pmatrix} \quad (1.1)$$

In the paper [11] for modeling a GA with dynamical adjustment of a search space size, the one-max problem was considered where each site contributed a different amount to the cost of the solution. Every individual of population was defined by L binary variables,

$x_i \in \{0,1\}$ with weight J_i, and it was mapped to the interval $[u_l, u_u]$, where u_l is a lower-bound and u_u is an upper-bound of a specified interval.

The problem was to optimize a cost function E over the x's,

$$E_\alpha = u_l + (u_u - u_l) \cdot \sum_{i=0}^{L-1} J_i \cdot x_i^\alpha \qquad (1.2)$$

where J_i are fixed weights at each site.

However, the paper [11] focused only on mathematical description of a size adjustment operator by considering for large selection rates and/or large mutation rate that the population converges rapidly towards its asymptotic limit, in which the average fitness and the variance fitness are not changeable by selection, mutation and crossover operators. That is, the following equations:

$$\begin{pmatrix} \mu(t) \\ \sigma(t) \end{pmatrix} \xrightarrow{size\ adjustment} \begin{pmatrix} \mu_a(t) \\ \sigma_a(t) \end{pmatrix} = \begin{pmatrix} \mu(t+1) \\ \sigma(t+1) \end{pmatrix} \qquad (1.3)$$

were considered to describe an effect of an adjustment of a search space size on the macroscopic statistical properties of population. This assumption greatly simplified the mathematical modeling of a CRGA and allowed for a detailed assessment of the role of the size adjustment operator in GA; however, to describe completely the behavior of a CRGA, the equation (1.1) needs to be derived.

This paper focuses on deriving the equations that describe the effect of all CRGAs operators on the macroscopic statistical properties of population that change in time. The full dynamics will be calculated by iterating the sequence in (1.1) starting from the initial population.

2. THE GENETIC ALGORITHM MODEL

The goal of the modeling is to assess the evolution in the distribution of population fitness, i.e., for each GAs operator, the distribution of fitness should be calculated after that operator is applied. In this paper the simplest model will be considered. First, we assume the population is infinite; in this case, the evolution of a single population can be considered rather than an ensemble. Second, we assume the distribution of fitness is Gaussian [12]. As a result, it is sufficient to consider only $\mu(t)$ and $\sigma^2(t)$.

2.1. Selection Dynamics

Selection is the operation whereby more fit strings are increased in the population at the expense of less fit ones. The effect of selection on the distribution of phenotypes within the population is independent of the genotype to phenotype mapping for a particular problem.

This is a consequence of the fitness being a function of the phenotype only; therefore, it is possible to model selection without reference to a specific problem.

There are two basic types of selection schemes commonly used: proportionate-based selection and ordinal-based selection [13]. Proportionate-based selection selects individuals on the basis of their fitness values relative to the fitness of the other individuals in the population. Proportionate selection [14], stochastic remainder selection [1], and stochastic universal selection [15, 16] are examples of proportionate-based selection schemes. Ordinal-based selection schemes select individuals not according to their fitness, but on the basis of their rank within population. In this case, the selection pressure is independent of the fitness distribution of population and solely based on the relative ordering (ranking) of the population. Tournament selection [1], truncation selection [17], and linear ranking selection [15, 16, 18] are examples of ordinal-based selection schemes. Ordinal-based selection schemes are often preferred over proportionate-based selection schemes because they are less sensitive to the shape of distribution or particular choice of fitness measure, and do not need a scaling procedure to keep appropriate levels of competition between individuals in the population [19].

In this paper, the binary tournament selection is used as a GA selection operator. In binary tournament selection, two members are randomly drawn from the population and the fitter member is copied into the mating pool. Blickle and Thiele [20] obtained the following expressions for infinite population to assess the effect of selection on the average and the variance fitness of population:

$$\mu_s = \mu + \frac{\sigma}{\sqrt{\pi}} \tag{2.1}$$

$$\sigma_s^2 = \left(1 - \frac{1}{\pi}\right) \cdot \sigma^2 \tag{2.2}$$

The expressions (2.1, 2.2) were obtained for an infinite population; however, the macroscopic properties of any finite population drawn from the ensemble will differ slightly due to well-known sampling effects. Expectation values for the average and variance fitness of a finite population (μ_f, σ_f^2) sampled from an infinite population were derived in [12]:

$$\mu_f = \mu \tag{2.3}$$

$$\sigma_f^2 = \left(1 - \frac{1}{P}\right) \cdot \sigma^2 \tag{2.4}$$

It can be seen that the mean is unchanged, but the variance of a finite population is reduced. Because of a finite population correction the formulae (2.1, 2.2) can be rewritten as follows:

$$\mu_s = \mu + \frac{\sigma}{\sqrt{\pi}} \tag{2.5}$$

$$\sigma_s^2 = \left(1 - \frac{1}{\pi}\right) \cdot \left(1 - \frac{1}{P}\right) \cdot \sigma^2 \tag{2.6}$$

where P is the number of individuals in finite population.

The increase in the mean is proportional to the variance in the population, and the variance is reduced through finite population sampling.

2.2. Mutation Dynamics

The mutation operator alters a string locally to hopefully create a better string. Mutation acts on each member of the population independently. The effect of mutation (and crossover) on a string will not depend only on the fitness distribution, it will also depend on the configuration of strings. Mutation operator m changes a site i of a string α with probability p_m [21], that is:

$$m_i^\alpha = \begin{cases} x_i^\alpha & \text{with probability } 1 - p_m \\ 1 - x_i^\alpha & \text{with probability } p_m \end{cases} \tag{2.7}$$

To calculate the average effect of mutation on a site of the string α we need to average overall possible mutations. Denoting the average overall possible mutations by $\langle \cdots \rangle_m$ the following can be obtained:

$$\langle x_i^{m\alpha} \rangle_m = (1 - p_m) \cdot x_i^\alpha + p_m \cdot (1 - x_i^\alpha) = p_m + (1 - 2 \cdot p_m) \cdot x_i^\alpha \tag{2.8}$$

The average fitness of a string after mutation can be calculated by combining the expressions (1.2) and (2.8):

$$\langle E_\alpha^m \rangle_m = \left\langle u_l + (u_u - u_l) \cdot \sum_{i=0}^{L-1} J_i \cdot x_i^{m\alpha} \right\rangle = u_l + p_m \cdot (u_u - u_l) \cdot \sum_{i=0}^{L-1} J_i + (1 - 2 \cdot p_m) \cdot (u_u - u_l) \cdot \sum_{i=0}^{L-1} J_i \cdot x_i^\alpha$$

Averaging over all members of population gives the average fitness after mutation:

$$\mu_m = \langle \langle E_\alpha^m \rangle_m \rangle_\alpha = 2 \cdot p_m \cdot u_l + p_m \cdot (u_u - u_l) \cdot \sum_{i=0}^{L-1} J_i + (1 - 2 \cdot p_m) \cdot \mu \tag{2.9}$$

For the weights $J_i = \dfrac{2^i}{2^L - 1}$, which represents a conversion of a binary code to a decimal code, the last formula can be simplified as follows:

$$\mu_m = p_m \cdot (u_u + u_l) + (1 - 2 \cdot p_m) \cdot \mu \tag{2.10}$$

The variance of the population after mutation can be calculated similarly by averaging a variance of a string over all members of population and all mutations. For the moment, the member label α will be dropped and the variance $\langle (E^m)^2 \rangle_m - \langle E^m \rangle_m^2$ over all mutation will be calculated:

$$\langle (E^m)^2 \rangle_m - \langle E^m \rangle_m^2 = (u_u - u_l)^2 \cdot \left(\left\langle \left(\sum_i J_i \cdot x_i^m \right)^2 \right\rangle_m - \left(\left\langle \sum_i J_i \cdot x_i^m \right\rangle_m \right)^2 \right)$$

where

$$\left(\left\langle \sum_i J_i \cdot x_i^m \right\rangle_m \right)^2 = \sum_i J_i^2 \cdot \langle x_i^m \rangle_m^2 + \sum_{i \neq j} J_i \cdot J_j \cdot \langle x_i^m \rangle_m \cdot \langle x_j^m \rangle_m$$

$$\left\langle \left(\sum_i J_i \cdot x_i^m \right)^2 \right\rangle_m = \sum_i J_i^2 \cdot \langle (x_i^m)^2 \rangle_m + \sum_{i \neq j} J_i \cdot J_j \cdot \langle x_i^m \cdot x_j^m \rangle_m = \sum_i J_i^2 \cdot \langle x_i^m \rangle_m + \sum_{i \neq j} J_i \cdot J_j \cdot \langle x_i^m \rangle_m \cdot \langle x_j^m \rangle_m$$

The last expression was simplified because $(x_i^m)^2 = x_i^m$, and $\langle x_i^m \cdot x_j^m \rangle_m = \langle x_i^m \rangle_m \cdot \langle x_j^m \rangle_m$ since mutation acts independently on each site of a string.

To find the variance of population after mutation, the last expression should be averaged over all members of population. After simplification and by using the formula (2.8) the variance fitness of population is:

$$\sigma_m^2 = p_m \cdot (1 - p_m) \cdot (u_u - u_l)^2 \cdot \sum_{i=0}^{L-1} J_i^2 + (1 - 2 \cdot p_m)^2 \cdot \sigma^2 \qquad (2.11)$$

The formula (2.11) should be corrected by coefficient $(1 - 1/P)$ because of a sampling effect of a finite population, like it was done for selection operator. If $J_i = \dfrac{2^i}{2^L - 1}$, which represents a conversion of a binary code to a decimal code, then:

$$\sigma_m^2 = \left(1 - \frac{1}{P}\right) \cdot \left(p_m \cdot (1 - p_m) \cdot (u_u - u_l)^2 \cdot \frac{2^L + 1}{3 \cdot (2^L - 1)} + (1 - 2 \cdot p_m)^2 \cdot \sigma^2 \right) \qquad (2.12)$$

Observation of (2.10) and (2.12) shows that the average and the variance fitness of population after mutation can be expressed in terms of the same properties of population before mutation.

2.3. Crossover Dynamics

The crossover operator allows the mixing of parental information when it is passed to their offspring. The result of crossover is a randomized exchange of genetic material between individuals, with the possibility that good solutions can generate even better ones. Like mutation, the effects of crossover depend on the configurations of strings, as well as the fitness distribution. There are many possible crossover schemes available, and the most appropriate depends on the problem under consideration [1]. The simplest way to calculate is "uniform crossover," in which a child is constructed from its two parents by choosing each site at random from either parent [12]. For problems with different weight at each site, it is often important to minimize disruption of the genotype. In this case, single-point crossover might be most appropriate, where parent genotypes are broken at one point and the segments on one side of this point are swapped.

For general form of the one-max problem with different weight at each site, a single-point crossover is most appropriate. The effect of mixing of alleles by a single-point crossover is estimated by assuming that it is similar to that for a uniform crossover [12].

For a uniform crossover, the alleles of a child γ produced by parents α and β are given in [22]:

$$x_i^\gamma = X_i^{\alpha\beta} \cdot x_i^\alpha + (1 - X_i^{\alpha\beta}) \cdot x_i^\beta, \quad \text{where} \quad X_i^{\alpha\beta} = \begin{cases} 1 & \text{with probability } p_c \\ 0 & \text{with probability } 1 - p_c \end{cases} \quad (2.13)$$

where p_c is a probability of crossover.

Prügel-Bennett and Shapiro [12] showed that with reasonable approximation a uniform crossover leaves the average fitness and the variance fitness of population unchanged; that is

$$\mu_c \approx \mu \quad (2.14)$$

$$\sigma_c^2 \approx \left(1 - \frac{1}{P}\right) \cdot \sigma^2 \quad (2.15)$$

In fact, a crossover involves the interaction of different population members, which can be measured by a correlation between the sites of two members. The correlation is a measure of the microscopic similarity of genotypes of population members. The changing of the mean correlation within the population indicates the changing of distribution of population, and, consequently, the macroscopic statistical properties of population. However, the mean correlation is unchanged by crossover, because although crossover changes the alleles within each population member, it conserves the mean number of alleles at each site within the population [22].

2.4. Dynamics of an Adjustment of a Search Space Size

To model an effect of an adjustment of a search space size (adjustment operation) on the macroscopic statistical properties, a law of a changing range of a mapped interval should be established. In this paper [11] a power law was used for a changing range of a mapped interval, that is

$$(u_{ua} - u_{la}) = (u_{u0} - u_{l0}) \cdot k^t, \quad (2.16)$$

where u_{la} and u_{ua} are a lower- and an upper–bounds, respectively, after the adjustment operation, u_{l0} and u_{u0} are the initial lower- and an upper-bounds, and t is a number of generations.

The new mapped interval is centered on the average fitness of population [11]; thus, according to (2.16), the following expressions can be established for calculating the lower- and upper-bounds of the new mapped interval:

$$u_{la} = \begin{cases} u_{l0} & \text{, if } u_{la} \leq u_{l0} \\ \mu - (u_{u0} - u_{l0}) \cdot k^t/2, & \text{otherwise} \end{cases} \quad (2.17)$$

$$u_{ua} = \begin{cases} u_{u0} & \text{, if } u_{ua} \geq u_{u0} \\ \mu + (u_{u0} - u_{l0}) \cdot k^t/2, & \text{otherwise} \end{cases} \quad (2.18)$$

Denoting the average over all individuals of the population by $\langle \cdots \rangle$ the average fitness μ_a after the adjustment operation can be expressed as follows:

$$\mu_a = \langle E_\alpha^a \rangle_\alpha = u_{la} + (u_{ua} - u_{la}) \cdot \sum_{i=0}^{L-1} J_i \cdot \langle x_i \rangle_\alpha \quad (2.19)$$

The average fitness of population in genotype space denoted by $\langle x_i \rangle_\alpha$ is not changed after the adjustment operation, and it can be expressed as follows [11]:

$$\sum_{i=0}^{L-1} J_i \cdot \langle x_i \rangle_\alpha = \frac{\mu - u_l}{u_u - u_l} \quad (2.20)$$

Consequently,

$$\mu_a = u_{la} + (u_{ua} - u_{la}) \cdot \frac{\mu - u_l}{u_u - u_l} \quad (2.21)$$

where u_l and u_u are a lower- and an upper-bounds, respectively, before the adjustment operation.

The variance σ_a^2 after the adjustment operation can be expressed in term of the variance σ^2 before the operation as follows:

$$\sigma_a^2 = \left\langle \left(E_\alpha^a - \mu_a\right)^2 \right\rangle_\alpha = \left(\frac{u_{ua} - u_{la}}{u_u - u_l}\right)^2 \cdot \sigma^2 \quad (2.22)$$

3. MODELING THE FULL DYNAMICS

For modeling the full dynamics of the GA according to (1.1), the iteration process of calculating macroscopic statistical properties of population should be applied, i.e., the changes of the macroscopic statistical properties of population caused by selection, mutation, crossover, and size adjustment operators need to be computed. This can be accomplished by applying the equations (2.5, 2.6) to initial the population for modeling the selection operator, then by applying the equations (2.10, 2.12) to model the mutation operator, then by applying the equations (2.14, 2.15) to model the crossover operator, and last by applying the equations (2.21, 2.22) to model the adjustment operator. Before executing the equations (2.21, 2.22), the lower- and upper-bounds of the size space must be calculated by using the equations (2.17, 2.18). This procedure is then iterated an arbitrary number of generations to get from the initial distribution to the final one.

For modeling the full dynamics of the GA, the initial macroscopic statistical properties of population should be calculated. The initial strings are generated by randomly setting each site to 0 or 1 with probability 0.5. The equations for mutation (2.10, 2.12) can be used to compute the initial macroscopic statistical properties of population.

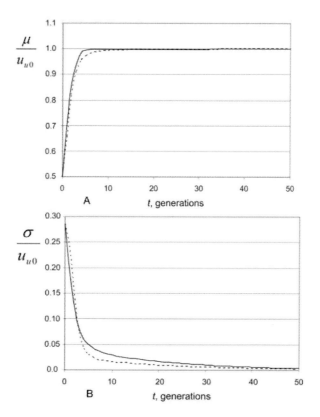

Figure 1. Simulation (dashed line) and Theory (solid line) for A) the average fitness and B) the standard deviation, *after [9-11]*.
The curves are for $P = 50$, $L = 100$, $p_m = 0.01$, $p_c = 0.65$ and $k = 0.95$; the simulations are averaged more than 6000 runs.

Placing $p_m = 0.5$ in equations (2.10) and (2.12) gives for the initial average fitness (μ_0) and the initial variance fitness of population (σ_0^2) the following expressions:

$$\mu_0 = 0.5 \cdot (u_{u0} + u_{l0}) \tag{3.1}$$

$$\sigma_0^2 = \left(1 - \frac{1}{P}\right) \cdot \frac{2^L + 1}{12 \cdot (2^L - 1)} \cdot (u_{u0} - u_{l0})^2 \approx \left(1 - \frac{1}{P}\right) \cdot \frac{(u_{u0} - u_{l0})^2}{12} \tag{3.2}$$

Figure 1 shows the result of computing the changes of the average fitness (μ) and the standard deviation (σ) of population caused by selection, mutation, crossover, and size adjustment operators more than 50 generations starting from initial population. Figure 1 represents a comparison between the theory (solid line) and simulations (dashed line) averaged more than 6000 runs.

Analysis of the curves in Figure 1 shows the theoretical model predicts accurately enough the behavior of a GA, although there are clearly systematic errors in the predictions of the theory. This discrepancy arises from a number of assumptions made to capture the full dynamics of the GA, but not at the expense of an over-complex model. The first assumption made was that the distribution of fitness is Gaussian; as a result, only $\mu(t)$ and $\sigma(t)$ were considered. This assumption in the number of macroscopic variables is significant, as it enables closed-form expressions to be derived for the evolution dynamics. The other source of error is in sample-to-sample fluctuations. The transformation that predicts the macroscopic statistical properties at a given time in terms of those in the previous time is nonlinear. Although the macroscopic statistical properties are self-averaging quantities, of which sample-to-sample fluctuations should be small, nonlinear functions of the properties are not necessarily self-averaging [12].

Figure 2 shows the influence of a coefficient of reduction of a mapped interval k on the macroscopic statistical properties of population ($\mu(t)$ and $\sigma(t)$), representing only the results of simulations. It is clear from Figure 2 that with the adjustment operation, the average fitness of population at the end of the run is significantly closer to the maximum than without an adjustment of a search space size.

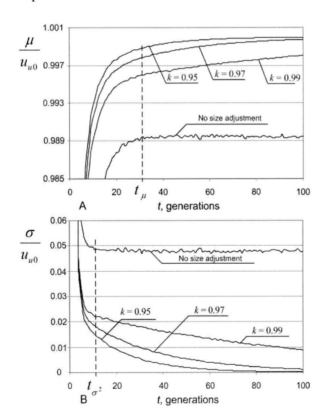

Figure 2. Simulations for A) the average fitness and B) the standard deviation with different values of coefficient k, *after [9-11]*.
The curves are for P = 50, L = 100, p_m = 0.01, p_c = 0.65; the simulations are averaged more than 6000 runs.

It can be seen that the curve without an adjustment of a search space size and all curves with any level of a coefficient k have almost the same transient period to reach an equilibrium point (the vertical dashed lines). At an equilibrium point, the improvement due to selection is balanced by a loss of fitness caused by mutation and crossover operators; at that point, the population becomes homogeneous. From that point, the improvement of the average fitness of population will be made only by the adjustment operator. The adjustment operator does not change the average fitness and the standard deviation of population in genotype space (a diversity of population), which is proved by the expressions (2.19, 2.22), but moves the population to the maximum by tightening a search space size.

For assessment, the transient period to reach an equilibrium point the model that consists of mutation, selection, and crossover operators can be considered. Applying the selection (2.5, 2.6), mutation (2.10, 2.12), and the crossover (2.14, 2.15) operators produce the following expressions for the average fitness and the variance fitness of population:

$$\mu(t+1) = p_m \cdot (u_u + u_l) + (1 - 2 \cdot p_m) \cdot \left(\mu(t) + \frac{\sigma(t)}{\sqrt{\pi}} \right) \tag{3.3}$$

$$\sigma^2(t+1) = \frac{1}{3} \cdot p_m \cdot (1 - p_m) \cdot (u_u - u_l)^2 \cdot \left(1 - \frac{1}{P}\right) + (1 - 2 \cdot p_m)^2 \cdot \left(1 - \frac{1}{\pi}\right) \cdot \left(1 - \frac{1}{P}\right)^3 \cdot \sigma^2(t) \tag{3.4}$$

The equation (3.4) is a first-order linear difference equation of the form [23]:

$$x(t+1) = \alpha \cdot x(t) + \beta \tag{3.5}$$

with the solution

$$x(t) = \alpha^t \cdot x(0) + \beta \cdot \frac{1 - \alpha^t}{1 - \alpha} \tag{3.6}$$

Thus, for the equation (3.4) the following solution can be derived (for simplification, a sampling effect of a finite population is not considered):

$$\sigma^2(t) = G_1 + G_2 \cdot ((1 - 2 \cdot p_m)^2 \cdot (1 - 1/\pi))^t \tag{3.7}$$

where

$$G_1 = \frac{p_m \cdot (1 - p_m) \cdot (u_u - u_l)^2}{3 \cdot (1 - (1 - 2 \cdot p_m)^2 \cdot (1 - 1/\pi))} \quad \text{and}$$

$$G_2 = \frac{(u_u - u_l)^2}{12} - \frac{p_m \cdot (1 - p_m) \cdot (u_u - u_l)^2}{3 \cdot (1 - (1 - 2 \cdot p_m)^2 \cdot (1 - 1/\pi))}$$

The variance fitness of population with $p_m \ll 1$ at equilibrium is

$$\sigma^2(t_{\sigma^2}) \approx \sigma^2(\infty) \approx \frac{\pi \cdot p_m \cdot (u_u - u_l)^2}{3} \tag{3.8}$$

where t_{σ^2} is a transient period of the variance to converge to its equilibrium value.

The characteristic time decay, at which $\sigma^2(t)$ converges to its equilibrium value, is $\tau_{\sigma^2} = -\log((1-2 \cdot p_m)^2 \cdot (1-1/\pi))$. This convergence rate depends on two factors: the convergence due to mutation and the convergence due to selection, and it does not depend on the size of a search space. But an equilibrium value of the variance depends on the size of a search space; the reduction of a search space size diminishes the variance fitness of population (see 3.8).

To find a solution for the average fitness $\mu(t)$ the transient period of the variance to converge to its equilibrium value can be ignored, then $\sigma^2(t)$ remains constant with a value calculated according to (3.8). In this case, the equation (3.3) is a linear difference equation with the solution

$$\mu(t) = K_1 + K_2 \cdot (1-2 \cdot p_m)^t \qquad (3.9)$$

where

$$K_1 = \frac{(u_u + u_l) \cdot \left(p_m + (1-2 \cdot p_m) \cdot \sqrt{p_m/3}\right)}{2 \cdot p_m} \quad \text{and} \quad K_2 = \frac{(u_u + u_l) \cdot (1-2 \cdot p_m) \cdot \sqrt{p_m/3}}{2 \cdot p_m}$$

The average fitness of population with $p_m \ll 1$ at equilibrium is

$$\mu(t_\mu) \approx \mu(\infty) \approx \frac{(u_u + u_l) \cdot \left(p_m + \sqrt{p_m/3}\right)}{2 \cdot p_m} \qquad (3.10)$$

where t_μ is a transient period of the average fitness to converge to its equilibrium value.

The characteristic time decay at which $\mu(t)$ converges to its equilibrium value is $\tau_\mu = -\log(1-2 \cdot p_m)$. The convergence of $\mu(t)$ is slower, as it converges first through the variance and then at a rate τ_μ, which is controlled by the mutation rate alone. The convergence rate of the average fitness does not depend on the size of a search space, but its equilibrium value depends on it; the reduction of a search space size increases the average fitness of population for the first scenario [11].

From the equilibrium point the average fitness of population continues to increase (see Figure 2), but now only because of the adjustment operator, i.e., the adjustment operator follows the first scenario by incrementing a lower-bound, and fixing an upper-bound of a mapped interval to u_{u0} because $u_{ua} > u_{u0}$. Thus, the equation (1.1) from the equilibrium point and up to the end of the run can be substituted by the equation (1.3).

The equation (2.20) can be rewritten for the first scenario

$$\sum_{i=0}^{L-1} J_i \cdot \langle x_i \rangle_\alpha = \frac{\mu(t_\mu) - u_l(t_\mu)}{u_{u0} - u_l(t_\mu)} = C \qquad (3.11)$$

C is a constant up to the end of the run, because only the adjustment operator is operational and it does not change the average fitness of population in genotype space denoted by $\langle x_i \rangle_\alpha$. The equation (2.21) that expresses the average fitness of population after the adjustment operator can be rewritten as follows

$$\mu(t) = u_{u0} \cdot C + u_l(t) \cdot (1-C) \qquad (3.12)$$

By moving an origin for t (a number of generations) to the equilibrium point the expression (2.17) of $u_l(t)$ can be rewritten (it was assumed that $u_{l0} = 0$) as follows

$$u_l(t) = \mu(t-1) - u_{u0} \cdot k^t / 2 \qquad (3.13)$$

Replacing $u_l(t)$ in (3.12) by expression (3.13), the following equation for the average fitness of population can be obtained

$$\mu(t) = (1-C) \cdot \mu(t-1) + C \cdot u_{u0} - \frac{u_{u0} \cdot (1-C)}{2} \cdot k^t \qquad (3.14)$$

The equation (3.14) is a first-order linear difference equation of the form [23]:

$$x(t) = \alpha \cdot x(t-1) + \beta + \gamma \cdot k^t \qquad (3.15)$$

with homogeneous solution

$$x(t) = x(0) \cdot \alpha^t \qquad (3.16)$$

and a particular solution of the form

$$x(t) = m + n \cdot k^t \qquad (3.17)$$

After simplification the following solution of the equation (3.14) can be obtained:

$$\mu(t) = \mu(t_\mu) \cdot (1-C)^t + C_1 + C_2 \cdot k^t \qquad (3.18)$$

where

$$C_1 = u_{u0} \quad \text{and} \quad C_2 = \frac{u_{u0} \cdot k \cdot (1-C)}{2 \cdot (1-k-C)}$$

Analysis of solution (3.18) shows that after a few generations (t) the term $\mu(t_\mu) \cdot (1-C)^t \approx 0$, because $(1 - C) < 0.5$ for the first scenario, i.e., the average fitness of population μ for the first scenario will reach the maximum with a characteristic time decay $\tau_\mu = -\log k$, and it comes to an upper-bound of the mapped interval, that is $\mu \to u_{u0}$ for $t \to \infty$.

Analysis of the expression (3.18) shows that the average fitness of population μ depends on a coefficient of shrinking of a mapped interval size k. The decrementing of a coefficient k reduces a number of generations to reach the maximum average fitness of population, but at the same time it reduces a level of the maximum that can be reached by GA. Comparisons of simulation results with theory prediction for the maximal level of the average fitness of the population show the discrepancies with any level of a coefficient k do not exceed 1%.

The variance fitness of population (or the standard deviation of fitness) is reduced according to the expression [11]:

$$\sigma(t) = \sigma(t_\mu) \cdot \left(a^t + \frac{k \cdot (k^t - a^t)}{2 \cdot (k - a)} \right) \quad (3.19)$$

where $a = 1 - \dfrac{\mu(t_\mu)}{u_{u0}}$.

The variance fitness of population reaches a zero for $t \to \infty$.

4. IMPLEMENTATION OF CHANGING RANGE GENETIC ALGORITHMS

4.1. Optimization of Multi-Component Binders

Concrete – is a complex nano-structured, multi-phase, multi-scale composite material that evolves over time. Comprehensive models predicting the behavior and properties of contemporary concrete involving a large number of components and, therefore, a large range of variables are under development [24-26]. The realization of these systems needs comprehensive computer models based on extensive experimental data and also on new design approaches which could predict the behavior of material saving time and research resources.

The application of chemical admixtures and mineral additives has become one of the most important developments in modern concrete technology. Added to the concrete mixture, relatively small amounts of chemical admixtures radically alter the behavior of fresh and hardened concrete [25, 27]. The performance of concrete can be improved by the application of selected mineral additives, especially industrial by-products like granulated blast furnace slag (GBFS), fly ash (FA), and silica fume (SF) [25, 28-30]. The replacement of portland cement (NPC) with mineral additives brings considerable economical savings and also helps to conserve natural resources. The relatively large number of components makes the problem of concrete mixture design more complicated than ever before; and the significant differences in the cost of the components makes the problem more complicated. The full-scale research of the behavior of chemical admixtures and mineral additives in concrete is time-consuming and expensive; therefore the application of an expert system based on existing knowledge and research data is essential for the proportioning of a competitive concrete mixture [24-26].

It was demonstrated that reliable models in the form of 2nd order polynomial equations can be obtained using factorial experiment [25, 31]. The optimization of cost is a numerical optimization problem with a nonlinear objective function and nonlinear constraints. Usually this kind of nonlinear programming problem cannot be solved by developing a deterministic method in the global optimization category [4, 5]. Therefore, the use of specially developed Changing–Range Genetic Algorithms (CRGA) which does not require consideration of the landscape of a search space nor the shape of an optimized function is attractive for the solution of cost optimization problem [9-11].

4.1.1. Modified Multi-Component Binders

The properties of concrete with GBFS, FA and SF including ternary mixtures of NPC-FA-SF or NPC-GBFS-SF have been discussed in the literature [25, 28-30]. Less information is available regarding the performance of ponded fly ash (PA) in concrete. It is suggested that the behavior of this type of concrete can be significantly affected by the fineness of the mineral additive and also by the application of an effective superplasticizer (SP) [25, 32]. The concept of a modified multi-component binder (MMCB) was proposed to describe this system. MMCB includes a binder composed of portland cement, finely ground mineral additive (fly ash, ponded ash or granulated blast furnace slag), and a highly reactive powder component (usually silica fume or metakaolin), modified by a superplasticizer (SP). The main idea of MMCB is to improve the reaction ability of the mineral additives by fine grinding. Consequently, the mineral additives react quicker, avoiding the delay of the development of concrete strength at an early age. It was hypothesized that the application of finely ground mineral additives (FGMA), as a component of the binder, provides better packing in the NPC-FGMA system, especially in combination with SF and SP. The resulting MMCB demonstrates a compressive strength in a range of 75-135 MPa, a significant increase over 68.0 MPa demonstrated by reference NPC. The improved range of strength and especially the increased number of components constituting MMCB led to the development of a special procedure for the proportioning of the MMCB based concrete mixtures [25, 32, 33]. An effective optimization of the performance characteristics at the level of MMCB (involving fewer components) was proposed to minimize the associated tests of concrete.

4.1.2. Strength and Cost Optimization Problem

In materials research, the development and exploration of the models is very important. Unlike actual tests, mathematical models describing concrete give a quick and inexpensive evaluation of the material. However, because of the typical inconsistency in the properties of the component materials, there is the risk of a possible discrepancy between the actual tests and the results of the model. Nevertheless, these results are important estimates which save the time and resources needed for research. It was demonstrated that 2nd order polynomial equations are appropriate for modeling the strength and rheological properties of MMCB systems [25]. Models were developed as a function of the composition for various MMCB systems including:

- NPC-SF-SP system;
- NPC-SF-FGPA-SP system;
- NPC-SF-FGBFS-SP system.

The models of MMCB compressive strength (f_c) were processed as 2nd order polynomial equations:

$$f_c = \sum_{i=0}^{n}\sum_{j=0}^{n} b_{ij} x_i x_j$$

where n - the total amount of variable factors ($n = 2$ for NPC-SF-SP system and $n = 3$ for NPC-SF-FGPA-SP or NPC-SF-FGBFS-SP systems);
b_{ij} - the coefficients of polynomial equation;

x_i, x_j - the values of variable factors; and $x_0 = 1$.

The coefficients of polynomial equations representing the developed models of compressive strength were reported by [25, 33]. The graphical representation of the strength of the NPC-SF-SP system is given in Figure 3. In developed models, the SP parameter is taken as a percentage of SF (on a dry basis); therefore SP - SF parameters are dependent. All the other parameters are considered as a percentage of the total content of MMCB; the remaining part is made up of NPC. The range of the variable factors used in the models is summarized in Table 1. Cost optimization is another important application for the developed models (Figure 3.): it helps to estimate the proportions of the concrete and minimizes the testing costs by omitting non-feasible compositions.

The design of the MMCB mixture of a specific strength and at a minimal cost comprises the global optimization problem (GOP) which could be resolved by finding an optimizer x^* such that

$$\varphi(x^*) = min\ \varphi(X),\ where\ X = [x_1,....,x_n] \in R^n.$$

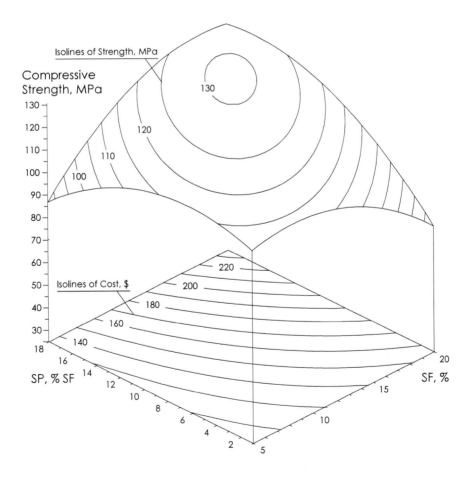

Figure 3. The Compressive Strength and Cost of the NPC-SF-SP Binders, *after [33]*.

The objective function φ is defined for the search space $S \subseteq R^n$, which is the finite interval region in n-dimensional Euclidean space. The lower and upper bounds define the domains of variables:

$l_i \leq x_i \leq u_i$, where $1 \leq i \leq n$

The search space in GOP is restricted to a feasible region (F, where $F \subseteq S$) by a set of constraints:

$g_j(X) \leq 0$, where $1 \leq j \leq p$
$f_r(X) = 0$, where $1 \leq r \leq q$

Usually the equality constraints can be substituted by pair of inequalities such as:

$f_r(X) \geq -\delta$ and $f_r(X) \leq \delta$

where δ is a small value to cover a tiny region. This case was considered in the research program and, consequently, the set of constraints consisted only of inequalities.

The cost of MMCB, C (as a function for optimization) is calculated by using the general formula:

$$C = \frac{1}{100} \sum_{i=1}^{m} x_i c_i$$

where m - the total amount of components ($m = 3$ for NPC-SF-SP system and $m = 4$ for NPC-SF-FGPA-SP or NPC-SF-FGBFS-SP systems);
x_i - the dosage/proportioning of i- component (NPC, SF, SP, and FGPA or FGBFS);
c_i - the cost of i- component.

The non-linearity of this equation is based on the dependency of the SP parameter from SF; therefore the dosage of the SP component (to be used in this formula as % of MMCB) was calculated using the following expression:

$x_{SP} = SF * SP / 100$

The polynomial equation describing the strength of the MMCB mixture represents the single equality constraint for the specific design that is substituted by a pair of inequality constraints as mentioned before.

Table 1. The Input Characteristics of MMCB Components

Component	Units	Bounds		Cost, $
		Lower	Upper	
NPC	%	20	95	100
SF	%	5	20	350
FGPA	%	5	60	40
FGBFS	%	5	60	80
SP	% SF	1	15	2500

4.1.3. The Application of CRGA to the Cost-Optimization

In CRGA, the separation of constraints and objectives method is used to handle the optimization problem [2]; and the fitness function $G(\vec{x})$ is calculated for two groups for individuals

- within the feasible region (i.e. satisfying constraints)
- outside the feasible region

The CRGA implementation uses a stochastic sampling remainder without replacing the selection procedure [1, 9]. A single-point crossover with probability $p_c=0.8$ between the first and last position of a binary string and the bitwise mutation with rate (per bit) $p_m= 0.02$ is used in CRGA implementation. The outline of CRGA is schematically presented in Figure 4 and differs from the conventional GA [1] only by implementing the *changerange()* function.

```
Input_Data();
Initial_Population();
for g:=1 to N_generation do
    for t:=1 to N_individuals do
        f_t := G(x);
    endfor
    Selection();
    Reproduction();
    Mutation();
    if g mod FIXGEN = 0 then
        changerange();
    endif
endfor
Report_Data();
```

Figure 4. Outline of CRGA Implementation.

The *Input_Data()* function establishes CRGA parameters and the input data of the problem. The CRGA parameters include:

- the number of generation ($N_{generation}$),
- the number of individuals in population ($N_{individuals}$),
- the *FIXGEN* is a number of generations passed to activate the *changerange()* function.

The *changerange()* function was introduced in order to provide a self-adaptive mechanism of CRGA. Figure 5 illustrates the core of the *changerange()* function including the steps of self-adaptive mechanism of CRGA, where every rectangle represents the range of the variables and the vertical lines indicate the binary representation scheme with constant resolution (equal to 10).

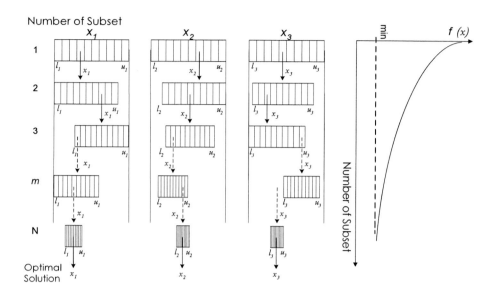

Figure 5. Shifting and Shrinking Mechanism (SSM) of CRGA, *after [33]*.

At the beginning, the lower and the upper bounds of the variables express the domain of the variables. After obtaining h_s individuals in the feasible region (one subset of the surviving individuals), the lower and upper bounds of each variable are changed because of the decreasing size of the search space relatively to the previous one. For example, the lower and upper bounds of variable X_1, X_2, X_3 at the beginning of subset 2 are calculated to embrace the new search space with equal distances to the left side (lower bound) and the right side (upper bound) according to values x_1, x_2, x_3 (where x_1, x_2, x_3 are the value of variables designated by *1, 2, 3* respectively at reference point). Then the new values of the lower and upper bounds of variable X_1, X_2, X_3 are limited to the lower and upper bounds of the domain of the variables (the dotted vertical lines in Figure 5).

As generations progress, the density of vertical lines inside the rectangles is increased; this means that the optimal solution is obtained with better precision. The changing range of the variables can be considered as an additional mutation rate that explores more precisely the search space and speeds up convergence towards the optimal solution. The right side of Figure 4 shows diagrammatically the changing $\varphi(x^*)$ at reference point versus set of generations in progress. It is predicted that the fluctuation of the $\varphi(x^*)$ will be significantly reduced as more sets of generations are developed [9-11].

A software package utilizing a CRGA was developed and applied to the optimization of MMCB. The target compressive strength levels varied within the range of 60-130 MPa with 10 MPa increment. The margin of accuracy was considered at the level of up to +1 MPa (that corresponds to $\delta = 1$) for all the strength levels.

Thus the equality constraint was converted to inequalities as:

$f_c \geq 0$ and $f_c \leq \delta$

These limitations are applied because the cost function has a convex shape depending on strength; and the minimum cost is targeted by the optimization procedure. Tables 2-3 and

Figures 6-8 summarize the research results which are based on the best values obtained from 10 program runs with a standard deviation of less than 1%.

Two cases were considered for cost optimization:

- The effect of SP and SF (using NPC-SF-SP system);
- The effect of FGMA type (i.e. comparison of NPC-SF-FGPA-SP and NPC-SF-FGBFS-SP systems).

Prior to its full-scale application in the research program, the performance of the developed CRGA was compared with the conventional GA. The example of the trial runs evaluating the performance of CRGA and GA is presented for NPC-SF-FGPA-SP in Figure 6. The obtained results clearly illustrate the advantage of CRGA over a conventional GA in finding the global optimum for the MMCB cost problem. Comparison of both curves shows that the shifting and shrinking mechanisms lead to the global optimum. Otherwise the GA population becomes homogenous and an additional mutation is needed to explore a more feasible region. This role is performed by the "shifting and shrinking" mechanism.

Table 2. The Effect of SP Cost on Optimum Composition of MMCB

Cost of SP, $	Compressive Strength, MPa	SF, %	SP, %SF	Cost, $
2500	130.0	12.7	10.2	162.7
	120.0	7.0	4.7	125.2
	110.5	5.0	1.0	113.7

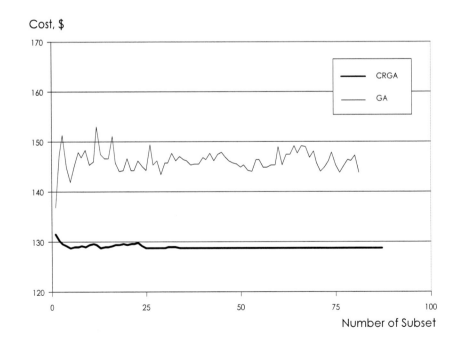

Figure 6. The Comparison of CRGA and Conventional GA, *after [33]*.

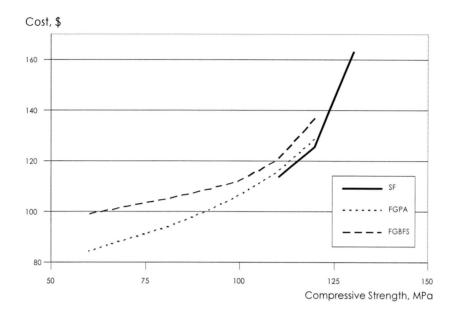

Figure 7. Strength – Cost Relationship for MMCB, *after [33]*.

Table 3. The Effect of FGMA Type on Optimum Composition of MMCB

Type of FGMA	Compressive Strength, MPa	SF, %	SP, %SF	FGMA, %	Cost, $
FGPA	120.0	7.9	6.1	5.1	128.7
	110.0	6.1	4.6	11.9	115.9
	100.0	5.0	3.7	20.7	106.6
	90.0	5.0	2.6	31.9	99.6
	80.0	5.0	1.6	41.1	93.9
	70.0	5.0	1.0	49.5	89.0
	60.0	5.0	1.0	58.1	84.6
FGBFS	120.0	9.0	7.3	5.0	137.0
	110.0	5.9	5.2	5.0	120.5
	100.0	5.0	1.2	5.1	112.4
	90.0	5.0	1.0	17.3	108.5
	80.0	5.0	1.0	28.8	105.1
	70.0	5.0	1.0	39.2	101.9
	60.0	5.0	1.0	48.8	99.1

The MMCB with a compressive strength of 110 MPa represents the composition with the lowest possible content of SF and SP. Increasing strength to 130 MPa required a rise in SF and SP dosage by 7.6 and 9.3% respectively, at a subsequent 43% increase in cost; Figure 7. The effect of FGMA type on the optimum compositions of MMCB is summarized in Table 3 and Figures 7-8. It is clear that the application of FGPA in MMCB is more cost-effective when compared with FGBFS.

This is due to the better strength properties of MMCB containing FGPA and also because of the lower cost of FGPA. For example, a MMCB with a compressive strength of 120 MPa was designed with almost the same volume of FGMA at its minimum level of 5% (actually, 5.1 and 5% for FGPA and FGBFS, respectively as per Table 3).

MMCBs containing FGBFS require an increased dosage of SF and SP (by more than 1% each) adding up to about 5% of additional costs above the already more expensive compositions with FGBFS. This difference in cost increases at lower strength levels (when the design strength is less than 100 MPa) with a subsequent increase in FGBFS content, reaching 17% for 60 MPa binders (Figures 7-8).

To obtain MMCB with strength of 100 MPa only 5% SF is required. Only 5.1% of FGBFS is allowed in this case at a SP dosage of 1.2%. Considerably higher volumes, i.e. 20.7% of FGPA can be used in this composition at a SP dosage of 3.7%. On the other hand, higher costs associated with the application of FGBFS could be offset by the superior corrosion resistance of this type of binder. Consequently, the maximum FGPA content is 58.1% for an MMCB with strength of 60 MPa; by contrast, only 48.8% of FGBFS is use for the same strength level (Figure 8). The models of MMCB containing FGMA are more conservative in the high strength range of 100-120 MPa; therefore high strength NPC-SF-SP binders can be designed at a slightly lower cost (Figure 7).

This binder optimization work demonstrated that CRGA performs better than a conventional GA in finding the global optimum in the case of an MMCB problem with non-linear constraints. This is achieved by the application of the "shifting and shrinking" mechanism. The proposed CRGA is highly accurate in locating the global optimum. It is also very quick and very efficient. The optimization of the strength characteristics helps to minimize the related tests of concrete. It was demonstrated that the 2nd order polynomial equations are suitable for modeling the strength of MMCB systems. The tabulated cost – optimized compositions can be used for the proportioning of the high performance concrete mixtures. Using virtual optimization helps to minimize the costs associated with laboratory research by omitting the non-feasible compositions.

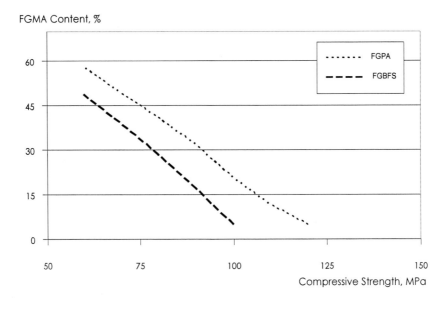

Figure 8. The Effect of FGMA on Strength of Cost-Optimized MMCB, *after [33]*.

4.2. Fractal Properties of Apollonian Packing of Spherical Particles

The quest to obtain the densest possible randomly packed arrangement for particles of varying shapes and sizes is an ongoing and challenging problem; it also has exceptionally wide applications in science and engineering. The hard-sphere randomly packing model is one of the simplest representations of particulate composite materials, colloids, amorphous metals, etc. The packing density of spheres is characterized either by the packing fraction η, or by porosity $\varepsilon = 1 - \eta$, which is a fraction of the unoccupied volume. As Kepler conjectured - and Hales proved [35] - the optimal packing of equal hard spheres is the face-centered cubic (fcc) arrangement with a maximal density of $\pi/\sqrt{18} \approx 0.7405$.

In order to improve the packing density for particles, it is necessary to fill the interstices between the larger particles with smaller particles without disrupting the original packing [36-40]. This kind of packing is known as "packing-limited growth" (PLG) or Apollonian packing. A PLG mechanism seeds the initial particles randomly which then dynamically grow according to a certain rule (an algorithm) [38].

Dodds and Weitz [38] have investigated PLG mechanism and demonstrated that some dynamic packing algorithms (where the growing spheres are seeded by a random injection in time and space) could be portrayed as a particular case of static packing. They examined a static packing model of Manna [41], which they referred as the "random Apollonian packing" (RAP) since it can be considered as a variation of the Apollonian packing.

The "random Apollonian packing" (RAP) is realized by a process that starts with an initial population of hard-spheres of a specified radius with new spheres added one at a time into the packing's unoccupied space. Therefore the center of the newly packed sphere is randomly selected and fixed, and then the sphere's size is determined by extending the radius of a sphere until it touches its closest sphere [38]. At the beginning, the Apollonian packing (AP) is realized by the same process that starts with an initial population of hard-spheres of a specified radius with new spheres added one at a time into the packing's unoccupied space, but the center of the newly packed sphere is arranged to place the sphere with a maximal radius possible into the unoccupied space.

For the Apollonian packing two particular cases can be described that are based on a configuration of the initial spheres:

1. The initial spheres prepacked according to Platonic Solids (tetrahedron, cube, octahedron, dodecahedron and icosahedron) configuration;
2. The random distribution of initially prepacked spheres.

The first case is named as specific Apollonian packing (SAP), and the second as Apollonian packing with the randomly prepacked spheres (APR) [42, 43].

As shown by Schaertl and Sillescu [44], increasing polydispersity raises the maximum packing fraction η of a hard-sphere system. The particle size distribution is a basic parameter which describes a polydispersity of a hard-sphere system that, for a given η, depends on the selected sphere-packing algorithm. However, for static polydisperse space-filling packings, Aste [37] has analytically proved that the size distribution for the spheres of small radius r ($r < r_{max}/5$) follows the power law: $N(r) \propto r^{-\alpha}$, where $N(r)$ is a number of particles of a particular radius. Further, Aste [37] demonstrated that in the case of static polydisperse packings there is an upper and lower bound of α, which depend on a certain sphere-packing algorithm. The SAP algorithm as the densest one provides the lowest bound for α [41].

Dodds and Weitz [38] explored the RAP algorithm as a case of the "packing-limited growth," PLG mechanism with various scenarios of the dynamics of particle's growth (heterogeneous, exponential, or linear). With scaling theory and a numerical simulation they analytically obtained $\alpha \approx 3.8$ for $d = 3$, where d is a number of dimensions.

Apollonian packings are self-similar [41] and are characterized by a fractal dimension D, which is related to α as $D = \alpha - 1$ [41]. Borkovec et al. constructed two fundamentally different numerical methods and computed the fractal dimension of the 3-dimensional SAP to be 2.4739465 with a tetrahedron configuration of initial spheres [45]. More recently Baram and Herrmann developed an algorithm to construct a 3-dimensional SAP with various configurations of initial spheres [46]. Investigating all possible configurations corresponding to Platonic Solids (tetrahedron, cube, octahedron, dodecahedron and icosahedron) and computing the fractal dimensions for these solids, they demonstrated that the fractal dimension D depends on the packing configuration of initial spheres. According to the reported results, D varies from 2.474 for the tetrahedron up to 2.588 for the bichromatic (octahedron-based) configuration of initial spheres [46].

A few articles report the fractal properties of 3-dimensional Apollonian packing with a random distribution of initially prepacked spheres (APR). Soppe used a conventional Monte Carlo method for computer simulation to get η value less than 0.68 [47]. It means that the Monte Carlo method does not provide a high packing density, because it produces configurations in which either the majority or all of the particles are not in contact with one another [48]. Anishchik and Medvedev [49] proposed a computer model of 3-dimensional Apollonian packing of the hard spheres based on the Voronoi-Delaunay method that was applied earlier to study the dense packing of equal spheres [50]. Anishchik and Medvedev used the Voronoi-Delaunay approach to determine the center of the largest sphere inscribed between the previously packed spheres. The largest inscribed spheres ($r > r_c$, where r_c is cutoff radius of sphere) were selected on every step and were turned into new particles first. Thereafter the current value of the cutoff r_c was decreased, and the packing was continued with a new value of r_c.

GAs can be used in the computer simulation models for 3-dimensional RAP and APR to evaluate the bounds of α. The proposed approach [42, 43] is based on a computer model described by Manna [41] for a 3-dimensional RAP of hard spheres. A 3-dimensional APR was realized using the same model enhanced by a genetic algorithm (GA). A GA searches the free space to inscribe the maximum-sized spheres among the previously packed spheres. Only few articles deal with the sphere-packing problem using GA. Franck-Oberaspach et al. employed a GA for the solution of a two-dimensional packing problem of different rigid objects [51]. In their work an arbitrary number of points is arranged within a given two-dimensional connected region in a such way that their mutual distances and the distance from the region boundary reaches a maximum [51]. Cornforth applied a GA for the placement of overlapping grids for the input space quantization in machine-learning algorithms, which is often visualized in a three-dimensional space with the task of fitting the maximum number of oranges (represented by equal spheres) into a box [52]. Therefore, Cornforth applied a GA for a sphere-packing problem to achieve the maximum density of equal spheres in a given space

[52]. However, the existing GA models cannot be employed in the case of APR because of their lack of dimensionality and the diversity of packing objects.

4.2.1. Computer Simulation Models

The RAP model is described as a predecessor of APR model. The developed algorithm of RAP starts with the random placement of the hard spheres of an initially specified configuration in a cube with periodic boundaries. The particle size distribution of an initial configuration of spheres is set by the Gaussian distribution. Further, the packing is provided by placing new spheres (one at a time) into the cube by a random selection of a fixed point (as the center of a new sphere) within the free space and extending its radius r_l to meet the closest sphere.

In the case of the APR model, the latter step includes the search for the center of a new sphere so that the new sphere can occupy the maximum volume within the available free space left between the previously packed spheres, i.e. providing an osculatory packing. An osculatory packing of the unit sphere is presented by four pairwise externally tangent spheres that are all internally tangent to the first one. This step represents a global numerical optimization problem, where $max(r_l)$ is an objective function defined by the search space $S \subseteq R^3$, which is the finite internal region in the 3-dimensional Euclidean space. This global numerical optimization problem is solved by GA (see the details below) which is nothing more but an evolutionary population-based search method that compares many solutions (the spheres packed) to find best one with maximum radius. This global numerical optimization problem imposes the additional constraints that is the search space is restricted to a feasible region (F, where $F \subseteq S$) by a set of constraints with the restriction for the spheres' overlapping:

$$(x_i - x_l)^2 + (y_i - y_l)^2 + (z_i - z_l)^2 \geq (r_i + r_l)^2,$$

where x_i, y_i, z_i are the coordinates of packed (i) and new (l) spheres,

$(x, y, z) \subseteq R^3$, $i = 1, ..., N-1$, $l = i+1, ..., N$ and

N is a total number of the spheres.

To realize a GA for an arrangement of a new sphere with maximum radius within the available free space, a population of solutions (N_{pop}), that is, the spheres with various radii, is initially generated. Every sphere is represented by a binary string of length L containing the coordinates of the center and the radius of the sphere. Next, the pair of spheres is selected randomly as parents to produce the new spheres (children) for the next generation. This reproduction procedure is accomplished by a crossover operation and by a mutation operation. A crossover operation is used to exchange partially the binary code between two parents with a probability p_c, but a mutation operation is applied to change the bit position of a binary string from 0 to 1 or vice versa with a probability p_m. New spheres are then evaluated by a "fitness function" (an objective function), in such a way that only the best spheres "survive" and "organize" a next population of spheres (N_{pop}). This procedure is applied for a certain number of generations (N_{gen}) at the end of which the best sphere (i.e. with maximum radius) will be packed within the free available space. The quality of a solution with a reasonable computational cost that is directly proportional to a number of generations is usually a trade-off. Coello, comparing several constraint-handling approaches of GAs, showed that the best results are generally obtained with high computational costs [2]. For the packing problem a computational cost (T) can be represented as $T \propto N^2$, where N is the number of packed spheres; therefore, a constraint-handling method that provides a minimum computational cost, but with reasonably high quality of a solution, should be applied. For APR, the constraint-handling method named CRGA (changing range GA), proposed by Amirjanov [9-11] was selected.

4.2.2. RAP and APR Packings: Computational Results

Both RAP and APR packings are simulated within a unit cube with periodic boundary conditions. An initial configuration consisting of 100 spheres is seeded according to Gaussian distribution with a mean value $\bar{r} = 0.5$ and standard deviation $\sigma = 0.05$ (first experiment) and $\sigma = 0.15$ (second experiment). All experiments were repeated 10 times (N_s). The following values were established using the trial runs for the best performance of GA to simulate APR: $N_{pop}=50$, $p_c=0.85$, $p_m=0.02$, $L=15$ bits, $P_f=0.42$, $h_s=0.2$, $k_r=0.99$, $t_r=0.0025$, $N_{gen}=150$. The best performance of GA intends to provide a minimum time (a computational cost) to reach a reasonably high quality solution. The parameters N_{pop} and N_{gen} which have the most influence on the computational cost were established by the preliminary experiments. To assess the quality of the GA the latter was applied to solve the optimization problem with a known solution. In these preliminary experiments, a rather simple optimization problem was constructed; that is, to inscribe a single sphere with known radius and known coordinates of its center between initially prepacked spheres. In this experiment after 150 generations, the GA found the coordinates of the center with an error less than 0.2%; for example, after 250 generations the error was reduced to 0.1%. In order to reduce the computational cost the first option was selected for the main experiment. The values of parameters p_c and p_m were selected from the range established in [1]. The parameters L, h_s, k_r and t_r also influence the quality of the solution and their optimal values were adopted from [9-11]. The value of parameter P_f was set according to [53].

The experimental details and the results of a simulation for both models are summarized in Table 4. Figure 9a illustrates the initial configuration of spheres seeded according to Gaussian distribution, where spheres are placed in the cube one at a time by randomly choosing the center of a new sphere in the matrix's free space and expanding its radius r_l to touch the closest neighbor (as per the 3-dimensional RAP approach).

Table 4. Experimental details and results of the simulation

Parameters	Experiment 1 $\sigma = 0.05$		Experiment 2 $\sigma = 0.15$	
	RAP	APR	RAP	APR
N_s	10	10	10	10
N_{total}	10^7	10^6	10^7	10^6
η_{ini}	0.3524	0.3524	0.5023	0.5023
α @ $N(r)$	3.710 ± 0.009	3.51 ± 0.01	3.712 ± 0.009	3.49 ± 0.01
α @ $\varepsilon(n)$	3.727 ± 0.001	3.4518 ± 0.0006	3.729 ± 0.002	3.4237 ± 0.0004
η	0.803	0.909	0.839	0.923

Figure 9b demonstrates the sequence of the 3-dimensional APR of the spherical particles. The packing sequence shown in Figure 9b demonstrates how the GA searches the free space within the already packed spheres in order to place a new sphere with the largest possible radius. Here, the packing sequence is visualized starting with the random placement and initial displacement of a new sphere (shown in a light color) at the 1st generation of GA, followed by the location and size change of the sphere at the 50th, 100th and 150th (final) generation. As a result of such procedure each new sphere is tightly packed within the neighboring matrix.

Figure 10 represents the frequency distribution $N(r)$ of the first experiment (double-logarithmic plot) in order to demonstrate the differences in α for both RAP and APR models (where mean values are presented). The straight lines specify the exponent α for RAP and APR. The RAP and APR simulations involved 10^7 and 10^6 spheres, respectively (N_{total}).

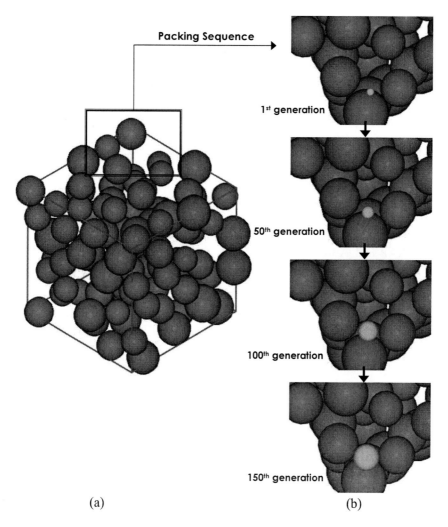

Figure 9. Apollonian Packing with the Random Distribution of Initially Prepacked Spheres: (a) the initial configuration of spheres seeded according to a Gaussian distribution and (b) GA search for 3-dimensional APR, *after [43]*.

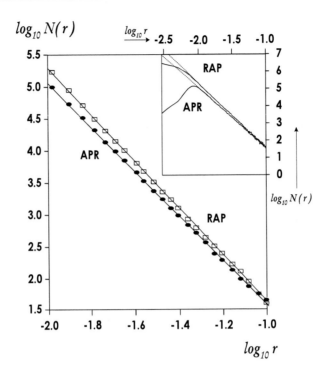

Figure 10. Frequency Distribution Function $N(R)$ vs. Radius r for the Experiment 1 ($\sigma = 0.05$), *after* [43].

The inset of Figure 10 shows the frequency distribution $N(r)$ vs. r at a wider range, up to a cutoff radius $r_c = 0.002$; and, accordingly, the radii of packed spheres differ by the factor 250 (for comparison, this scale factor was as low as 10 in reference [48]). The number of spheres packed with RAP and APR simulations, N_{total} was restricted by the computer and a time resources. As it was mentioned above, a computational cost (T) is evaluated as $T \propto N^2$, where N is the number of packed spheres. Therefore, for an RAP simulation with 10^7 spheres about 6 hours and about 200 MB MB of RAM are required on a Pentium 4 computer with 3.2 GHz processor. For an APR simulation the converging to the global optimum (to set the sphere with maximal radius) would require the $N_{pop}* N_{gen}$ solutions, each similar to the RAP packing step. Consequently, for APR simulation with 10^6 spheres some 96 hours and about 20 MB of RAM is needed on the same type of computer.

From the inset of Figure 10 it can be observed that a significantly larger number of spheres is required to occupy the packing's free space when the sphere's size is approaching the cutoff radius r_c. The dashed straight lines indicate the characteristics of the frequency distribution $N(r)$ vs. r for RAP and APR as more spheres would be packed and the additional simulation experiments for RAP with $2*10^7$ spheres confirmed this assumption.

The best-fit estimation for the exponent α is given with a 95% confidence interval. The estimation of α using a frequency distribution $N(r)$ for RAP closely agrees with the scaling theory proposed by Dodds and Weitz [38]. The curves of a frequency distribution $N(r)$ for Experiment 2 are shifted to the left as can be seen in Figure 11 (RAP1 vs. RAP2) since the volume fraction of the initial spheres (η_{ini} in Table 1) is greater in Experiment 2.

The estimation of α obtained with a frequency distribution $N(r)$ for APR matches the range described by Baram and Herrmann [46]. The estimated lower bound of α for APR in both experiments is found to vary from 3.48 to 3.52 (which is within the limits specified by Baram and Herrmann [46] for 3-dimensional SAP with various configurations of initial spheres). So, the minimum value of α for APR is greater than that for the densest 3-dimensional SAP with a tetrahedron configuration of the initial spheres. This agreement between APR and SAP implies that the APR cannot result in denser packing than SAP. For APR the bound of α reported by Anishchik and Medvedev [49] is less than the obtained values (3.45 vs. 3.48 - 3.52) and even less than that for the densest 3-dimensional SAP with a tetrahedron configuration of the initial spheres. This contradiction can be attributed to a specific strategy used in [49] to add new spheres within the available free space left between the previously packed ones.

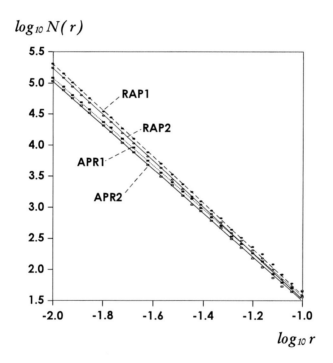

Figure 11. Function of $N(r)$ vs. Radius r for the Experiments 1 and 2, *after [43]*.

For 3-dimensional SAP, Baram and Herrmann [46] proved that the fractal dimension D (or α) depends on a configuration of initial spheres. For 3-dimensional APR based on random distributions and various configurations of the initial spheres we found no significant difference in the value of exponent α. However, more experiments may be required to support this assumption.

The effect of porosity on the fractal properties of the models was also studied following our approach. The exponent α corresponding to porosity is given in Table 4. The relative volume of porous space (or porosity) for both models follows the power law [38, 54]: $\varepsilon(n) \propto n^{-\beta}$, where n is a number of particles. For RAP Dodds and Weitz [38] developed the scaling theory of the distribution of spheres; they determined that $\alpha = (d + 1 + \beta)/(1 + \beta)$.

Figure 12 represents the porosity function of $\varepsilon(n)$ versus n for the Experiment 1 and 2 (RAP1 and RAP2). Both curves extend in alignment within the zone corresponding to the particles of small radii. This zone is located next to the points 1 and 2 representing the porosity of the cube after the placement of the initial spheres. It means that an RAP algorithm quickly produces the particles of small radii. The agreement in α values calculated from $\varepsilon(n)$ and $N(r)$ is considered to be satisfactory (the difference is less than 3%). This similarity in values of α governing the functions of $\varepsilon(n)$ and $N(r)$ can be presented as an additional proof of the scaling theory. Based on the relation described by Herrmann et al. [54]: $\varepsilon(r) \propto r^{-D+3}$, for a 3-dimensional APR the exponent α is specified as $\alpha \approx 4/(1+\beta)$. For the spheres of small radii the curves corresponding to APR1 and APR2 are also aligned (Figure 12). But the zone corresponding to the particles of small radii for APR originates at the left side of the dashed vertical line a-b. It can be noted that an APR algorithm requires more particles to start the generation of the particles of small radii.

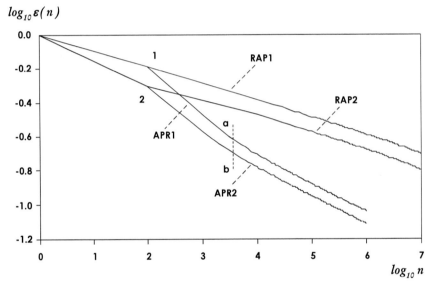

Figure 12. The Double-Logarithmic Function of Porosity $\varepsilon(n)$ vs. n, after [43].

Computer models of the fractal properties of 3-dimensional RAP and APR were used in extensive numeric simulations to estimate the bounds of an exponent α for RAP and APR, the values of which are represented in Table 5.

For RAP the bound of α was numerically estimated for $N(r)$ and $\varepsilon(n)$. Further, these new experimental results have provided an additional support for the scaling theory proposed by Dodds and Weitz [38]. For APR the bound of α was found to be within the range from 3.48 to 3.52, and therefore within the limits reported by Baram and Herrmann [46] for 3-dimensional SAP with various configurations of initial spheres. For APR the bound of α reported by Anishchik and Medvedev [49] is less than the obtained values (3.45 vs. 3.48 - 3.52) and even less than that for the densest 3-dimensional SAP with a tetrahedron configuration of the initial spheres.

This contradiction can be attributed to a specific strategy used in [49] to add new spheres within the available free space left between the previously packed ones. The results of the experiment confirmed that the exponent α was not significantly affected by the configuration of the initial spheres used in both models (RAP and APR).

Table 5. Comparison of the experimental results and reported data

Packing Method	Fractal Dimension α		Reference
	Results of Experiment	Reported	
RAP	3.70-3.72	3.70	[38]
APR	3.48-3.52	3.45	[49]
SAP	-	3.474-3.588	[46]

4.3. Three-Dimensional Apollonian Packing with Randomly Prepacked Spheres

A better understanding the general laws related to the densest possible packings in particulate assemblies is important for modeling of non-crystalline, polydisperse and composite systems. The static polydisperse space-filling packing can be divided in two cases:

- Apollonian packing (AP), where the spheres of random sizes and positions are packed tangent to each other.
- Disordered packing, where the spheres have random sizes and positions, and are packed with at least one contact to each other. Dodds and Weitz [38] have referred to this kind of packing as the 'random Apollonian packing' (RAP) since it can be considered to be a variant of the AP.

By using PLG algorithm for two-dimensional packing Andrienko et al. [55] reported that $D \approx 1.75$ and hence $\alpha \approx 2.75$ based on numerical evidence, where D is a fractal dimension which is related to α as $D = \alpha - 1$ [41]. Elsewhere, the same authors [56] determined numerically that $\alpha \approx 2.53$. A few articles report an assessment of the fractal properties of three-dimensional AP. Anishchik and Medvedev [49] proposed a computer model of three-dimensional AP of the hard spheres based on the Voronoi–Delaunay method that was applied earlier only to study the dense packing of equal spheres [38]. For investigated AP they reported the fractal dimension of $D \approx 2.45$.

4.3.1. Computer Simulation Model

A genetic algorithm was used for computer simulation of three-dimensional RAP and AP to evaluate the bounds of α. For RAP the upper bound of α was numerically estimated to be within the range 3.70-3.72. For AP the lower bound of α was found to be within the range 3.48–3.52 [43].

Figure 9 demonstrates the generalized sequence of the 3-dimensional Apollonian packing of the spherical particle by using GA. The GA searches the available free space within the cell by reallocating the sphere's center in order to arrange a new sphere with the largest possible radius.

Because of the stochastic nature of a GA, this approach did not provide the packing of spheres that all are tangent to each other, and, consequently, it assessed the lower bound of α with low precision. Therefore, a GA possesses the following drawbacks:

- A GA may stick to the local maximum which means that the smaller radius that the possible maximum will be inscribed to the hollow space. It means that the inscribed sphere will not be tangent to four closed spheres.
- A GA converges to the global maximum, but may not reach exact maximal solution; it means that a GA provides the optimal solution with a certain precision.

There are two possible approaches to reduce the influences of the drawbacks:

- increase a number of generations for converging of a GA
- explicitly set and explicitly decrement a cut-off radius of inscribed spheres

First approach is directly applicable to previously described algorithm (referred as an AP model). The approach improves the precision of a GA, but it will significantly increase a computational cost of packing process.

Second approach explicitly sets a cut-off radius and gradually reduces this radius of inscribed spheres. For implementing this approach, additional step of the algorithm is needed that reduces a cut-off radius of inscribed spheres. Packing process starts filling the available space with the spheres of maximum radius R_{max}. Once the interstitial space between these spheres become too small to contain spheres with maximum radius the procedure reduces the radius of the new spheres to insert $R_1 = R_{max}/k_{red}$. The filling continues iterating this process gradually reducing the radii $R_\lambda = k_{red}^{-\lambda} \cdot R_{max}$ in order to insert spheres of biggest radii compatibly with the interstitial sizes. Placing the centers in appropriate positions by using the optimization feature of GA, this procedure generates the Apollonian packing of tangent spheres. For implementing the approach, a number of trials N_{tr} to make a reduction of a cut-off radius should be established. N_{tr} is a number of unsuccessful attempts to fill the interstitial space with the spheres that have radii in range $R_\lambda \div R_{max}$. This approach (referred as an APC model) significantly reduces the influence of the drawbacks of a GA described above, and reasonably increases the computational cost that depends on the reduction coefficient k_{red} and a number of trials N_{tr}.

4.3.2. APC Model: Computational Results

The APC model like AP model is simulated within a unit cube with periodic boundary conditions. An initial configuration consisting of 100 spheres is seeded according to a Gaussian distribution with a mean value $\bar{r} = 0.5$ and standard deviation $\sigma = 0.05$. To investigate the lower limit for α the experiments with the different reduction coefficient k_{red} were run. For all experiments the following values (adopted from [43]) were established for the best performance of GA to simulate Apollonian packing: $N_{pop} = 50$, $p_c = 0.85$,

$p_m = 0.02$, $L = 15$ bits, $P_f = 0.42$, $h_s = 0.2$, $k_r = 0.99$, $t_r = 0.0025$, $N_{gen} = 150$. All experiments were repeated 10 times (N_s). The experimental details and the results of a simulation for all experiments are summarized in Table 6.

Table 6. Experimental details and results of the simulation of Apollonian packing

Parameters	Results reported [43]	Reduction coefficient k_{red}		
		1.1	1.05	1.01
N_s	10	10	10	10
N_{tr}	-	10^4	10^4	10^4
N_{total}	10^6	10^6	10^6	10^6
η_{ini}	0.3524	0.3524	0.3524	0.3524
$\alpha @ N(r)$	3.51 ± 0.01	3.389 ± 0.016	3.341 ± 0.018	3.292 ± 0.009
$\alpha @ \varepsilon(n)$	3.452 ± 0.002	3.387 ± 0.012	3.347 ± 0.011	3.311 ± 0.006
η	0.909	0.9161	0.9182	0.9241

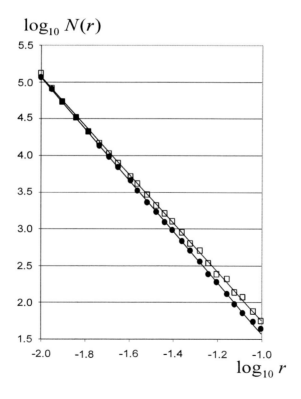

Figure 13. Frequency Distribution Function $N(r)$ Versus Radius r: AP model reported in [43] (circles), APC model with $k_{red} = 1.01$ (squares), *after [42]*.

Figure 13 represents the frequency distribution $N(r)$ (double logarithmic plot) for the APC model with a reduction coefficient $k_{red} = 1.01$ and the frequency distribution $N(r)$ (double logarithmic plot) for the AP model in order to demonstrate the differences in α for both models (where mean values are presented).

The straight lines specify the exponent α for AP and APC. The best-fit estimation for the exponent α is given with a 95% confidence interval. It can be observed from Figure 13 that the APC with $k_{red} = 1.01$ accommodated significantly higher number of spheres with radius $r = 0.05$ (related to -1.0 in logarithmic scale) vs. AP. However, the curves for both models strive to the same point related to the filling of all interstitial space between the spheres. As a result, the curves have a different exponent α. APC model provides the densest packing than any reported before. APC model with reduction coefficient $k_{red} = 1.01$ reaches the packing fraction $\eta = 0.856$ by packing $N_{total} = 11200$ spheres. In comparison, the model described by Anishchik and Medvedev [49] provided the same packing fraction by packing $N_{total} = 40000$ spheres.

Analysis of results presented in Table 6 shows that decrementing of a coefficient k_{red} reduces the exponent α that may reach the analytical limit ($\alpha_{min} \approx 3.2$) calculated by Aste [36, 37]. Aste proved that for Apollonian packing, where a dense sphere-packing is generated by filling the interstitial spaces with non-overlapping spheres of maximum sizes, the only relevant metric parameter is the ratio between the external radii of the spheres that generate an interstice and the internal radius of the sphere that fill this interstice. By incrementing the ratio between the external radii and the internal radius of the spheres one can obtain more dense Apollonian packing. Decrementing of a coefficient k_{red} increases the ratio between the external radii and the internal radius of the spheres and, consequently, reduces the exponent α.

The effect of porosity on the fractal properties of the APC model was also studied. The exponent α corresponding to porosity is given in Table 6. The relative volume of porous space (or porosity) for Apollonian packing follows the power law [54]: $\varepsilon(n) \propto n^{-\beta}$, where n is a number of particles. The fractal dimension D or an exponent α can also be estimated from the porosity [54]. Based on the relation described by Herrmann et al [54]: $\varepsilon(r) \propto r^{-D+3}$, for a three-dimensional Apollonian packing the exponent α can be estimated as $\alpha \approx 4/(1+\beta)$. The similarity in values of α governing the functions of $\varepsilon(n)$ and $N(r)$ can be presented as additional proof of the theory developed by Aste [37].

These new results related to estimate of frequency distribution $N(r)$ and the porosity distribution $\varepsilon(n)$ have provided an additional support for the theory of Apollonian packing proposed by Aste [37]. It was shown that the APC model can be used as a simulation tool for estimation of an exponent α with high precision levels, but at an increased computational cost.

CONCLUSION

This article outlines the modeling of the dynamics of GA with an adjustment of a search space size for one-max problem that follows a power law and derived the equations that describe the effect of GA's operators on the macroscopic statistical properties of the population, such as an average fitness and a variance fitness of population.

The formalism described for modeling GA with an adjustment of a search space size assumes the environment and the population form a unique system. The genetic operators (selection, mutation, and crossover) move the population to the region where the maximum is located, and at the same time, parameter-space size is adaptively reduced, which is based on the idea that shrinking the solution space can help to obtain better and accurate fitness value by individuals of population.

It was revealed that the conventional GAs operators (selection, mutation, and crossover) drive the population to the equilibrium point where the improvement due to selection is balanced by a loss of fitness caused by mutation and crossover operators. A number of generations (a transient period) to reach that point depends on the selection rate and the mutation rate of GA, but does not depend on a coefficient of reduction of a search space size k. However, a coefficient k influences the levels of the macroscopic statistical properties of population obtained at the end of the transient period. At the end of the transient period the population becomes homogeneous, which means the conventional GA operators do not change the macroscopic statistical properties of population, and only the adjustment operator continues to drive the population to the maximum. A coefficient of shrinking of a mapped size k exponentially changes the macroscopic statistical properties of population and significantly influences on the computational cost. It was shown that a coefficient k is a trade-off procedure: to reduce the computational cost, the value of k should be reduced; however, the value of k should be increased to obtain a better optimum.

The derived equations that express the effect of GAs operators on the macroscopic statistical properties of the population showed the theoretical model predicts accurately enough the behavior of GA, although there are clearly systematic errors in the predictions of the theory arising from numerous assumptions made to capture the full dynamics of GA.

Implementation of Changing Range Genetic Algorithms - CRGAs was effective for solving some problems related to the science and design of engineering materials.

In optimization of multi-component binders the evaluation tests demonstrated that CRGA performs better than a conventional GA in finding the global optimum in the case of an MMCB problem with non-linear constraints. This is achieved by the application of the "shifting and shrinking" mechanism. The proposed CRGA is highly accurate in locating the global optimum. It is also very quick and very efficient (i.e. it does not require additional parameters or additional computational efforts). By implementing CRGA it was found that SP cost has a little effect on the optimal composition of MMCB: the proportioning of components with minimum cost is virtually the same for given strength levels. The "virtual lab" optimization of the strength characteristics at the level of MMCB helps to minimize the full-scale tests related to binder or concrete.

To investigate the fractal properties of three-dimensional RAP and APR the computer models with CRGA were developed. APR was represented as a numerical optimization problem resolved by a CRGA. These models were used in extensive numeric simulations to estimate the bounds of an exponent α for RAP and APR. The lower bound of α was numerically estimated for the frequency distribution $N(r)$ and the porosity distribution $\varepsilon(n)$. These new results have provided an additional support for the theory of Apollonian packing proposed by Aste [36]. It was shown that the Apollonian packing model which includes a CRGA can be used as a simulation tool for estimation of an exponent α with high level of precision, however, at an additional computational cost.

REFERENCES

[1] Goldberg, D. E., *Genetic Algorithms in Search, Optimization and Machine Learning*, 1989 (Addison-Wesley: Reading, Mass).
[2] Coello Coello, C. A., Theoretical and numerical constraint-handling techniques used with evolutionary algorithms: a survey of the state of the art, Comput. *Methods Appl. Mech. Eng.*, 2002, 191, 1245 –1287.
[3] Gen, M. and Cheng, R., *Genetic Algorithms and Engineering Design*, 1997 (Wiley: New York).
[4] Michalewicz, Z., *Genetic Algorithms + Data Structures = Evolution Programs*, 1996 (Springer-Verlag: New York).
[5] Michalewicz, Z. and Schoenauer, M., Evolutionary algorithms for constrained parameter optimization problems, *Evolutionary Computation*, 1996, 4(1), 1 –32.
[6] Djurisic, A. B., Elazar, J. M. and Rakic, A. D., Genetic algorithms for continuous optimization problems - a concept of parameter-space size adjustment, *J. Phys. A.*, 1997, 30, 7849-7861.
[7] Chelouah, R. and Siarry, P., A Continuous Genetic Algorithm Designed for the Global Optimization of Multimodal Functions, *Journal of Heuristics*, 2000, 6, 191-213.
[8] Hernández-Aguirre, A., Botello-Rionda, S., Coello Coello, C. A., Lizárraga- Lizárraga, G. and Mezura-Montes, E., Handling constraints using multiobjective optimization concepts, *Int. J. Numer. Meth. Eng.*, 2004, 59 (15), 1989-2017.
[9] Amirjanov, A., A changing range genetic algorithm, *Int. J. Numer. Meth. Eng.*, 2004, 61(15), 2660-2674.
[10] Amirjanov, A., The development a changing range genetic algorithm, *Comput. Methods Appl. Mech. Eng.*, 2006, 195, 2495-2508.
[11] Amirjanov, A., Modelling the dynamics of an adjustment of a search space size in a genetic algorithm, *Int. J. Mod. Phys. C*, 2008, 19, 1047-1062.
[12] Prügel-Bennett, A. and Shapiro, J. L., The dynamics of a genetic algorithm for simple random Ising systems, *Physica D*, 1997, 104, 75-114.
[13] Miller B, Goldberg D. Genetic algorithms, selection schemes, and the varying effects of noise. *Evolutionary Computation* 1997; 4(2): 113-131.
[14] Holland J. *Adaptation in natural and artificial systems.* University of Michigan Press: Ann Arbor, MA, 1975.
[15] Baker J. Reducing bias and inefficiency in the selection algorithm. In J. J. Grefenstette (Ed.), *Proceedings of the Second International Conference on Genetic Algorithms*, Hillsdale, NJ. Lawrence Erlbaum, 1987; 14-21.

[16] Grefenstette J, Baker J. How genetic algorithms work: A critical look at implicit parallelism. In J. D. Shaffer (Ed.), *Proceedings of the Third International Conference on Genetic Algorithms*, San Mateo, CA. Morgan Kaufmann: New York, 1989; 20-27.

[17] Mühlenbein H, Schlierkamp-Voosen D. Predictive models for the breeder genetic algorithm: I. Continuous parameter optimization. *Evolutionary Computation* 1993; 1(1): 25-49.

[18] Baker J. Adaptive selection methods for genetic algorithms. In J. J. Grefenstette (Ed.), *Proceedings of an International Conference on Genetic Algorithms and Their Applications*, Hillsdale, NJ. Lawrence Erlbaum, 1985; 101-111.

[19] Whitley D. The Genitor algorithm and selective pressure: Why rank-based allocation of reproductive trials is best. In J. D. Schaffer (Ed.), *Proceeding of the Third International Conference on Genetic Algorithms*, San Mateo, CA. Morgan Kaufmann: New York, 1989; 116-123.

[20] Blickle T, Thiele L. A comparison of selection schemes used in evolutionary algorithms. *Evolutionary Computation* 1997; 4(4): 361-394.

[21] Rogers A, Prügel-Bennett A. The dynamics of a genetic algorithm on a model hard optimization problem. *Complex Systems* 2000; 11(6): 437–464.

[22] Rattray L. *Modelling the Dynamics of Genetic Algorithms Using Statistic Mechanics* Ph.D. Thesis, University of Manchester, Manchester, UK. 1996.

[23] Spiegel M. *Calculus of finite differences and difference equations.* McGraw-Hill, 1971.

[24] Dewar, J.D. (1995), A Concrete Laboratory in a Computer-Case-Studies of Simulation of Laboratory Trial Mixes; ERMCO-95, *Proceedings of the XIth European Ready Mixed Concrete Congress,* Istanbul, pp. 185-193.

[25] Sobolev K., The Development of a New Method for the Proportioning of High-Performance Concrete Mixtures. *Cement and Concrete Composites,* Vol. 26, No. 7, 2004, pp. 901-907.

[26] de Larrard, F. and Sedran, T. (2002), *Mixture-Proportioning of High- Performance Concrete, Cement and Concrete Research,* Vol. 32, pp. 1699-1704.

[27] Ramachandran, V.S. (1995), *Concrete Admixtures Handbook, Second Edition,* Noyes Publications, New Jersey.

[28] Malhotra, V.M. (1998), *High- Performance High- Volume Fly Ash Concrete, High Performance High- Strength Concrete: Material properties, Structural Behavior and Field Application,* Perth, Australia, pp. 97-122.

[29] Mehta, P.K. (1989), Pozzolanic and Cementitious By-Products in Concrete. Another Look, CANMET/ACI International *Conference on Fly Ash, Silica Fume, Slag, and Natural Pozzolans in Concrete,* ACI SP-114, pp. 1-14.

[30] Neville, A.M. (2000), *Properties of Concrete*, Prentice Hall, 844 pp.

[31] Simon, M., Snyder, K.A. and Frohnsdorff, G.J.C. (1999), *Advances in Concrete Mixture Optimisation, Concrete Durability and Repair Technology Conference.* Proceedings. Scotland, UK, Thomas Telford Publishing, London, England, Dhir, R. K.; McCarthy, M. J., Editors, pp. 21-32.

[32] Sobolev K., and Arikan M., *High-Volume Mineral Additive ECO- Cement. American Ceramic Society Bulletin,* USA, Vol. 81, No. 1, 2002, pp. 39-43.

[33] Amirjanov A. and Sobolev K., Genetic Algorithm for Cost Optimization of Modified Multi-Component Binders. *Building and Environment* Vol. 41, No. 2, 2006, pp. 195-203.

[34] Koziel S., and Michalewicz Z., "Evolutionary Algorithms, Homomorphous Mapping, and Constrained Parameter Optimization Problems", *IEEE Trans. Evolutionary Computation*, 7/1 (1999) 19-44.
[35] Hales, T.C., *Sphere Packings* I, *Discrete* and Computational *Geometry* 18 (*1997*), pp. 135–149.
[36] Aste T. and Weaire D. *The Pursuit of Perfect Packing*. Bristol: Institute of Physics Publishing; 2000.
[37] Aste T. Circles, spheres and drops packings. *Phys. Rev. E,* 1996; 53: pp. 2571-2579.
[38] Dodds PS and Weitz JS. Packing-limited growth. *Phys. Rev. E,* 2002; 65: pp. 056108.
[39] Sobolev K. and Amirjanov A., The Simulation of Particulate Materials Packing Based on the Solid Suspension Model. *Advanced Powder Technology,* Vol. 18, No. 3, 2007, pp. 261–271.
[40] Sobolev K., and Amirjanov A., The Development of a Simulation Model of the Dense Packing of Large Particulate Assemblies. *Powder Technology,* Vol. 141, No. 1-2, 2004, pp. 155-160.
[41] Manna SS. Space filling tiling by random packing of discs. *Physica A,* 1992; 187: pp. 373-377.
[42] Sobolev K. and Amirjanov A. Application of Genetic Algorithm for Modeling of Dense Packing of Concrete Aggregates. *Construction and Building Materials,* Vol. 24, No. 8, 2010, pp. 1449-1455.
[43] Amirjanov A. and Sobolev K., Fractal Properties of Dense Packing of Spherical Particles. *Modelling and Simulation in Materials Science and Engineering,* Vol. 14, 2006, pp. 789-798.
[44] Schaertl W and Sillescu H. Brownian dynamics of polydisperse colloidal hard spheres: Equilibrium structures and random close packings. *J. Stat. Phys.,* 1994; 77: pp. 1007-1025.
[45] Borkovec, M.; De Paris, W.; Peikert, R. The Fractal Dimension of the Apollonian Sphere Packing", *Fractals,* 2 (4): 1994, pp. 521–526.
[46] Baram R. M. and Herrmann H.J. Self-Similar Space-Filling Packings in Three Dimensions, *Fractals,* 12, 2004, 293-301.
[47] Soppe, W. Computer simulation of random packings of hard spheres, *Powder Technology,* 62, 1990, pp. 189– 196.
[48] Torquato S., Truskett T. M. and Debenedetti P. G., Is Random Close Packing of Spheres Well Defined? *Phys. Rev. Lett.* 84 (10), 2000, pp. 2064-2067.
[49] Anishchik SV and Medvedev NN. Three-dimensional Apollonian packing as a model for dense granular systems.*Phys. Rev. Lett.,* 75, 1995; pp.4314-4317.
[50] Bryant S and Blunt M. Prediction of relative permeability in simple porous media.*Phys. Rev. A,* 46, 1992; pp.2004-2011.
[51] Franck-Oberaspach G., Schweiger D. B. and Svozil K., A Packing Problem, Solved by Genetic Algorithms, *Journal of Universal Computer* Science (Springer) 5, 1999, pp. 464-470.
[52] Cornforth D., Evolution in the orange box - A new approach to the sphere-packing problem in CMAC-based neural networks. In Mckay and Slaney (eds.), AI 2002: *Advances in Artificial Intelligence, Lecture notes in Artificial Intelligence* 2557, Springer, 2002, pp. 333-343.

[53] Runarsson T. P. and Yao X., Stochastic Ranking for Constrained Evolutionary Optimization, *IEEE T Evolut Comput,* 4 (3), 2000, pp. 284-294.
[54] Herrmann HJ, Mantica G and Bessis D. Space Filling Bearings. *Phys. Rev. Lett.,* 65, 1990, pp.3223-3226.
[55] Andrienko YA, Brilliantov NV and Krapivsky PL. Pattern formation by growing droplets: The touch-and-stop model of growth. *J. Stat. Phys.,* 75, 1994, pp. 507-523.
[56] Brilliantov NV, Krapivsky PL and Andrienkov YA. Random space-filling-tiling: fractal properties and kinetics. *J. Phys. A: Math. Gen.,* 27, 1994, pp. L381-L386.

In: Handbook of Genetic Algorithms: New Research
Editors: A. Ramirez Muñoz and I. Garza Rodriguez
ISBN: 978-1-62081-158-0
© 2012 Nova Science Publishers, Inc.

Chapter 2

MODEL-FREE DECONVOLUTION OF TRANSIENT SIGNALS USING GENETIC ALGORITHMS

László Bengi, Balázs Kovács, Máté Bezdek and Ernő Keszei
Eötvös University Budapest, Department of Physical Chemistry,
and Reaction Kinetics Laboratory, Hungary

ABSTRACT

Model-free deconvolution of transient signals usually suffers from a wavy small-frequency component as well as an increase of high frequency noise. The more these features are damped, the more the deconvolved signal becomes distorted. If we can include as much information concerning the genuine deconvolved signal as possible, the quality of the deconvolved transient can be greatly improved.

We have developed a deconvolution procedure which does not make use of a presupposed (often arbitrary) functional form of the transient signal. Instead, it is based on the inversion of the distortion effects of convolution, which means temporal compression, amplitude enhancement, increasing of the steepness of rise and decay of the measured convolved signal, and cutting initial data to zero to reproduce an eventual sudden jump. As our results have proved that the choice of a fairly good initial population is crucial for a successful deconvolution, we use additional genetic algorithms to generate an initial population with a relatively high fitness.

The initial population is generated from the measured convolved signal in two subsequent stages, each consisting of a genetic algorithm and the check of the potential to improve the fitness of the population during further breeding, as a candidate for the deconvolved data set. This generation can be done "automatically" so that the user does not need to experiment much to get a good initial population and satisfactory final results.

The final genetic algorithm performing the main iteration to get the estimate of the deconvolved data is constructed in such a way that it does not enhance either the low-frequency wavy behaviour, or the high frequency noise. It is based on a smooth mutation in a range containing several data, and a dynamic adjustment of the mutation.

The method is described in connection with the deconvolution of ultrafast laser kinetic (or femtosecond chemistry) data. The advantage of this method to find an underlying molecular mechanism via statistical inference is also discussed.

Keywords: deconvolution, genetic algorithm, transient signals, femtochemistry, fluorescence decay.

1. INTRODUCTION

Deconvolution is a process to reverse the result of – usually unwanted – convolution which distorts measured signals.[1-3] In many cases, measured signals are not those we are interested in, but – due to the limited capacities of the measuring apparatus – they are distorted by a phenomenon that can be interpreted as collecting data by moving averages. If we consider a continuous time-dependent signal $o(t)$ to be measured, the effect of *convolution* means that instead of this function, we can only measure

$$i(\tau) = \int_{-\infty}^{\infty} o(t)\, s(\tau-t)\, dt \; . \tag{1}$$

Using terms of image processing, the *measured image function* $i(\tau)$ is the integral with respect to time of the product of the true *object function* $o(t)$ and the distorting *spread function* $s(\tau-t)$. Signals are typically measured as a discrete time-series of digital recordings. The discrete equivalent of the convolution equation (1) is the sum

$$i_k = \sum_{k-M}^{k+M} o_i s_{k-i} \; , \tag{2}$$

where i_k is the k-th recording of the time-series of the image function measurements, o_i is the actual value of the object function between the recordings $k-M$ and $k+M$, and s_{k-i} is the integral of the spread function within one measured interval between $k-M$ and $k+M$. This latter equation illustrates more transparently the moving average over $2M+1$ discrete (non-zero) values of the spread function. (We get a genuine average without dividing by the sum of all the nonzero s_{k-i} values if the spread function is defined so that the sum of its values between $k-M$ and $k+M$ is exactly unity. In this case, we can also consider the spread function as a kind of probability distribution characterizing the detection device. This is also why it is called the *instrument response function*.) As a result, the measured time series will be a distorted version of the original object function time series o_i. Integral (1) or the sum (2) is usually written in a shorthand-type notation as

$$i = o \otimes s \; , \tag{3}$$

which is fully equivalent to either of the two equations. The symbol $\otimes$ denotes the operation of convolution.

To get the true object function o from the known i and s, we have to solve the convolution equation (1) or (2). This procedure is called *deconvolution*. As we can see, the (analytical) task is to solve an integral equation; thus the operation of convolution does not have a simple inverse. This problem can be circumvented with the help of an integral transformation. The *Fourier transform* (FT) of the image function (I) is simply the product of the Fourier transforms of the object (O) and the spread (S) functions: [3]

$$I = O \cdot S \qquad (4)$$

As the product does have an inverse, it is easy to get the FT of the object function O when dividing I by S. The inverse Fourier transform then readily provides the original object function o.

If the detection of the measured signal is free of any errors, the solution can be unique. However, if the measured signal contains some error (even if it is only a rounding-off error due to digital truncation of the recorded values), we get an infinite number of solutions. This means that no unique solution exists for any measured convolved data sets, which can easily be shown if we write the additional noise term explicitly:

$$i = o \otimes s + n, \qquad (5)$$

where n is the noise term. Due to the distributivity and associativity of convolution [3], this can be rewritten as

$$i = o \otimes s + n = o \otimes s + n' \otimes s = (o + n') \otimes s . \qquad (6)$$

Accordingly, the convolved noisy signal is equivalent to the convolution of the sum of the true object function and the "deconvolved noise" n'. Due to the additivity, *any* function n' added to the true object will give the measured image whose convolution with the spread function results in zero. Obviously, there exist infinite such functions with a high-frequency oscillation, as such oscillations disappear when their weighted average is calculated by convolution. Solution in the frequency domain – which makes use of the simple division of the Fourier transforms – does not help either. Both the image and the spread are relatively slowly varying functions (compared to the width of the sampling intervals), thus the high-frequency components in the FT in both cases are very much close to zero. Due to experimental errors, this "zero amplitude" fluctuates quite substantially, which is further amplified by the division of the close-to-zero image FT by the close-to-zero spread FT. The resulting inverse FT – the estimate of the original object – typically contains much more noise than signal.

This is a major problem during the deconvolution of measured signals, which inevitably contain some noise. The consequence is that a simple-minded deconvolution usually results in a rather noisy deconvolved data set, along with a lower frequency wavy behavior somewhere in the dominant frequency range of the spread function. There are several methods used that get rid of these unwanted features, which are more or less efficient. Most of them are based on the fact that the noise content results in a high-frequency oscillation, thus the elimination (or at least a severe limitation) of high-frequency components avoids this oscillation. However, if there are sudden changes in the true object function – which is typically the case in transient signals –, this damping of high-frequency components leads to a distortion of the reconstructed object function. Keeping in mind that the effect of noise is a kind of artifact in the deconvolved function, information about the true shape of the original object function helps a lot to drive the solution towards a physically sound deconvolved. The more information we can enter into the deconvolution procedure, the better estimates of the true object we can get.

In our previous research concerning ultrafast laser chemistry (or *femtochemistry*) measurements, we have tried to use deconvolution methods described in the literature. We have challenged several time-domain iterative methods, and many frequency-domain methods, including digital filtering and regularization.[4-6] We have found that proper adaptation of these methods leads to quite reasonable deconvolution results. However, none of them provide a genuine noise-free, unbiased deconvolved result, unless we know the functional form of the object function and search only for the parameters of this function so that the *reconvolved* – the calculated equivalent of the image function – best fits the measured image. However, this *reconvolution* is not a true deconvolution method; it is more of a fitting procedure that estimates the parameters of an already known function. In many physical problems, no model of the measured object is known *a priori*, thus it is necessary to calculate a *model-free* (*or nonparametric*) *deconvolution* of the measured signal. The nature of this task necessitates so called *soft methods* to search for a physically sound solution. One of the soft methods that has been found to be very effective in solving many problems is a genetic algorithm. [7] In addition to its great power to find a solution, a genetic algorithm also makes use of constraints, "herding" the solution by specifying several constraints derived from the *a priori* knowledge about the shape of the desired solution. This is the reason we have developed a method based on genetic algorithms to perform model-free deconvolution of transient signals.

Genetic algorithms are not widely used for deconvolution purposes yet, but – as it turned out from our research presented here – they are rather promising candidates to be used more widely in the near future. We have only found relatively few applications in the literature including image processing [8-10], spectroscopy [11-12], chromatography [13-15], and pharmacokinetics [16]. Most of the applications are not using model-free deconvolution but reconvolution methods, where the fit of the reconvolved model function to the measured data is performed, thus only parameters of this fit are determined using a genetic algorithm. It should also be noted that the term "deconvolution" is often used for the process of decomposing a set of overlapping peaks into separate components via curve fitting of peak models to the measured data set. This is conceptually distinct from deconvolution, because in deconvolution the original peak shape is unknown, and the measured function is distorted by the spread function. In the decomposition procedure, there is no distortion of the composite peak resulting from the superposition of the components, only a resolution of the components is necessary.

In this chapter we describe the use of genetic algorithms (GAs) that we successfully implemented for model-free (nonparametric) deconvolution of a special kind of transient signals: femtosecond kinetic traces. We do not deal with blind deconvolution where only the image is known and both the spread and the object are determined during the deconvolution procedure; only with deconvolution in case of known image and spread. After a brief description of the physical background concerning unwanted and unavoidable convolution in femtochemistry experiments, we describe synthetic functions to test deconvolution. The main part of the chapter deals with the description of the particular genetic algorithm used to estimate the deconvolved data set. This also includes a careful selection of the initial population, which is automatically generated using two subsequent steps also based on genetic algorithms. Following the discussion of the performance of this procedure tested on synthetic images, we also show deconvolution of real-life experimental data. The chapter ends with a conclusion, discussing an outlook to use the described deconvolution procedure for other problems where convolved transient signals occur.

2. CONVOLUTION OF THE MEASURED SIGNALS IN FEMTOCHEMISTRY

Femtochemistry is a new branch of molecular science dealing with direct experimental observation of molecular events during the breaking of chemical bonds and the formation of new bonds. [17-18] These events typically happen at timescales ranging from a few tens to a few hundreds of femtoseconds (fs), *i. e.*, between about 20×10^{-15} and 500×10^{-15} seconds. As the clocking speed of any available electronic devices is several orders of magnitudes lower, real-time measurements cannot be made at this timescale. To circumvent this limitation, the speed of light is used for clocking. First, an ultrashort laser pulse excites the molecules and thus starts the reaction, followed by a second laser pulse that tests some optical properties of the reacting system. The reaction time is coded into the delay between the two pulses. As 1 µm difference in the optical path length is equivalent to roughly 3 fs delay, a mechanical or electromechanical device is convenient to control the path length so that 1 fs delays can be adjusted easily.

Ultrafast laser pulses can be as short as 10 fs. However, excitation and detection are expected to be selective, which necessitates that their spectral width be quite small. Due to the uncertainty relation between time and energy, the typical pulse width is usually limited to about 100 fs or more.[19] Thus, characteristic times to be measured are about the same as the width of the laser pulses used to measure them; which results in convolved measured signals. A detailed description of how the convolved signal emerges is given elsewhere.[7] Here we only recall the result: the detected signal is a convolution of the *effective pulse* (spread, or the instrument response function, which is the correlation of the exciting and measuring pulses) and the *object*, as given in Eq. (3). As a consequence, femtosecond kinetic traces are usually heavily distorted by the convolution described above. Consequently, it is necessary to deconvolve femtochemical transient signals when performing a kinetic interpretation of the observed data. Typical data sets can be seen in Figures 4-6.

3. DECONVOLUTION USING GENETIC ALGORITHMS

In the "classical" version of a GA, individuals are represented as a binary string, which codes either the complete solution, or parameters to be optimized. These strings are considered the genetic material of individuals, or *chromosomes*, while the bits of each string are regarded as *genes*, having two different *alleles*, 0 or 1. It is often problematic to represent a solution in the form of binary strings. Thus, for numerical optimization purposes, *floating point* coding is more convenient, which allows for a virtually infinite number of alleles. Classical (binary) genetic operators have also been replaced accordingly by *arithmetic* operators. The "art" of using GAs is in finding a suitable representation or data structure for the solution, and using genetic operators that can explore the *solution space* in an efficient way, avoiding local optima and converging quickly to the global optimum. [20-21]

Deconvolution methods described in the literature use different representations and a variety of genetic operators. While binary coding and classical binary crossover and mutation are appropriate for processing black and white images [9-10], quite different encoding and operators should be used to deconvolve measured kinetic signals.[21] Generation of the initial population may also be critical. One method is the completely random generation of the first individuals, while a careful generation of already fit individuals is sometimes important.

When deconvolving femtosecond kinetic data using GAs, potential problems that may arise are similar to those of other methods: we should avoid the amplification of experimental noise and the oversmoothing of the image that results in – among other effects – a low frequency "wavy" distortion. The non-periodic nature and sudden stepwise changes of the transient signals should also be reconstructed without distortion. We have found that the above needs cannot be fulfilled if we start the GA with a randomly generated initial population. Therefore, the first task is to create individuals who already display some useful properties of a good solution.

3.1. Data Structure

The solution of the convolution equation (2) is a data set containing the undistorted object (instantaneous kinetic response) at the same time instants as the measured image function. Conveniently, the coded solution is this data set, which is an array containing floating point elements, with each element representing a measured value of the undistorted object. In terms of GAs, this is a single haploid chromosome containing as many genes as there are data points measured. As each parent and each offspring is a haploid, there is a haploid mechanism of reproduction to implement, in which two parent chromosomes are combined to form a single haploid chromosome of the offspring. There is no need to "express" the genes as phenotypes; the chromosome already represents the solution itself.

Convolution – since it is a kind of weighted moving average – *widens* the signal temporally, *diminishes its amplitude*, makes its *rise and descent less steep*, and *smoothes out* its sudden steplike jumps. Accordingly, to create a fairly fit initial population, we started from the image itself and implemented an operator to *compress* the image temporally, another to *enhance its amplitude*, third and fourth ones to *steepen its rise and decay*, and finally, a fifth operator to *restitute the stepwise jump* by setting some leading elements of the data to zero. All five operators – which we may call *creation operators* – are constructed to conduct a random search in a prescribed modification range. To this purpose, normally distributed random numbers are generated with given expectation and standard deviation for the factor of temporal compression, of the enhancement of amplitude, for increasing the steepness of rise and decay, and for the number of data points to be cut to zero at the leading edge of the data set. The effect of these operators is schematically represented in Figure 1.

The first implementation of the GA for deconvolution started with a user-defined set of creation operators, and the resulting initial population was displayed graphically, along with the original image. The best individual of this population and the result of the convolution of this individual with the spread function (the *reconvolved*), as well as the deviation of this reconvolved from the image were also displayed.

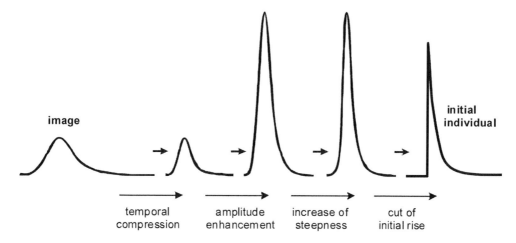

Figure 1. Schematic representation of the creation of initial individuals from the measured image function. The four steps depicted here show the effects of five different creation operators applied subsequently.

If the user found that the reconvolved data set was too different from the image, or if there were spurious oscillations present in the best individual, the user initiated another selection of the initial population with different expectation and/or standard deviation parameters of the creation operators. The procedure was repeated until the user considered the selected initial population satisfactory.

In order to arrive at fairly good estimates of the deconvolved data set that contained no spurious oscillations and had the desired shape, this trial and error procedure of finding satisfactory creation operators required somewhat tedious experimentation. The procedure also required good insight on the part of the user, not only in the details of the kinetic background involved, but also in the way the procedure responded to changes in the creation parameters. This was the reason that motivated the development of another method, also based on genetic algorithms, to replace user experimentation, which is explained following the description of the applied breeding procedure.

3.2. Parent Selection and Crossover

The quality of the deconvolved data set – an individual of the population – can be measured readily by the mean square error (*MSE*) of the reconvolution, given by

$$MSE = \sqrt{\frac{\sum_{k=1}^{N}[(\hat{o} \otimes s)_k - i_k]^2}{N-1}}, \qquad (7)$$

where $\hat{o}$ is the estimate of the deconvolved, *i. e.*, the actual individual of the population, and N is the number of data in the image data set. GA literature suggests that some kind of *inverse* of this error should be used so that the resulting fitness function is normalized. We implemented a dynamic scaling of fitness by adding the minimal *MSE* of the population in the denominator:

$$fitness_j = \frac{1}{\min(MSE) + MSE_j},\qquad(8)$$

which maintains the fitness values in the range from 1 / [min(*MSE*) + max(*MSE*)] to 1 / [2 min(*MSE*)].

To choose parents for mating, we use stochastic sampling with replacement, implemented as a *roulette-wheel selection*. [20-21] Offsprings are produced by *arithmetic crossover* of the parents, resulting in an offspring whose data points are the average of the parents' data. (Fitness-weighted averages did not result in an important difference concerning convergence and the quality of the winner.) This procedure of parent selection and crossover is made until the number of produced offspring becomes the same as in the previous population.

3.3. Mutation and Selection of the New Generation

Arithmetic crossover explores the potential solutions solely within the range represented by the initial population; it cannot drive the population out of this region. Mutation is therefore needed to further explore the fitness landscape of the solution space. In addition, a properly chosen mutation operator can also be used to avoid the usual noise amplification and low-frequency wavy behavior. A brute-force mutation by randomly changing the values of individual data points has the consequence of increasing the noise content of the signal, as reconvolution prior to the computation of fitness smoothes out even large differences in neighboring points, thus the population would evolve to a better fit by increasing high-frequency noise. In contrast, a "smooth" mutation of neighboring data points results in an effective smoothing of the mutated individuals after a few crossovers. It has been implemented as an addition of a randomly generated Gaussian to the actual data set. The expectation (centre), the standard deviation (width) and the amplitude of the additive Gaussian correction is randomly selected within a specified range, including both positive and negative amplitudes. (Leading zeros of the initial population get never changed by mutation, which is equivalent to a semi-finite-support constraint. [1,6]) If the kinetic response function has a long tail (*e. g.* due to largely different characteristic times involved in the reaction mechanism), its slow decrease can also be reconstructed by this mutation, even if the initial population had a much sharper decrease without a long tail.

There is another feature which proved to be useful in finding the fittest possible individuals; non-uniform mutation. [20] This is responsible for fine-tuning the mutations in such a way that it drives even a rather uniform and close to optimal population further towards the global optimum. If the mutation amplitude is small, the convergence at the start of the iteration is also small. A larger amplitude results in a faster convergence but makes the improvement of individuals quite improbable after the deviation of the solution from the optimum is much less than the mutation amplitude, as mutations typically result in a less fit individual compared to the one prior to mutation. To effectively get closer to the optimum, it is necessary to diminish the amplitude of the mutation as the number of generations increases, or with decreasing deviation from the optimum. We perform this adjustment by estimating the experimental error as the standard deviation of measured image data in the range where its values are more or less constant. (*E. g.*, in the leading zero level of the signal, or in an almost constant tail.) Comparing this experimental error to the difference between the *MSE* of the fittest and the least fit individuals, the amplitude parameter of the Gaussian mutation is multiplied by the factor

$$f = 1 - e^{1-\left(\frac{MSE \text{ difference}}{\text{experimental error}}\right)^p}. \qquad (9)$$

The factor *f* approaches zero as the ratio of the *MSE* difference between the most and least fit individuals to the experimental error approaches one. This correction avoids too large modifications, thus resulting also in a less noisy deconvolved data set. The higher the power *p*, the more enhanced the tuning effect of the factor. A value of the exponent $p = 1.5$ proved to be efficient when deconvolving the data shown in this chapter. To avoid problems arising from a possible overestimation of the experimental error, the factor *f* can be checked and set equal to a prescribed smallest value if the ratio in the exponent is too close to one, or even less.

When the number of newly generated offsprings equals the population number minus one, selection of the new generation is done. All the parents die out except for the fittest one, which also becomes member of the next generation. This selection method is called *single elitism* and guarantees a monotonous improvement of the best individual.

3.4. Termination and the Choice of the Winner

After each generation, the quality of individuals is evaluated by the *MSE* between the image and the reconvolved, as this quantity is used to calculate the fitness. In addition, the differences between the reconvolved fittest individual (the *winner*) and the image – the residual errors – are calculated. For this residual data set, Durbin-Watson (DW) statistics indicate whether subsequent differences are correlated – pointing to systematic error. [22-23] The equivalent of the experimental error – the standard deviation of a few data points of the image data set, which can be considered constant – can be calculated similarly from the deconvolved data.

As a termination criterion, we can check if the *MSE* between the image and the reconvolved set of the winner is less than or equal to the experimental error. However, this does not guarantee that the reconvolved solution closely matches the image; there might be some bias present in the form of low-frequency waviness. It is the Durbin-Watson statistics that sensitively indicates such misfits. For the large number of data in a kinetic trace (typically more than 200), its critical value for a test of random differences is around 2.0 [22-23], and it is typically much lower than that for a wavy set of differences. Thus we may either use a DW value close to 2.0 as a criterion, or combine the experimental error criterion with the DW criterion so that both of them should be fulfilled.

There are less specific GA properties that can also be used to stop the iteration. If the *MSE* of the best individual does not change for a prescribed number of generations, the algorithm might have converged. Similarly, if the difference between the *MSE* of the best fit and the least fit individual becomes less than the experimental error, we cannot expect too much change in the population due to mutations. However, the use of these criteria only indicates that the GA itself has converged but it does not guarantee a satisfactory solution.

3.5. Automatic Generation of a Fairly Good Initial Population

Having explained the breeding procedure of the genetic algorithm, let us return to the problem of creating a fairly good initial population which can evolve to give a physically sound deconvolved data set, having all the properties the user expects. The first step of the creation is to get an estimate of the deconvolved signal by using inverse filtering. We use a modified adaptive Wiener filter for this purpose. [4,6] This filtering usually results in a wavy and somewhat noisy deconvolved data set, but its onset – the first nonzero data point – can be approximately calculated in the following way. The data point between the first nonzero value and the maximum of the Wiener-filtered deconvolved is determined. The onset of the object function is in the vicinity of this location. To explore the fitness landscape, typically 50 individuals are created at each location within the range, considering it as the actual onset. Values of the other creation parameters are also generated with a uniform probability distribution, except for the maximum enhancement of the signal which is the same for all individuals. The fittest initial individual is chosen from all the initial populations, and the location of its onset will become the next estimate of the onset of the object function.

The procedure then continues in a second stage with the creation of other initial generations, this time within a narrowed range of equally distributed onsets of the individuals around the new location. Values of the other creation parameters are now generated with a normal distribution using their expectation and standard deviation given in the project description file. Breeding is done usually for 50 generations in this stage. (This number can, of course, be different.) However, the search in this stage is not for the fittest individual within a generation, but for the initial population in which the fitness increases most, compared to that of the fittest individual at the end of the first stage, as it has the greatest potential to improve. To determine this, each initial generation is bred several times (we typically use 20 breedings) with a different series of random numbers. After each of the 20 different breeding runs, the one that contained the fittest individual is noted. The "winner of the creation" will be the population that proved to increase in fitness most, with respect to the fittest individual at the end of the first stage, most frequently out of the 20 breedings. Its location of the onset and the rate of enhancement of amplitude will be used to generate the initial population of the final genetic algorithm to search for the optimal deconvolved data set.

In some cases – when the transient signal does not have a steplike onset but a slower rise – the two stages described above do not always give a satisfactory result. They often miss the true onset of the object function (which is known for synthetic data) by a few data points. However, the user is able to see the mismatch and can manually correct for the onset which gives the best result visually. Therefore, there is also a possibility to fix the location of the onset, in which case the first two stages are skipped and the procedure begins with the third stage by creating an initial population using the input values of the creation parameters.

3.6. Implementation of the Deconvolution Procedure

We have implemented the procedure described above as a package of user defined Matlab functions and scripts. The control of the deconvolution procedure is done by a graphical user interface (GUI). All the input data including filenames and operator parameters can be entered via this interface. As these data are then all written into a project descriptor text file, there is also a possibility to simply enter the name of a previously saved project descriptor file to reload all its data. The output file contains the entire project descriptor, statistical evaluations, and all relevant results. In addition, there is a four-panel figure displayed, containing most of the results for immediate graphical evaluation after each stage of the procedure.

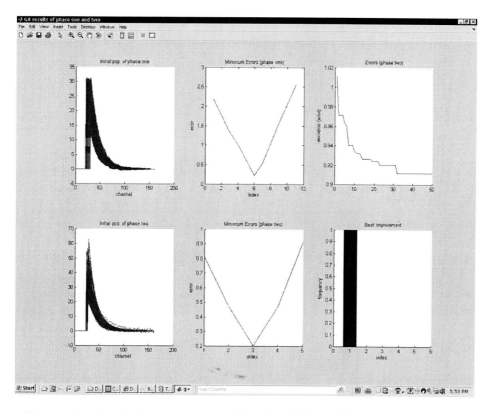

Figure 2. Screenshot after the first two stages searching for a fairly good initial population. For the description of the panels, see text.

In case the value of the input parameter specifying the expectation of the location of the first nonzero element is zero, the procedure starts with the first two stages searching for the location of the onset of the deconvolved signal. At the end of this procedure, the results of the two stages are displayed graphically in a six-panel figure (see Figure 2). The upper left panel shows the initial population generated with onsets around the location obtained with Wiener filtering. The upper middle panel is the most important here; it shows the MSE errors of the best individuals at each location after the first breeding stage, as a function of the channel index. As we can see, in Figure 2 it is channel No. 6 which gives the fittest individual. The upper right panel shows the evolution of the MSE errors of the fittest individual in each generation as a function of the number of iteration steps (*i. e.* the number of subsequent generations). In the figure we can see that an improvement of fitness (decrease of the MSE error) occurred until step No. 32, but there was no improvement from then up to step No. 50.

The lower left panel shows the initial population generated with onsets around the location obtained as that of the fittest individual at the end of the first stage. It is clearly seen that the variety of the individuals in this population is greater than in the first stage, but in a narrower interval of the location of the onset (five channels instead of twelve). The lower middle panel shows again the *MSE* of the best individuals at each location after the second breeding stage, as a function of the channel index. However, this time it is not this statistics that is used to choose the winner, but the frequency at which an initial population becomes the fittest one after 20 breedings with different sets of random numbers to control the procedure of generating the initial population as well as the breeding procedure, which is shown in the lower right panel. In this particular case, it is the initial population generated with the onset at channel No. 1 which has the greatest improvement 20 times out of the 20 runs, thus it had been chosen as the winner. (If there are more than one placements of the onset resulting in a few best scores, it is the one with the highest frequency which is chosen as the winner.) The winner of this second stage determines the placement of the onset, and this value is used as the expectation of this parameter to generate the initial population used in the "main" GA to search for the optimal deconvolved data set.

After the first two stages, the user can observe the graphically displayed results described above, and can decide whether to launch the main GA, or halt the procedure and restart with a different set of creation parameters. If the decision is to continue the procedure, the number of iteration steps (the number of subsequent generations) can be determined before actually launching the main iteration.

In case the value of the input parameter specifying the expectation of the location of the first nonzero element at the beginning of the procedure is different from zero, the procedure skips the first two stages and uses the location of the onset of the deconvolved signal entered to generate the initial population for the main GA iteration.

At the end of the prescribed number of iterations, the program halts and a four-panel figure is displayed (see Figure 3). The upper left panel shows the image (measured) data set, the spread function (called "pulse" in the figure), the object function in case of a synthetic data set (in which case the object is known), along with the initial population. The upper right panel shows the evolution of the *MSE* of the best individual in subsequent generations, as a function of the iteration number (or generation number). The lower left panel shows the results; *i. e.* the deconvolved data set (the "winner"), its reconvolved with the spread function, along with the original image for comparison. In the case of synthetic data, the object function is also shown for comparison with the deconvolved result. (In the figure, due to the screen resolution, the differences between the image and reconvolved, as well as the differences between the object and winner cannot be seen.) The lower right panel helps to judge the nature of differences between the data sets shown in the lower left panel by displaying their (discrete) Fourier transforms. These transforms show the frequency behavior of the corresponding time series, which sensitively detect mismatches and contain useful information on the success of the deconvolution. We do not deal with the details of the Fourier transforms here; the interested reader is advised to consult an earlier publication. [5] If the user is content with the final results, the procedure is finished. If the results are not satisfactory, a new run with or without the first two stages can be made, starting with a different set of input parameters.

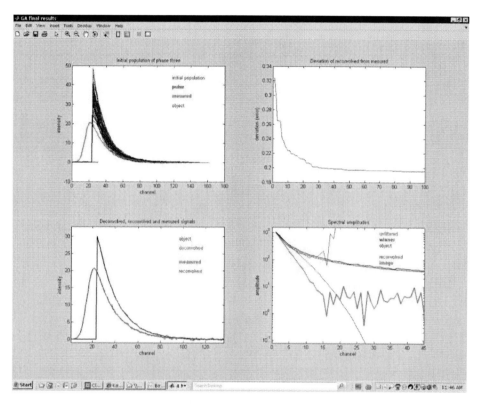

Figure 3. Screenshot showing the graphical display of the final results of deconvolution, after the main iteration. For the description of the panels, see text.

4. RESULTS

To test the performance of the algorithm, we used synthetic data calculated for a simple consecutive reaction containing two first-order steps; the first with a 200 fs, the second with a 500 fs characteristic time, as described in previous publications. [4-6] They mimic different typical transient absorbance signals. The first one is recorded at a wavelength where a completely decomposing reactant species can be detected with no remaining absorptivity ("reactant"), the second one at another wavelength with an absorbing intermediate species and also a product with positive remaining absorptivity, while the third one is similar but with a negative remaining absorptivity ("bleaching"). Calculated object data have been convolved with a 255 fs fwhm Gaussian spread function, and an "experimental noise" of 2 % of the maximum of the signals was added to the convolved data set to become the model of measured data. As recent measurements in fluorescence detection explored extremely short characteristic times with simultaneous long-time components [24-25], we also tested a fourth synthetic data set mimicking this situation, with time constants of 100 and 500 fs, and a spread function of 310 fs fwhm. While the absorption data set with bleaching necessitates the most careful transformations using the creation operators, the fluorescence decay challenges the power of the algorithm to reconstruct an extremely large stepwise jump followed immediately by a steep decay and ending in a long tail.

4.1. Test Results for Synthetic Data

Figure 3 shows the best result obtained for the reactant transient absorption signal. As it can be seen, there is no visible difference between the deconvolved set and the synthetic object data, except for the "experimental error" added to the synthetic image. However, fluctuations in the deconvolved data set are inferior to that of the image data set which means that an effective smoothing of the "experimental errors" has also been achieved by the genetic algorithm performing deconvolution. Due to the creation operator that renders the leading data prior to the onset of the signal to zero, the steplike jump of the signal is completely reproduced, without any wavy behavior. (Leading zero values are not modified by mutations either.) Thus, we can conclude that a transient with a sudden jump and fast decay that undergoes a severe distortion due to convolution can be deconvolved to an unprecedented level.

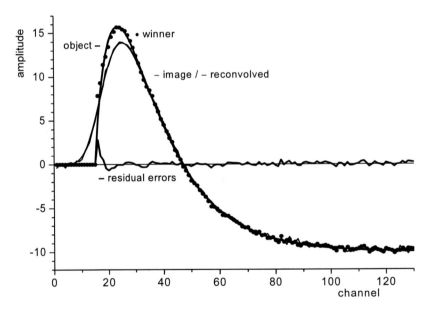

Figure 4. Deconvolution results for a non-periodic transient signal with a negative final level (in femtochemistry, it is an intermediate absorbing species with a residual bleaching). The two smoothly rising overlapping curves (solid lines) show the initial image data set to be deconvolved, and the final estimate of the deconvolved signal reconvolved with the spread function (labelled as reconvolved). The solid curve with a sudden jumplike rise is the object function without errors. Closed circles indicate the values obtained for the deconvolved at the end of the final iteration (labelled as winner). Residual errors are shown as a solid curve around zero signal level, calculated as the difference between the winner and the object.

Figure 4 shows the results obtained for the transient signal with bleaching. This signal is not periodic, as its initial value is not identical to the last one; there is a large difference between the first and last data points. In the context of the genetic algorithms used, this problem can be treated with a careful design of the creation operators in a way that the beginning and the end of the data set should not change while applying the changes schematically illustrated in Figure 1. The success of this procedure can be illustrated by the final result of the deconvolution, shown in Figure 4. The matching smooth continuous curves are again the "measured" image and the reconvolved data set, which can hardly be distinguished at the resolution of the figure. The object (continuous curve with a steplike rise) is almost completely fit by the winner of the GA iteration (circles). Residual errors are displayed as another continuous curve close to zero signal level, which indicate a randomly distributed small fluctuation of the winner around the true object, except for a few initial points following the onset of the signal. However, this initial part of the signal is always very hard to reproduce by deconvolution; it contains typically a few initial mismatches in a wavy form. This initial wavy bias is usually present even in the case of reconvolution using a known model function.

It should also be added that the deconvolution procedure described in this chapter has been primarily optimized to the steplike transient function (reactant-like absorptivity or fluorescence decay). To illustrate the success of this optimization, deconvolution results for the simulation of the highly distorted fluorescence signal mentioned above are also shown in Figure 5. The notation of this figure is similar to that of Figure 4, with the matching "measured" and reconvolved data sets, the object and the winner, along with the residual errors concerning these latter data. The challenge of deconvolution in this case is to reconstruct the large maximum of the steplike jump which is more than four times the maximum of the image, and the long tail which begins with a much steeper decay than the image, and goes over into a shallower final part than the image. As we can see from the residual errors, there are no marked wavy mismatches at the very beginning of the signal, thus the maximum is also excellently recovered by the deconvolution.

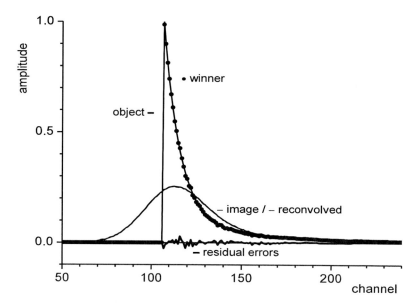

Figure 5. Deconvolution results for a heavily distorted reactant signal with a sudden initial jump followed by a steep decrease and ending in a rather shallow tail (in femtochemistry, it can be a reactant-like absorption signal, or a fluorescence decay). The description of the signals and the notation used is the same as in Figure 4.

An interesting feature of the procedure applied has been explored during the deconvolution of this transient signal. While the success of the deconvolution depends largely on a fairly close value of the signal maximum to that of the true object of the initial population, even a quite large mismatch in the tail part can successfully be eliminated during the final iteration. This is a much desired property of the procedure used if we want to deconvolve signals with largely different decay time constants. As this feature is discussed in a previous paper [7], we do not deal with it in details here.

We have also conducted another test of the deconvolution procedure. Using the best deconvolved of the three synthetic data sets calculated for a simple consecutive reaction containing two first-order steps, determined with the described procedure, we have made a global estimation of the kinetic and photophysical parameters involved. (For the description of the model function and the global estimation, see ref. 7.) The results indicate that most of the parameters have a narrower confidence interval and a smaller (or nonexistent) bias than in the case of parameter estimation from the results of a reconvolution procedure. These results support that the procedure using GAs described here leads to deconvolved data sets whose analysis provides parameters that are less correlated with each other, and are thus also less biased. However, we consider these results only preliminary, for the deconvolution of the product species with a small transitory peak and a remaining positive absorptivity is not yet satisfactory. (As it is mentioned above, optimization of the deconvolution procedure focused on the reactant-like data set.).

4.2. Test Results for Experimental Data

Model-free deconvolution is necessary if we do not know even the functional form of the object function, only some of the qualitative properties concerning its expected shape. In a real-life experiment, the image and the spread functions are only known as measured data sets. Therefore, we cannot use the comparison of the object and the winner to monitor the development and the success of the iteration procedure. (For the same reason, it has not been used either as a criterion when testing the procedure with synthetic data.) In Figure 6, we show deconvolution results of a recent experiment still unpublished, where the fluorescence decay of adenosine monophosphate in aqueous solution was measured by femtosecond fluorescence upconversion (excited at 267 nm, observed at 310 nm).

Data are collected at 33.33 fs intervals per channel, and the experimentally determined effective pulse has a width of 270 fs fwhm, *i. e.*, 8 channels. [26] As we can see, the sudden jump characteristic of fluorescence data is completely recovered by deconvolution. Residual errors (the difference between the image and the reconvolved data sets) show an almost completely random distribution with no markedly wavy feature. Fluctuations of the residual error do not exceed the experimental noise contained in the measured image. The only seemingly systematic behavior can be seen between channels 31 and 36, where there is a humplike feature in the residual error. Observing the measured image, we can state that there is a marked shoulderlike feature in this data set as well, indicating that the deconvolution procedure truly reconstructs the behavior of the measured data. Though it might be an artifact, it might also be a kinetic effect. It is up to the user to take it seriously as an effect, or consider it an erroneous result. Apart from this small feature, the deconvolved curve has all the properties we expect from an instantaneous fluorescence response, thus we can consider the deconvolution highly successful.

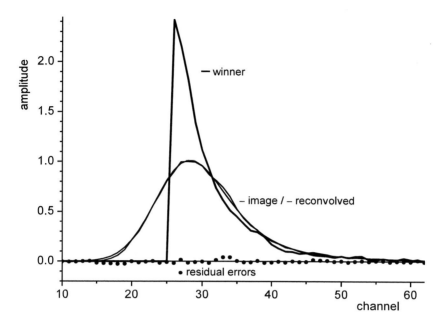

Figure 6. Deconvolution results for experimental transient fluorescence data. The two smoothly rising overlapping curves (thinner solid lines) show the initial image data set to be deconvolved, and the final estimate of the deconvolved signal reconvolved with the spread function (labelled as reconvolved). The solid curve with a sudden jumplike rise is the deconvolved data set at the end of the final iteration (labelled as winner). Residual errors (solid circles) are calculated as differences between the image and the reconvolved data sets.

CONCLUSION

Model-free deconvolution methods previously used to treat femtosecond transient kinetic data had two major shortcomings. Due to noise suppression, they resulted in deconvolved signals having a low-frequency wavy behavior, and the greater the noise suppression was, the more biased the overall shape of the reconstructed signal became. To overcome these shortcomings, we used GAs to achieve better deconvolution results. In order to adapt GAs for the purpose of deconvolving time-dependent transient signals heavily distorted by convolution, we have modified the classical GA and introduced a careful generation of the initial population and a special mutation changing neighbouring genes simultaneously.

Modification of the data structure included the use of floating-point arrays as chromosomes, representing the solution of the deconvolution problem at the chromosome level. Accordingly, genetic operators performed floating point arithmetical operations, not binary ones. To create a fairly good approximation of the deconvolved data set, we have transformed the measured image function to invert the effects of convolution. Due mainly to the operator which cuts a series of leading data of the signal to zero, and to the mutation procedure that maintains these leading zeros, the high-frequency part of transient signals with a sudden jump-like start can be fully reconstructed. Waviness of the solution is avoided by applying smooth arithmetic mutation of neighboring genes instead of single point mutations, along with a dynamically changing mutation that fine-tunes the amplitude of the changes according to the closeness of the population to the optimal solution.

Deconvolution using this modified GA outperforms existing algorithms and produces an unprecedented quality of the reconstructed original signal for highly nonperiodic transient absorption as well as highly distorted fluorescence decay traces with a sudden initial steplike rise. Results on real-life experimental data also support the applicability of the proposed deconvolution method. Even if a few thousand generations are needed to reach an optimal solution, the iterations only take a couple of minutes on a moderately fast desktop PC. Further work is in progress to explore the full power of the applied genetic algorithm, and to develop an optimized procedure also for transients with a less sharp initial rise.

It should be noted that deconvolution is a necessary tool in many applications where it is only possible to measure a convolved signal, independently from the actual timescale of the measurement. Examples are, among others, pharmacokinetic recordings at a few hours' timescale, or transient electric signals of common devices at a few microseconds' timescale.

ACKNOWLEDGMENTS

Thanks are due to Thomas Gustavsson and Ákos Bányász for experimental data and detailed information of the experimental conditions. Péter Pataki is acknowledged for his contribution to the Matlab code.

REFERENCES

[1] Jansson PA (editor). Deconvolution of Images and Spectra, 2nd edition; Academic Press: San Diego, 1997.
[2] A. Meister: *Deconvolution problems in nonparametric statistics*, 2009 Springer, Berlin Heidelberg.
[3] Press WH, Flannary BP, Teukolsky SA, Vetterling WT. Numerical Recipees – The Art of Scientific Computing; Cambridge University Press: Cambridge, 1986.
[4] Bányász Á, Mátyus E, Keszei E. Deconvolution of ultrafast kinetic data with inverse filtering. Radiat. Phys. Chem. 2005; 72: 235-242.
[5] Bányász Á, Dancs G, Keszei E. Optimisation of digital noise filtering in the deconvolution of ultrafast kinetic data. Radiat. Phys. Chem. 2005; 74: 139-145.
[6] Bányász Á, Keszei E. Nonparametric deconvolution of femtosecond kinetic data. J. Phys. Chem. A 2006; 110: 6192-6207.
[7] Keszei E. Efficient model-free deconvolution of measured femtosecond kinetic data using genetic algorithm, *J. Chemometrics*, 23, 188-196 (2009).
[8] Johnson EG, Abushagur MAG. Image deconvolution using a micro genetic algorithm. Opt. Commun. 1997; 140: 6-10.
[9] Chen YW, Nakao Z, Arakaki K, Tamura S. Blind deconvolution based on genetic algorithms. IEICE T. Fund. Elect. 1997; E80A: 2603-2607.

[10] Yin HJ, Hussain I. Independent component analysis and nongaussianity for blind image deconvolution and deblurring. *Integr. Comput-Aid E.* 2008; 15: 219-228.

[11] Sprzechak P, Moravski RZ. Calibration of a spectrometer using a genetic algorithm. *IEEE T. Instrum. Meas.* 2000; 49: 449-454.

[12] Tripathy SP, Sunil C, Nandy M, Sarkar PK, Sharma DN, Mukherjee B. Activation foils unfolding for neutron spectrometry: Comparison of different deconvolution methods. *Nucl. Instrum. Meth. A.* 2007; 583: 421-425.

[13] Vivo-Truyols G, Torres-Lapasio JR, Garrido-Frenich A, Garcia-Alvarez-Coque MC. A hybrid genetic algorithm with local search I. Discrete variables: optimisation of complementary mobile phases. *Chemometr. Intell. Lab.* 2001; 59: 89-106; *ibid.* A hybrid genetic algorithm with local search II. Continuous variables: multibatch peak deconvolution. *Chemometr. Intell. Lab.* 2001; 59: 107-120.

[14] Wasim M, Brereton RG. Hard modeling methods for the curve resolution of data from liquid chromatography with a diode array detector and on-flow liquid chromatography with nuclear magnetic resonance. *J. Chem. Inf. Model.* 2004; 46: 1143-1153.

[15] Mico-Tormos A, Collado-Soriano C, Torres-Lapasio JR, Simo-Alfonso E, Ramis-Ramos G. Determination of fatty alcohol ethoxylates by derivatisation with maleic anhydride followed by liquid chromatography with UV-vis detection. *J. Chromatogr.* 2008; 1180: 32-41.

[16] Madden FN, Godfrey KR, Chappell MJ, Hovorka R, Bates RA. A comparison of six deconvolution techniques. *J. Pharmacokinet. Biop.* 1996; 24: 283-299.

[17] Zewail AH, Laser femtochemistry. *Science* 1988; 242: 1645-1653.

[18] Simon JD (editor): *Ultrafast Dynamics of Chemical Systems*, Kluwer Academic Publishers, Dordrecht (1994).

[19] Donoho DL, Stark PB. Uncertainty principle and signal recovery. SIAM J. Appl. Math. 1989; 49: 906-931.

[20] Michalewicz Z. *Genetic Algorithms + Data Structures = Evolution Programs.* Springer: Berlin, 1992.

[21] Mitchell M. *An Introduction to Genetic Algorithms.* MIT Press: Cambridge, Mass., 1996.

[22] Durbin J, Watson GS. Testing for serial correlation in least squares regression I. *Biometrika* 1950; 37: 409-428.

[23] Durbin J, Watson GS. Testing for serial correlation in least squares regression II. *Biometrika* 1950; 38: 159-178.

[24] Bányász Á, Gustavsson T, Keszei E, Improta R, Markovitsi D. Effect of amino substitution on the excited state dynamics of uracil, *Photoch. Photobio. Sci.* 2008; 7: 765-768.

[25] Gustavsson T, Bányász Á, Lazzarotto E Markovitsi D, Scalmani G, Frisch MJ, Barone V, Improta R. Singlet Excited-State Behavior of Uracil and Thymine in Aqueous Solution: A Combined Experimental and Computational Study of 11 Uracil Derivatives, *J. Am. Chem. Soc.* 2006; 128: 607-619.

[26] Bányász Á. Gustavsson T. unpublished results.

In: Handbook of Genetic Algorithms: New Research
Editors: A. Ramirez Muñoz and I. Garza Rodriguez
ISBN: 978-1-62081-158-0
© 2012 Nova Science Publishers, Inc.

Chapter 3

APPLICATION OF GENETIC ALGORITHM OPTIMISATION TECHNIQUE IN BEAM STEERING OF CIRCULAR ARRAY ANTENNA

A. D. Adebola, R. A. Abd-Alhameed and M. M. Abusitta

Mobile and Satellite Communications Research Centre,
Bradford University, Bradford, UK

ABSTRACT

Beam steering entails changing the direction of the mainlobe of the radiation pattern of an antenna in order to achieve spectrum efficiency enhancement and mitigation of multipath propagation. Beams formed by antenna arrays are categorised as either switched array beams or adaptive array beams. While switched array beams are fixed, finite and predetermined, adaptive array beams possess infinite number of patterns and can adapt in real time to RF signal environment hence adaptive beamforming arrays are always preferred. Adaptive arrays have the capability of directing or steering in real time the main beam in a desired direction or towards signal of interest (SOI) and suppressing interference or multipath signals.

A reactive loading and time modulated switching techniques are applied to steer the beam of a circular uniformly spaced six element antenna array having a source element at its centre. Genetic algorithm (GA) process is used to calculate the optimal values of reactances loading the parasitic elements for which the gain can be optimized in a desired direction. For time switching, GA is also used to determine the optimal on and off times of the parasitic elements for which the difference in currents induced optimizes the gain and steers the beam in a desired direction. These methods are demonstrated by vertically polarised antenna configuration operated at frequency of 2.45GHz. Simulations results showed that near optimal solutions for gain optimization, sidelobe level reduction, VSWR 3 over a 100MHz bandwidth and beam steering is achievable by Genetic Algorithm as optimisation techniques.

1. Introduction

Over the last decade, there has been great advancement in wireless technology bringing about new and improved services at low cost to consumers. This has led to a massive increase in the number of consumers utilising wireless technologies hence a corresponding growth in the wireless communications industry. The challenge of coping with this growth resulted in the need to increase and improve capacity and efficiency as well as reduce interference.

Since antennas play a fundamentally critical role in wireless communications as there can be no wireless communication without antennas, there has been lots of research into efficient methods of designing antennas to meet these needs. Antennas having basic characteristics of high directivity and beam steering in real time are known to be capable of meeting these needs as they increase spectral efficiency and gain and mitigate multipath propagation [1-6].

Beam steering entails changing the direction of the mainlobe of the radiation pattern of an antenna in order to achieve spectrum efficiency enhancement and mitigation of multipath propagation [6]. Beams formed by antenna arrays are categorised as either switched array beams or adaptive array beams. While switched array beams are fixed, finite and predetermined, adaptive array beams possess infinite number of patterns and can adapt in real time to RF signal environment hence adaptive beamforming arrays are always preferred. Adaptive arrays have the capability of directing or steering in real time the main beam in a desired direction or towards signal of interest (SOI) and suppressing interference or multipath signals [7],[8],[9].

Optimisation is a process of obtaining the best solution amidst several solutions to the problem to be optimised. Classical optimisation and Heuristic optimisation are conventional techniques used for implementing optimisation. Because the classical optimisation techniques require objective functions that must be differentiable and have the likelihood of being limited to local minima they find limited application in practical implementations hence the heuristic optimisation technique is preferred in optimising antenna related problems. There are several heuristic optimisation algorithms in use today for implementing optimisation, they include differential evolution, simulated annealing, genetic algorithm, ant colony optimisation, and particle swarm optimisation.

2. Beam Steering Techniques

Beam steering is a popular technique utilised in changing the direction of the major lobe of the radiation pattern of an antenna so as to mitigate the problem of multipath propagation and to achieve improvement in spectrum efficiency [6]. Prominent techniques of achieving beam steering in array antennas are phase shifting, reactive loading and time modulated switching.

2.1. Phase Shifting

This is a fundamental concept in the steering of the mainlobe of array antennas and relies on a uniform, progressive phase change across the array. The phase distribution on radiating elements of an array is varied by varying the feed to each element such that the varied phase distribution corresponds to a desired scan angle. Varactor diode phase shifters, bit phase shifters, ferrite phase shifters at RF or digital signal processing at baseband are examples of phase shifters. The utilisation of phase shifters in beam steering is not cost effective as it is nearly half the cost of an entire electronically scanned phased array [10]. Also, phase shifters are known to have high insertion loss which results in degradation of the signal to noise ratio (SNR) in systems [11].

2.2. Reactive Loading

This approach of steering the beam of an array antenna involves the use of a single active RF chained active element and multiple reactance circuit terminated parasitic elements [12, 13-15]. Sufficient mutual coupling between the active and parasitic elements is utilised and beam steering is achieved by changing or varying the reactance values. An advantage of this approach is its small size, low power dissipation and low fabrication cost when compared to other techniques.[2, 12,16,17]

2.3. Time Modulated Switching

Time modulated switching was originally proposed in the late 1950's and it utilises time as an additional parameter so as to permit extra degree of freedom in the synthesis of array radiation characteristics. For beam steering, RF switches are used to control the 'on' and 'off' time of an antenna array element thereby making it possible to have at the fundamental frequency a time averaged radiation pattern with low sidelobe levels [18]. Although this technique has a major drawback in that it produces unwanted signals at multiples of the fundamental frequency resulting in energy losses and interference, however, simultaneous scan operation can be achieved using this technique with the unwanted sideband beams used to point at other directions [12, 16, 18-22].

3. OPTIMIZATION TECHNIQUES

Optimization is the process of obtaining the best solution amidst several solutions to a problem. By implication, what optimization does is rather than obtain a single solution to a problem, it finds many solutions and selects the best of these solutions [23-24]. This is achieved by finding the minimum or maximum of an objective or mathematical function to be optimised. In order words, optimization finds the combination of variable which brings about the highest or lowest possible values of an objective or mathematical function. Classical and heuristic optimizations are known optimization techniques. While classical optimization is an analytic method of optimisation which uses differential calculus to locate the optimum solution, heuristic optimization is an advanced numerical optimization method which models behaviour found in nature to obtain the optimum solution [25]. Classical optimization requires a starting point and has the likelihood of being limited to local minima, they find limited practical applications as some of them involve objective functions that are not continuous and or differentiable.

There are a couple of approaches to finding the minimum or maximum of an objective function depending on if the objective function has single or multi variables though some approaches do well for both.

3.1. Exhaustive Search

This approach checks all the possible combinations of the input variables of the objective function in order to find a minimum or maximum. It is only practicable when the number of points involved is very small since it is time consuming.

3.2. Random Search

Instead of checking all possible points, this approach randomly picks points across the interval of interest to find the minimum or maximum.

3.3. Line Search

With this approach, a line vector which originates from an arbitrary point and cuts across the cost surface is chosen. Steps are then taken along this line to reach a maximum or minimum. It is an approach best suited for only quadratic functions because with many minima or maxima, the line vectors could completely miss the area where the global minimum or maximum exists. This approach does not exploit any information about the cost function.

3.4. Simulated Annealing Algorithm

This algorithm is modelled after the annealing process in which a substance is heated above its melting temperature and then allowed to gradually cool so as to form crystalline structure. This approach begins with a single guess at the solution, and then finds the solution by working in a serial manner and as it converges, it gradually becomes less random.

3.5. Genetic Algorithm

Genetic algorithm is a stochastic numerical search algorithm introduced by John Holland in the 1960's and was inspired by the biological processes of natural genetics, selection and evolution [23-24, 26-27]. Based on Darwin's concept of natural evolution, GA begins with a random population of possible solutions called chromosomes spread across the search space and over a series of time steps or generations evolves towards a global optimal point. This process is achieved by subjecting the initial population of possible solutions to a selection process at which point the fitter possible solutions are picked and the weak ones eliminated using a fitness weight as selection criteria. The fit possible solutions then exchange information through recombination in a manner similar to biological organisms undergoing sexual reproduction. This is called crossover. Finally, the characteristics present in population after crossover is mutated. With the initial population gone through the selection, crossover and mutation processes, a new generation is formed and the generation counter is increased by one. The new generation becomes the initial population upon which selection, crossover and mutation is performed and the process continues until a fixed number of generations are attained or a convergence criterion is met [28-29].

The steps involved in the implementation of GA are

a. Creation of initial population: this is the initialization of the population to be used in the algorithm by random guess of possible solutions (chromosomes) to an optimization problem. The chromosomes contain variables constrained between zero and one. Binary number variables could also be used.
b. Fitness evaluation: the fitness of each chromosome in the population is evaluated by passing it through the cost function and a fitness value is assigned. The cost function used is a very important parameter in optimization and since it is used severally in evaluating chromosome fitness, a trade-off between the evaluation time and calculation accuracy will be needed.
c. Natural selection: this is a sorting phase where only fit chromosomes are passed to the next generation. Two common approaches utilised to accomplish natural selection. First approach is the retention of fit chromosomes and elimination of all others while the second approach called thresholding retains chromosomes with fitness below a threshold value.
d. Mate selection: this involves the assignment of highest probability to most fit chromosomes. The roulette wheel and tournament selection are two common means of performing mate selection though the latter is preferred because it is not recomputed after every generation, fitness does not have to be normalised in order to develop probabilities and works well with low mutation rate [24].
e. Offspring generation: portion of each parent is combined to create an offspring in ways similar to reproduction in biological organisms. There are several methods of generation offsprings [30-31].
f. Mutation: this process involves the introduction of random variations into the population and could take different forms depending on whether the variables involved are binary or continuous. While binary mutation converts a one to a zero or zero to one, a continuous variable mutation may completely replace selected mutated value with another random value.
g. Termination: steps one through six is repeated on a generational basis until termination criteria is achieved. Common conditions under which the algorithm is terminated include

- Defined number of iterations is attained.
- Set time is reached

- Cost function evaluation is reached.
- Value of the best solution has not changed after certain number of iterations.

These processes bring about new generation of chromosomes which differ from preceding generations and the average fitness increases for each generation with the retention of only best chromosomes from the preceding generation.

It is pertinent to note that genetic algorithm being a population based random search algorithm, utilise random transition rules rather than deterministic ones. It is not limited to local minima and is efficient in large scale optimisation problems hence its application in resolving and optimising antenna related applications [2, 6, 25, 32-33].

4. Overview of Numerical Electromagnetics Code (NEC)

Numerical Electromagnetics Code (NEC) is a computer code developed for analysing electromagnetic response of antennas and other metal structures. It is an advancement to the Antenna Modelling Program (AMP) developed in the 1970's and is based on using the Moment of Method to solve integral equations for currents induced on antenna structures by fields incident on them or from sources. It was sponsored by the Naval Ocean Systems Centre and the Air Force Weapons laboratory and developed at Lawrence Livermore National Laboratory in California [34-35]. NEC-2 is the highest version available for free in the public domain.

The approach utilised in NEC is known to have no theoretical limit. It is easy to use with scripting programs and its full source code can be modified without restrictions. NEC is generally applicable to thin wires which is modelled as short straight segments and closed conducting surface patch antenna structure modelled as flat surface patches. The major idea behind the use of Method of Moment is the breakdown of the problem into smaller independently bounded linear elements which can be solved utilising many functions called basis functions. A right weighting function is then applied to each of the resulting functions which are then integrated. Finally, the linear equation is solved using simple matrix form [36].

5. Vertically Polarised Antenna Model

5.1. Why Circular Array?

Single element antennas are known to have wide radiation pattern and low directivity making its use in the achievement of improved capacity, spectral efficiency and interference mitigation goals impracticable. High directivity with single element antennas can only be achieved by increasing its electrical size by means of enlarging the physical dimension of the single element or bringing together more than one single element whose physical dimensions has not been altered.

The combination of more than one single element to form a new antenna is termed an array. The type and number of elements in the array, their geometry, and the manner in which the elements are excited are parameters which determine how directive array antennas can be and there has been proposals of techniques in which these parameters can be optimised to achieve highly directional antennas with real time beam steering capabilities.

Linear, circular and planar arrays are array geometries which have been implemented to achieve adaptive beamforming [2, 12, 16, 19]. Although linear arrays are the simplest array geometry for which array processing can be easily applied, it is unable to perform scan in three dimensions (3-D) due to the presence of edge elements in its geometry. Also, as linear arrays are steered away from boresight, the beams formed are known to become significantly wide [7, 37-39]. An interim solution to achieving 3-D steering is to have several linear arrays arranged in a triangular or rectangular shape forming a planar array but this increases processing intensity and cost hence the use of linear arrays in adaptive beamforming is not attractive [40]. Uniform rectangular arrays having non-omnidirectional elements are also unable to achieve full 3-D steering and this makes circular arrays an array of choice for optimum adaptive beamforming. The beam of uniform circular arrays can be steered azimuthally since it has no edge elements and depending on the element's radiation pattern, they can provide a certain level of source-elevation information. Another advantage of circular array is its ability to overcome the effect of mutual coupling and prevent the occurrence of directional beams that have constant shape over broad bandwidths [37, 40].

5.2. Antenna Structure and Analysis

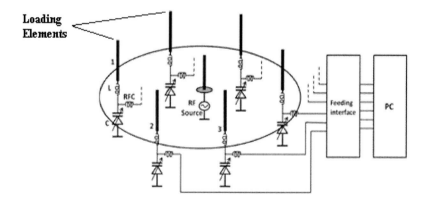

Figure 1. Reactive loading.

Figure 1 shows vertically polarised antenna geometry with capacitive reactive elements introduced at the feed point of each of the six parasitic loaded elements to implement reactive loading. Figure 2 shows a similar geometry but with simple time switches used to implement time modulated switching. The field pattern of both geometries can be given by:

$$E = E_m + \sum_{i=1}^{N}(E_i) \tag{1}$$

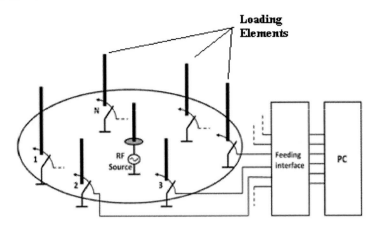

Figure 2. Time modulated switching.

For vertical polarization case, then $E_m = 0$. Taking the central element as the reference element and assuming all antenna elements are a quarter-wavelength in length, then equation 1 can be approximated by the following form:

$$E = I_m(\hat{d}_m \bullet \hat{u}_\theta \hat{a}_\theta + \hat{d}_m \bullet \hat{u}_\phi \hat{a}_\phi) + \sum_{i=1}^{N}\left\{I_i\begin{bmatrix}\hat{d}_{ei} \bullet \hat{u}_\theta \hat{a}_\theta + \\ \hat{d}_{ei} \bullet \hat{u}_\phi \hat{a}_\phi\end{bmatrix}e^{-jk\overline{p}_i \bullet \hat{r}}\right\} \qquad (2)$$

Where d_m and d_e are the unit vectors that describe the orientations of the main and outer elements. These can be simply defined by:

$$\hat{d} = \sin(\theta_d)\cos(\phi_d)\hat{u}_x + \sin(\theta_d)\sin(\phi_d)\hat{u}_y + \cos(\phi_d)\hat{u}_z \qquad (3)$$

θ_d and ϕ_d are the zenith and azimuth angles.

I_m and I_i are the current maxima of the main elements and the ith outer element, respectively.

The ith $\hat{r}$, $\hat{u}_\theta$ and $\hat{u}_\phi$; position vector of the ith element of the ith radiator is denoted by p_i. Subject to the ring array of radius r_a shown in Figure 1 and 2, this vector can be expressed by:

$$\overline{p}_i = r_a \cos(\phi_i)\hat{a}_x + r_a \cos(\phi_i)\hat{a}_y \qquad (4)$$

The unit vectors $\hat{r}$, $\hat{u}_\theta$ and $\hat{u}_\phi$ are given by the following:

$$\hat{r} = \sin(\theta)\cos(\phi)\hat{a}_x + \sin(\theta)\sin(\phi)\hat{a}_y + \cos(\theta)\hat{a}_z \qquad (5)$$

$$\hat{u}_\theta = \cos(\theta)\cos(\phi)\hat{a}_x + \cos(\theta)\sin(\phi)\hat{a}_y + \sin(\theta)\hat{a}_z \qquad (6)$$

$$\hat{u}_\phi = -\sin(\phi)\hat{a}_x + \cos(\phi)\hat{a}_y \qquad (7)$$

Using time switching process on these elements, the fields may still be expressed by equation 2. It should be noted that,

$$\frac{|E_i^{on}(\theta)|}{\max|E_i^{on}(\theta)|} \approx \frac{|E_i^{off}(\theta)|}{\max|E_i^{off}(\theta)|} \tag{8}$$

In other words, the normalized fields due to ON and OFF states of the ith vertically polarized element are equivalent. These assumptions based on the separation distance d_s of each element from the ground plane to be limited to $d_s \ll \lambda$.

The induced current on the ith element for ON and OFF states shown in Figure 3 are expressed by A_{on} and A_{off} respectively. The limited time constraint is applied on i^{th} element as $|\tau_{off} - \tau_{on}| < T$.

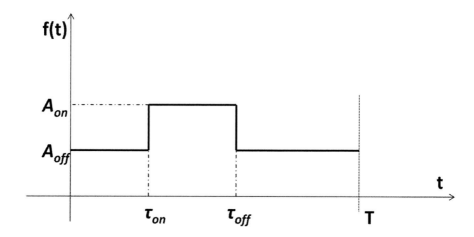

Figure 3. The variation of current over one time cycle.

The Fourier series coefficient for this pulse shape can be given (using [41]) by:

$$f_{mi} = A_{ioff} T \operatorname{sinc}(0.5\pi m T) e^{-j0.5\pi m T} + (A_{ion} - A_{ioff}) \frac{\sin(\pi m(\tau_{ioff} - \tau_{ion}))}{\pi m} e^{-j\pi m(\tau_{ion} + \tau_{ioff})} \tag{9}$$

Using the results in [41], the time switching can be found subject to the weighting induced currents (w_i) assumed over the antenna structure. This simply can be put into similar format as following:

$$\frac{\sin(\pi m(\tau_{ioff} - \tau_{ion}))}{\pi m} e^{-j\pi m(\tau_{ion} + \tau_{ioff})}$$

$$= \frac{w_i - A_{ioff} T \mathrm{sinc}(0.5\pi m T) e^{-j0.5\pi m T}}{(A_{ion} - A_{ioff})} = w_{di} \tag{10}$$

$$= w_{di} = |w_{di}| \, e^{j\gamma_i} \tag{11}$$

Assuming the same procedure used in [41], the switch-on and switch-off of the i_{th} element can be given as follows:

$$\tau_{ioff} = \frac{1}{2\pi m}\left(\frac{\gamma_i}{\pi m} + \frac{1}{\pi m}\sin^{-1}\pi m |w_{di}|\right) \tag{12}$$

$$\tau_{ion} = \frac{1}{2\pi m}\left(\frac{\gamma_i}{\pi m} - \frac{1}{\pi m}\sin^{-1}\pi m |w_{di}|\right) \tag{13}$$

The induced currents A_{ion} and A_{ioff} are known for all cases that can be considered within the antenna operation. It should also be noted for a given baseband bandwidth B_a the time period T might be predicted by $T > 1/B_a$. Thus, the i_{th} element times i.e., switch-on and switch-off can be easily determined.

Locus variation of the weighting coefficients: The boundary variations of the weighted coefficients can be estimated subject to Eqns. 11 and 12. This can be explained in following example when m equals 1. It can be noted that the coefficient w_{di} can be given by the following inequality:

$$\left|\frac{w_i - A_{off}\sin(0.5\pi)e^{-j0.5\pi}}{(A_{ion} - A_{ioff})}\right| \leq \frac{1}{\pi} \tag{14}$$

Equation 14 can be reduced further to the following:

$$\pi\left|\frac{w_i - jA_{off}}{A_{ion} - A_{off}}\right| \leq 1 \tag{15}$$

Substitute $w_i = x + jy$ into equation 15,

$$\left(x - \frac{\pi}{2}a_{if}\right)^2 + \left(y + \frac{\pi}{2}a_{rf}\right)^2 \leq \frac{1}{\pi^2}|A_{ion} - A_{ioff}|^2 \tag{16}$$

Where a_{rf} and a_{if} are the real and imaginary parts of the complex current at off state. Thus the locus variation can be simplified using polar co-ordinates as follows:

$$x = a + r_d \cos\phi \tag{17}$$

$$y = b + r_d \sin \phi \qquad (18)$$

Where

$$r_d = \frac{1}{\pi}|A_i^{on} - A_i^{off}|, \quad a = \frac{2}{\pi}a_{if}, \quad b = -\frac{2}{\pi}a_{rf}$$

5.3. GA Driver Implementation for Antenna Model

The GA driver was designed to maximize the cost function in the desired direction. Equation 19 is the cost function optimised by GA.

$$F = \frac{1}{1+\sqrt{w_1 S_1 + w_2 S_2}} \qquad (19)$$

$$S_1 = \left(\frac{R_{in}-50}{50}\right)^2 + \left(\frac{X_{in}-0}{50}\right)^2 \qquad (20)$$

$$S_2 = \sum_{i=1}^{K}\left(\frac{|E|-E_{min}}{E_{min}}\right)^2 \qquad (21)$$

$$= \sum_{i=K}^{L}\left(\frac{|E|-E_{ref}}{E_{ref}}\right)^2$$

$$= \sum_{i=L}^{360}\left(\frac{|E|-E_{min}}{E_{min}}\right)^2$$

E_{min} is the minimum value of the antenna gain located for the angles between 1 to K (i.e. 360° is equivalent to 360 points).

E_{ref} is the gain set to the steered angles

K is the number of angle to be minimised including a defined angle ranged from L to 360.

Points K to L is the angle that should maximise the gain to the reference value.

w_1 and w_2 are the weighting factor defined by the user. In this case, these values were set to 0.5.

R_{in} and x_{in} are the input resistance and reactance respectively.

Computer programs were written to optimise the cost function using FORTRAN to implement each algorithm. The program initializes the search space with chromosomes and particles by randomization method. Selection phase of the GA implementation uses the tournament selection method and this is realised in the program by shuffling technique. Binary coding is implemented as a subroutine for the uniform and non-uniform processes of single point crossovers. The GA parameters are imputed via an associated file. The output from the respective algorithm driver is then fed into the NEC source code.

GA parameters
Population size = 4
Number of parameters = 6
Probability of mutation = 0.02
Probability of crossovers = 0.5
Maximum generations = 450
Number of possibilities = 8192
Minima = -1000
Maxima = 1000

5.4. Results and Discussion

5.4.1. Reactive Loading Using GA

Using equations 19, GA maximizes the value of the fitness function obtained by the combination of all parameters of the antenna. Figures 4 and 5 show the best fitness and average fitness convergence curves of GA for steered angles with specified side lobe level of -10dB, -20dB and -30dB.

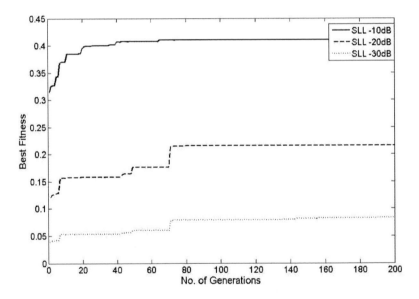

Figure 4. GA convergence curve for best fitness.

5.4.1.1. Speed and Level of Convergence of GA

With parameters of the genetic algorithm as specified in chapter three and setting the level of the main lobe to 0dB, the fitness function of GA for the design converged at value of about 0.42 for a predefined maximum SLL of -10dB. With the main lobe level kept constant and the maximum predefined SLL altered to -20dB and then -30dB, GA converged at a value of 0.21 and 0.07 respectively as depicted in the best fitness convergence curves of figure 4.1. Since optimization entails finding the maximum or minimum of a mathematical function i.e. fitness function, the best fitness convergence curve also shows that the fitness function convergence occurred after the 60th generation for SLL of -10dB and 70th generation for SLL of -20dB and -30dB.

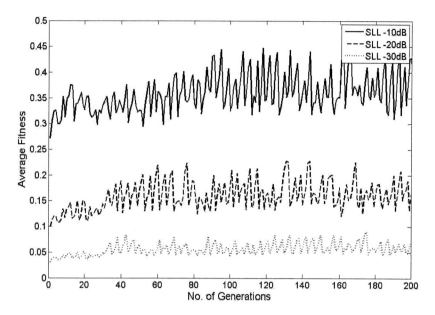

Figure 5. GA convergence curve for average fitness.

5.4.1.2. Stability Around Convergence Value pf GA

Figure 5 shows the results of the average fitness convergence curves for GA depicting the stability of the algorithm before and after convergence. Focussing on the stability after convergence has occurred, the figure shows that as the maximum SLL is increased the step size of the oscillation around the convergence value decreases hence the maximum SLL of -30dB has the highest stability about its convergence value.

Simulation results of figures 4 and 5 shows that 200 generations is more than sufficient for the algorithm to converge with acceptable optimised values obtained for each element of the antenna array at respective side lobe level.

5.4.1.3. Beam Steering Capability

The optimised values of the reactive loads of each element of the array antenna obtained from the algorithm is fed into the NEC2 code. Figure 6 shows the radiation pattern obtained for steered angles 70°, 128°, 230° and 350°.

The simulation results of the radiation pattern in figure 6 above justify the ability of GA to be used to optimize the reactive loads of a uniform circular array antenna to realize beam steering. The result shows the radiated power is concentrated in a particular direction which is then steered in the desired direction i.e. the radiation pattern response is directional having a peak at the steered angle. It is observed that the beamwidth and side lobes remain significantly the same at various steered angles. With the main beam adequately steered in the desired direction and very small side lobes as shown in the result, the array antenna is highly directional and hence capable of object tracking, attenuating multipath and interference rejection.

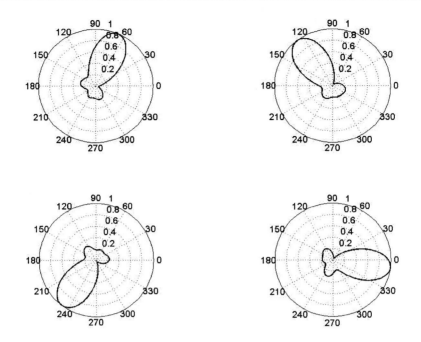

Figure 6. Radiation pattern for beam steering at angle 70°, 128°, 230° and 350° using GA.

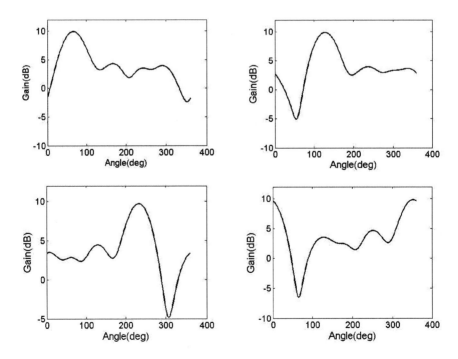

Figure 7. Gain of steered angle 70°, 128°, 230° and 350° with SLL of -10dB using GA.

Figure 7 above is the plot showing the result of the antenna gain at respective steered angle with SLL of -10dB. The gain of 9.92dB was obtained at the steered angle of 70°. This gain is also the maximum front gain of the antenna while the maximum back gain obtained was 4.29dB at an angle of 166° bringing about a front-to-back ratio of 5.63dB. The HPBW obtained was 65° and this was measured at -3dB points to the left and right of the point of maximum gain which corresponds to points at angles 35° and 100° on the plot. At steered angle 128°, a gain of 9.85dB (also maximum front gain) was obtained with back gain of 3.92dB at 128° hence front-to-back ratio of 5.93dB. The HPBW was 67° measured at angles 95° and 162°. For steered angle 230°, the front gain, back gain, front-to-back ratio, HPBW obtained was 9.79dB, 4.38dB, 5.41dB and 66° respectively. The back gain was obtained at angle 129° and HPBW measured at angles 198° and 264°. For steered angle 350°, the front gain, back gain, front-to-back ratio, HPBW obtained was 9.84dB, 4.70dB, 5.14dB and 65° respectively. The back gain was obtained at angle 250° and HPBW measured at angle 318° and 23°.

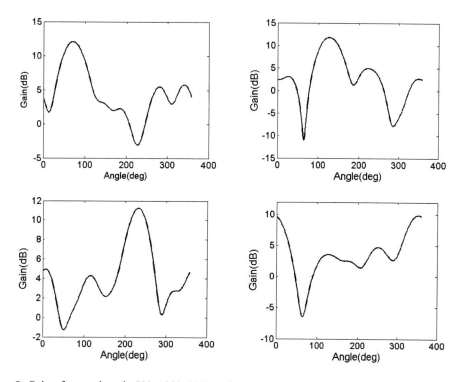

Figure 8. Gain of steered angle 70°, 128°, 230° and 350° with SLL of -20dB using GA.

Figure 8 above is the plot showing the result of the antenna gain at respective steered angle with SLL of -20dB. The gain of 12.04dB was obtained at the steered angle of 70°. This gain is also the maximum front gain of the antenna while the maximum back gain obtained was 5.74dB at an angle of 340° bringing about a front-to-back ratio of 6.30dB. The HPBW obtained was 59° and this was measured at -3dB point to the left and right of the point of maximum gain. These points correspond to angles 42° and 101° shown on the plot. At steered angle 128°, a gain of 11.76dB (also maximum front gain) was obtained with back gain of 4.95dB at 222° hence front-to-back ratio of 6.81dB. The HPBW was 58° measured at angles 99° and 157°. For steered angle 230°, the front gain, back gain, front-to-back ratio, HPBW obtained was 11.22dB, 4.95dB, 6.27dB and 59° respectively. The back gain was obtained at angle 7° and HPBW measured at angles 201° and 260°. For steered angle 350°, the front gain, back gain, front-to-back ratio, HPBW obtained was 11.20dB, 7.20dB, 4.00dB and 57°

respectively. The back gain was obtained at angle 128° and HPBW measured at angles 322° and 19°.

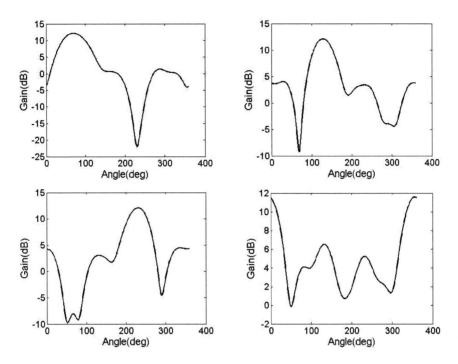

Figure 9. Gain of steered angle 70°, 128°, 230° and 350° with SLL of -30dB using GA.

Figure 9 above is the plot showing the result of the antenna gain at respective steered angle with SLL of -30dB. The gain of 12.16dB was obtained at the steered angle of 70°. This gain is also the maximum front gain of the antenna while the maximum back gain obtained was 1.39dB at an angle of 288° bringing about a front-to-back ratio of 10.77dB. The HPBW obtained was 66° and this was measured at -3dB point to the left and right of the point of maximum gain. These points correspond to angles 39° and 105° shown on the plot. At steered angle 128°, a gain of 12.12dB (also maximum front gain) was obtained with back gain of 4.08dB at 28° hence front-to-back ratio of 8.04dB. The HPBW was 56° measured at angles 101° and 157°. For steered angle 230°, the front gain, back gain, front-to-back ratio, HPBW obtained was 12.09dB, 4.46dB, 7.63dB and 57° respectively. The back gain was obtained at angle 334° and HPBW measured at angles 201° and 258°. For steered angle 350°, the front gain, back gain, front-to-back ratio, HPBW obtained was 11.57dB, 6.49dB, 5.08dB and 55° respectively. The back gain was obtained at angle 131° and HPBW measured at angles 327° and 22°.

Figure 10 is the graphical representation of the antenna's optimal VSWR at respective steered angle. At the operating frequency of 2.45GHz, the VSWR obtained was 1.98, 1.66, 2.22 and 1.78 for steered angles 70,128, 230 and 350 degrees respectively. Using the good rule of thumb, VSWR values of 1.98, 1.66, 2.22 and 1.78 indicate power transfer achieved is between 10dB and 14dB for all steered angles which is a very good range. It is obvious from this figure that the antenna can operate within a bandwidth of 100MHz having a maximum VSWR ≤ 3.

Application of Genetic Algorithm Optimisation Technique ... 77

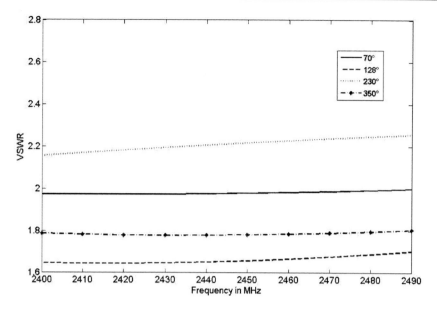

Figure 10. VSWR vs Frequency for beam steering angles 70°, 128°, 230° and 350° using GA.

5.4.2. Time Modulated Switching Using GA

Fig. 11 show the radiation pattern obtained for respective steered angles and their time switching sequence is presented in figure 12.

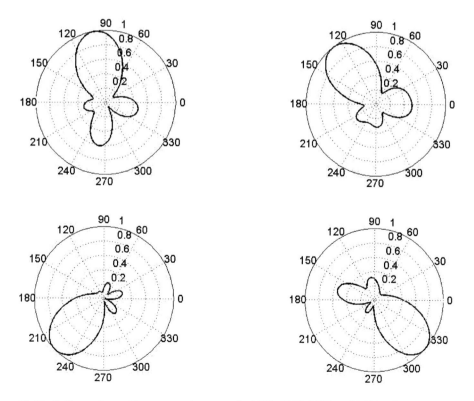

Figure 11. Radiation pattern of beam steering at angle 100°, 130°, 220° and 320° using GA.

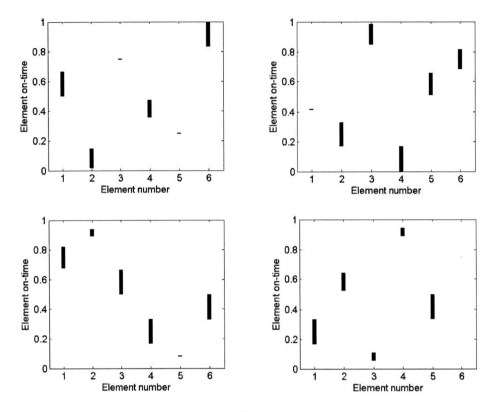

Figure 12. Normalised time – sequence corresponding to the antenna ring array steering angles in fig 11.

The results shown in figures 12 above depicts that with sequence of switching time for each steered angle, the antenna has a directional response with peak value at the desired direction. The sequence of on-time plot for each element show that no two elements have the same duration of on-time although more than one element can be off to steer the main beam in the desired direction as shown in the time sequence plot for steered angle 230 degrees. It is also observed from figure 12 that although the width of the main beam does not vary significantly with different steered angle, the amplitudes of the side lobes vary significantly.

5.5. Model Performance

The results obtained showed GA converged to an optimal value within the specified number of generations. Also, the converged optimal values of reactances for each element of the array antenna resulted in a VSWR ≤ 3 using GA. These results show that GA as an optimization algorithm is very efficient in determining the optimal reactances and on-time sequence to achieve beam steering was a novel idea.

The results of the radiation field patterns highlighted in section four justify the use of circular array antenna in the execution of this project. Not only was beam steering achieved over 360 degrees coverage in the azimuth unlike linear array antennas whose steering angle is limited to 180 degrees, the beams were highly directional and gain was optimum in the desired direction.

6. HORIZONTALLY POLARIZED ANTENNA MODEL

6.1. Design Motivation

The concept or understanding of antenna polarization is extremely important in their deployment in systems. Antenna polarization could fundamentally be either linear or circular. While linear polarization could be in the form of vertical or horizontal, circular polarization is either clockwise (right hand circular) or anticlockwise (left hand circular).

Although various antenna polarisation are utilised in different applications and systems, it is a common practice to find vertical polarization in applications where signal radiation is required over short to medium range. Horizontal polarization on the other hand is used over long distance and circular polarisation finds application mainly in satellite communications.

Since horizontally polarised antennas find application in critical systems such as unmanned aerial vehicles among others, the study of how GA can be utilised in steering the beam of a horizontally polarised array antenna system is very pertinent.

6.2. Antenna Design Structure and Analysis

In realizing the objectives of this section with respect to beam steering, a horizontally polarized circular array antenna operating at 2.4 GHz was used. Using the Ansoft HFSS program, a cylindrical Alford slot antenna operating at 2.4 GHz was modelled and its electromagnetic field pattern was analysed and exported to an optimisation process. This data was used to create the other elements of a six element array antenna, equally spaced over the circumference of a ring whose radius is a quarter of the wavelength ($r_a = \lambda/4$) as shown in Figures 13, 14 and 15 respectively.

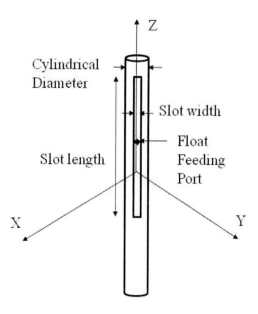

Figure 13. Basic antenna geometry, 3D of alford slot cylindrical antenna.

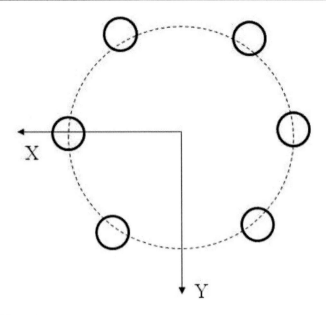

Figure 14. 2D of the antenna array.

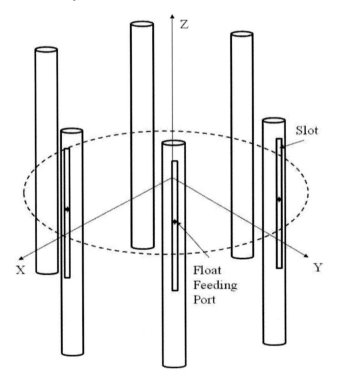

Figure 15. 3D of the antenna array.

The antenna dimensions are presented in Table 1 below. The lumped source dimensions for modelling purpose are length = 4 mm, width = 3 mm, located at the centre of the slot.

Table 1. Antenna dimensions

Antenna Dimensions	
Cylinder height	400 mm
Internal cylindrical / hollow radius	8.5 mm
External cylinder radius	11.5 mm
Slot length	200 mm
Slot width	3 mm

The radiation pattern was synthesized and the parameters which determine the beam steering of the array were then optimized by implementing a genetic algorithm (GA) [42]. Optimization is achieved by utilising the approach detailed below.

The overall antenna field pattern can be expressed by the relation,

$$E = \sum_{i=1}^{N} (E_i) \tag{22}$$

Since horizontal polarization is being considered and working under the assumption that the six elements are arranged in a circular array, equation 23 can be simplified to become:

$$E = \sum_{i=1}^{N} \left\{ \left[E_{\theta i} \hat{a}_\theta + E_{\phi i} \hat{a}_\phi \right] e^{-jk\overline{P}_i \bullet \hat{r}} \right\} \tag{23}$$

where $E_{\theta i}$ and $E_{\phi i}$ are the ith electric fields in a_θ and a_ϕ directions respectively of the ith element.

P_i and r can be given by the following:

$$\overline{P}_i = r_a \cos\phi_i \hat{a}_x + r_a \sin\phi_i \hat{a}_y \tag{24}$$

$$\hat{r} = \sin\theta\cos\varphi\hat{a}_x + \sin\theta\sin\varphi\hat{a}_y + \cos\theta\hat{a}_z \tag{25}$$

By adopting the same procedure used in [41], the switch-on and switch-off of the ith element can be given as follows

$$\tau_{ion} = \frac{1}{2\pi m}\left(\frac{\gamma_i}{\pi m} - \frac{1}{\pi m}\sin^{-1}\pi m |w_{di}|\right) \tag{26}$$

$$\tau_{ioff} = \frac{1}{2\pi m}\left(\frac{\gamma_i}{\pi m} + \frac{1}{\pi m}\sin^{-1}\pi m |w_{di}|\right) \tag{27}$$

where w_{di} ($=|w_{di}|\arg(\gamma_i)$) is the ith weighting coefficient for each element, which is related to the switching times of equations 22 and 23 as follows:

$$w_{di} = \frac{\sin(\pi m(\tau_{ioff} - \tau_{ion}))}{\pi m} e^{-j\pi m(\tau_{ion} + \tau_{ioff})} \qquad (28)$$

Taking the above weighting elements, equation 22 can now be written as:

$$E = \sum_{i=1}^{N} (w_{di} E_i) \qquad (29)$$

For beam steering to be achieved, the fitness function has to have a maximum value at the angle of intended steering. Achieving this maximum is carried out by the GA driver. In order to analyse the radiation pattern across the H Plane covering a range of angles, each shift of 1° is regarded as 1 discrete point. The fitness function applied for this process is defined by the following:

$$CF = \begin{cases} \sum_{i=H}^{Q} \left(\frac{|G| - G_{max}}{G_{max}}\right)^2 & H \leq \phi \leq Q \\ \sum_{i \neq [H \to Q]}^{n} \left(\frac{|G| - G_{min}}{G_{min}}\right)^2 & \phi \neq [H \to Q] \end{cases} \qquad (30)$$

where G is the antenna array gain. G_{max} and G_{min} are the requested gains for the beam steering angles [H→Q] and ($\phi \neq$[H→Q]) respectively.

Before initiating the process of optimization, the target objectives and number of parameters required for the whole process in order to achieve the optimum desired goal must be defined. Since in most cases, it is also required to define a threshold for the GA which enables it to evaluate the designed antenna performance and terminate where necessary, certain constraints inside the cost function to support the data processing at the point of nearly attaining the optimal design requirements was factored in.

For this design, the coupling between the elements was eliminated and the most important antenna parameters which are directly targeted were selected for optimization hence, the field data can be exported for optimisation process in which GA was characterized by the parameters listed below.

Maximum number of chromosomes (binary bits) per individual = 200
Maximum population size = 50
Maximum number of parameters which the chromosomes make up = 12
Maximum number of generations = 200
Likelihood/probability of a uniform crossover = 0.5.

6.3. Results and Discussion

A circular antenna array of 6 elements was considered for the above model using the Alford slot cylindrical antennas operated at 2.4 GHz.

Figure 16 illustrates the simulation and measured return loss. The simulated was generated by using HFSS software package the measured was obtained using a HP8510C ANA. The resultant bandwidth for -10 dB return loss at 2.4 GHz was found to be 123 MHz. The radiation pattern for single element is shown in Figure 17, and Figure 18. It is clearly noticed that the E_ϕ is the dominant field component.

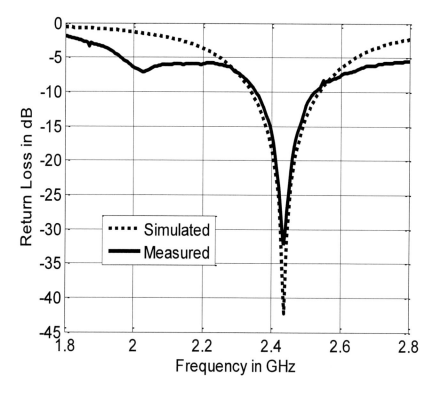

Figure 16. Simulated And Measured Return For Alford Slot Cylindrical Antenna.

Figure 19 shows the radiation pattern obtained for the respective angles 30°, 65°, 180°, and 330°; in which their time switching sequence is presented in Figures 20, 21, 22 and 23 respectively. The results show that with this sequence of switching time for each steered angle, the antenna has a directional response with peak value at the desired direction. The sequence of on-time plot for all elements might be existing for two or more elements. It is also observed from Figure 18 that although the width of the main beam does not vary significantly with different steered angle, the amplitudes of the side lobes vary significantly.

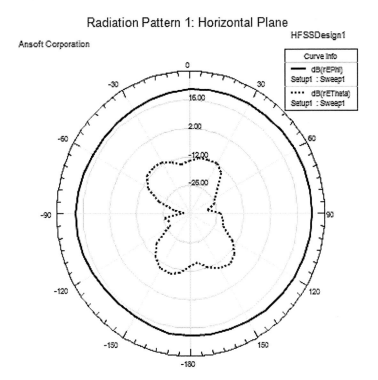

Figure 17. E_θ and E_ϕ distributions at xy plane.

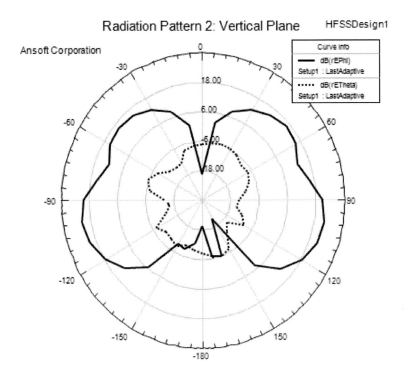

Figure 18. E_θ and E_ϕ distributions at vertical plane.

Application of Genetic Algorithm Optimisation Technique ... 85

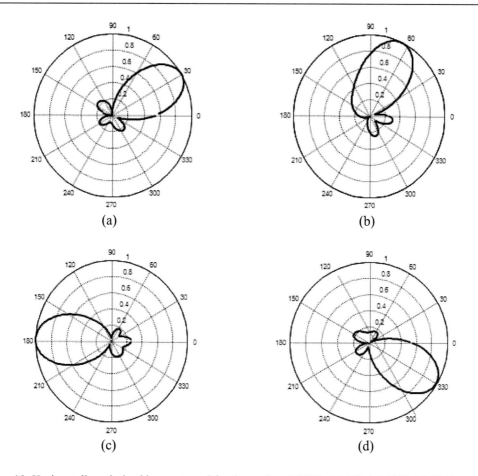

Figure 19. Horizontally polarized beam steered for the angles; (a) 30°, (b) 65°, (c) 180°, (d) 330°.

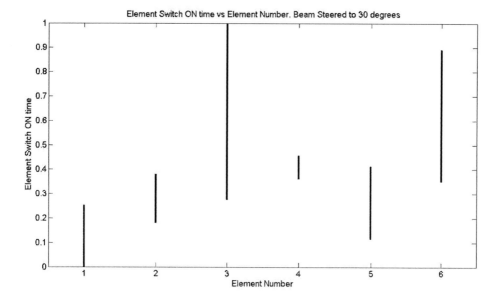

Figure 20. Normalized time sequence corresponding to steering angles; 30°.

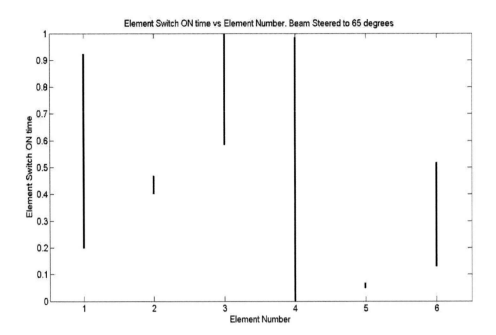

Figure 21. Normalized time sequence corresponding to steering angles; 65°.

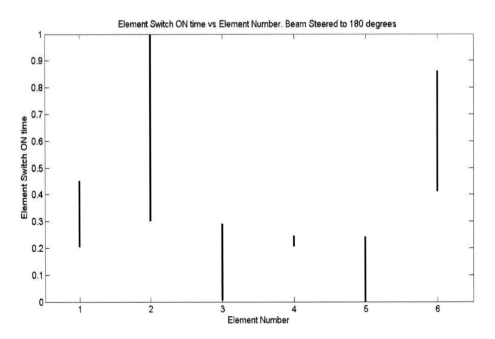

Figure 22. Normalized time sequence corresponding to steering angles; 180°.

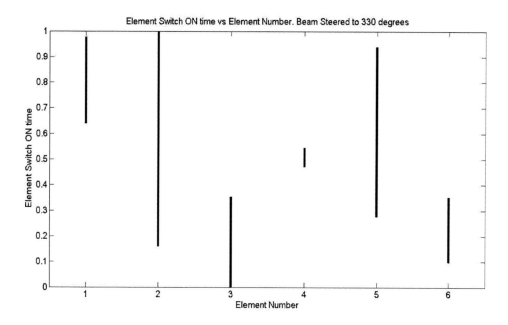

Figure 23. Normalized time sequence corresponding to steering angles; 330°.

7. SUMMARY

The application of genetic algorithm in steering the beam of circular array antenna has been presented in this chapter. The goal of this project was to design, simulate and implement a prototype polarised antenna whose main beam can be steered in a desired direction. After a careful review on various antenna types and their characteristics as well as methods by which beam steering can be achieved, a design procedure using reactive loading and time modulated switching methods of beam steering on a six element circular array antenna at an operating frequency of 2.45GHz was proposed. The use of genetic algorithm as optimization tool for determining optimal input value of reactance and optimal switch on time sequence of each element of the array for these beam steering methods was also proposed.

A reactive loading and time modulated switching techniques were applied to steer the beam of a circular uniformly spaced six element antenna array of vertical polarisation having a source element at its centre. Time modulated switching technique was applied to a second circular array antenna, horizontally polarised and without a centre element. Genetic algorithm (GA) process was used to calculate the optimal values of reactances loading the parasitic elements for which the gain can be optimized in a desired direction. For time switching, GA was also used in determining the optimal on and off times of the elements for which the difference in currents induced optimizes the gain and steered the beam in a desired direction.

Simulations results for vertically polarised antenna showed that near optimal solutions for gain optimization, sidelobe level reduction, VSWR ≤ 3 over a 100MHz bandwidth and beam steering is achievable by Genetic Algorithm as optimisation techniques. On the horizontally polarised antenna, resultant bandwidth of 123MHz at -10dB return loss was obtained.

The implementation of a six element circular array antenna whose beam was steered in the desired direction by utilising the proposed beam steering and optimization method provides a very good way of achieving object tracking, mitigation of multipath and interference effect as well as network capacity improvement. By using a circular array antenna, it was possible to steer the antenna's beam over a 360 degree range in the azimuth. The utilisation of genetic algorithm in determining optimal input values of reactance for the reactive loading method and on time sequence for the time modulated switching method resulted in having an optimal gain range of 9dB and 12dB in the desired steered direction and VSWR ≤ 3 over around 100MHz bandwidth. Hence the implemented antennas could be said to be very efficient.

In terms of convergence rate and the optimal value of convergence of the algorithms, GA proved efficient in realising optimal values within specified convergence parameters resulting in an antenna model having acceptable VWSR value over a frequency range of about 100MHz and 123MHz for vertical and horizontal polarisation respectively.

In general, the radiated power of the array source element and reactively loaded (or time-switched) array elements combined to give a directional beam having optimal gain at the desired steered direction. The reactively loaded array can therefore be said to have achieved beam steering by using different optimal value of reactance on each array element while the time modulated array achieved beam steering by setting the time at which each element of the array is connected. Hence, beam steering was achieved by both methods without using any phase shifter or time delay networks.

From results obtained, it can also be concluded that GA is an efficient optimization technique for realising optimal values and on-time for elements of an array antenna in order to maximize gain and steer the antenna's beam in desired direction. This optimization technique can be used in the searching and finding solutions quickly for complex design problems.

REFERENCES

[1] A. Tennant, "Experimental Two-Element Time-Modulated Direction Finding Array," *Antennas and Propagation, IEEE Transactions on,* vol. 58, pp. 986-988, 2010.

[2] M. M. Abusitta, *et al.*, "New approach for designing beam steering uniform antenna arrays using Genetic Algorithms," in *Antennas & Propagation Conference, 2009. LAPC 2009. Loughborough,* 2009, pp. 617-620.

[3] S. Bellofiore, *et al.*, "Smart antenna system analysis, integration and performance for mobile ad-hoc networks (MANETs)," *Antennas and Propagation, IEEE Transactions on,* vol. 50, pp. 571-581, 2002.

[4] M. Chryssomallis, "Smart antennas," *Antennas and Propagation Magazine, IEEE,* vol. 42, pp. 129-136, 2000.

[5] A. Tennant and B. Chambers, "A Two-Element Time-Modulated Array With Direction-Finding Properties," *Antennas and Wireless Propagation Letters, IEEE,* vol. 6, pp. 64-65, 2007.

[6]   P. K. Varlamos and C. N. Capsalis, "Electronic beam steering using switched parasitic smart antenna arrays - Abstract," *Journal of Electromagnetic Waves and Applications,* vol. 16, pp. 927-928, 2002.

[7]   C. A. Balanis, *Antenna Theory Analysis and Design,* 3rd ed. New Jersey: John Wiley and Sons, 2005.

[8]   K. R. Mahmoud, *et al.*, "Analysis of uniform circular arrays for adaptive beamforming applications using particle swarm optimization algorithm," *International Journal of Rf and Microwave Computer-Aided Engineering,* vol. 18, pp. 42-52, 2008.

[9]   *Smart Antenna Systems.* Available: www.iec.org

[10] Y. Yusuf and G. Xun, "A Low-Cost Patch Antenna Phased Array With Analog Beam Steering Using Mutual Coupling and Reactive Loading," *Antennas and Wireless Propagation Letters, IEEE,* vol. 7, pp. 81-84, 2008.

[11] K. No-Weon, *et al.*, "Feasibility study on beam-forming technique with 1-D mechanical beam steering antenna using niching genetic algorithm," *Microwave and Wireless Components Letters, IEEE,* vol. 12, pp. 494-496, 2002.

[12] S. Chen, *et al.*, "Fast beamforming of electronically steerable parasitic array radiator antennas: theory and experiment," *Antennas and Propagation, IEEE Transactions on,* vol. 52, pp. 1819-1832, 2004.

[13] R. Schlub, *et al.*, "Seven-element ground skirt monopole ESPAR antenna design from a genetic algorithm and the finite element method," *Antennas and Propagation, IEEE Transactions on,* vol. 51, pp. 3033-3039, 2003.

[14] J. Lu, *et al.*, "Dielectric Embedded ESPAR (DE-ESPAR) Antenna Array for Wireless Communications," *Antennas and Propagation, IEEE Transactions on,* vol. 53, pp. 2437-2443, 2005.

[15] C. Plapous, *et al.*, "Reactance domain MUSIC algorithm for electronically steerable parasitic array radiator," *Antennas and Propagation, IEEE Transactions on,* vol. 52, pp. 3257-3264, 2004.

[16] S. Sugiura and H. Iizuka, "Reactively Steered Ring Antenna Array for Automotive Application," *Antennas and Propagation, IEEE Transactions on,* vol. 55, pp. 1902-1908, 2007.

[17] S. Sugiura and H. Iizuka, "Study of reactively steered ring antenna array suitable for automobiles," in *Antennas and Propagation Society International Symposium 2006, IEEE,* 2006, pp. 2365-2368.

[18] T. Yizhen and T. Alan, "Beam steering and adaptive nulling of low sidelobe level time-modulated linear array," *Electronic Letters,* 2010.

[19] J. Fondevila, *et al.*, "Optimizing uniformly excited linear arrays through time modulation," *Antennas and Wireless Propagation Letters, IEEE,* vol. 3, pp. 298-301, 2004.

[20] J. C. Bregains, *et al.*, "Signal Radiation and Power Losses of Time-Modulated Arrays," *Antennas and Propagation, IEEE Transactions on,* vol. 56, pp. 1799-1804, 2008.

[21] Y. Shiwen, *et al.*, "A new technique for power-pattern synthesis in time-modulated linear arrays," *Antennas and Wireless Propagation Letters, IEEE,* vol. 2, pp. 285-287, 2003.

[22] T. Yizhen, *et al.*, "Beam steering techniques for time-switched arrays," in *Antennas & Propagation Conference, 2009. LAPC 2009. Loughborough*, 2009, pp. 233-236.
[23] E. M. Y. Rahmat-Samii, *Electromagnetic optimization by Genetic Algorithms*. Canada: John Wiley & Sons, 1999.
[24] R. L. Haupt and D. H. Werner, *Genetic Algorithms in Electromagnetics*. Hoboken, New Jersey: John Wiley & Sons, Inc, 2007.
[25] M. Gilli and P. Winker, "A Review of Heuristic Optimization Methods in Econometrics," *Swiss Finance Institute Research Paper No. 08-12,* 2008.
[26] J. H. Holland, *Adaptation in Natural and Artificial Systems*. Ann Arbor: Univ. Michigan Press, 1975.
[27] D. A. Coley, *An introduction to Genetic Algorithms for scientists and engineers*. Singapore: World Scientific, 1999.
[28] A. I. Ferrero, Ed., *Antennas Parameters, Models and Applications*. New York: Nova Science Publishers Inc, 2009, p.^pp. Pages.
[29] D. E. Goldberg, *Genetic Algorithms in search, optimization and machine learning*. Canada: Addison-Wesley Publishing Company, 1989.
[30] R. L. Haupt and S. E. Haupt, *Practical Genetic Algorithms* 2nd ed. New York: Wiley, 2004.
[31] A. Wright, *Genetic algorithms for real parameter optimization, in Foundations of Genetic Algorithms*, G. J. E. Rawlins ed. San Mateo, CA: organ Kaufmann, 1991.
[32] N. Karaboga, *et al.*, "Null steering of linear antenna Arrays with use of modified touring ant colony optimization algorithm," *International Journal of Rf and Microwave Computer-Aided Engineering,* vol. 12, pp. 375-383, 2002.
[33] D. Gies and Y. Rahmat-Samii, "Reconfigurable array design using parallel particle swarm optimization," in *Antennas and Propagation Society International Symposium, 2003. IEEE*, 2003, pp. 177-180 vol.1.
[34] G. T. Burge and A. J. Poggio, "*Numerical electromagnetic code (NEC): method of moments,"* U. N. O. S. Center, Ed., ed. San Diego, 1981, pp. 1-37.
[35] B. J. Strait, *Applications of method of moments to electromagnetic fields*: SCEEE Press, 1980.
[36] M. A. Mangoud, *et al.*, "Simulation of human interaction with mobile telephones using hybrid techniques over coupled domains," *Microwave Theory and Techniques, IEEE Transactions on,* vol. 48, pp. 2014-2021, 2000.
[37] P. Ioannides and C. A. Balanis, "Uniform circular arrays for smart antennas," *Antennas and Propagation Magazine, IEEE,* vol. 47, pp. 192-206, 2005.
[38] P. J. Bevelacqua and C. A. Balanis, "Minimum Sidelobe Levels for Linear Arrays," *Antennas and Propagation, IEEE Transactions on,* vol. 55, pp. 3442-3449, 2007.
[39] R. H. Morelos-Zaragoza and M. Ghavami, "Combined Beamforming and space-time block coding for high speed wireless indoor communications," presented at the *4th Int'l symposium on wireless personal multimedia communications,* Aalborg, Denmark, 2001.
[40]]C. M. Tan, *et al.*, "On the application of circular arrays in direction finding Part II: Experimental evaluation on sage with different circular arrays," in *1st Annual COST 273 workshop*, Finland, 2002, pp. 29-30.

[41] Y. Tong and A. Tenant, "Simultaneous control of sidelobe level and harmonic beam steering in time-modulated linear arrays," *Electronic Letters,* vol. 46, pp. 201-202, 2010.

[42] G. Golino, "A Genetic Algorithm for Optimizing the Segmentation in Sub Arrays of Planar Array Antenna Radars with Adaptive Digital Beam Forming", appears in: Phased Array Systems and Technology, *IEEE International Symposium*, 2003, pp. 211-216.

In: Handbook of Genetic Algorithms: New Research
Editors: A. Ramirez Muñoz and I. Garza Rodriguez
ISBN: 978-1-62081-158-0
© 2012 Nova Science Publishers, Inc.

Chapter 4

EXAMINATION OF TWO DIFFERENT MATHEMATICAL TECHNIQUES FOR DETERMINING THE IMPORTANT FACTORS AFFECTING INDOOR AIR QUALITY

Akhil Kadiyala and Ashok Kumar
Department of Civil Engineering, The University of Toledo,
Toledo, Ohio, US

ABSTRACT

This chapter provides a comprehensive discussion on the performance of regression and regression tree analyses in determining the important factors affecting in-vehicle air quality. Indoor contaminants of particulate matter, carbon dioxide, carbon monoxide, sulfur dioxide, nitric oxide, and nitrogen dioxide were monitored inside two public transport buses. One bus operated on 20% biodiesel and the other operated on ultra low sulfur diesel in the city of Toledo, Ohio. The independent variables considered in this study include meteorological variables (ambient temperature (temp.), ambient relative humidity (RH), wind speed, wind direction, precipitation, visibility, ambient PM2.5), indoor comfort parameters (indoor temp., indoor RH), and real-time on-road variables (passenger count, bus status (bus position/door position - Idle/Open, Idle/Close, Run/Close), number of cars and buses/trucks ahead). Regression analysis was performed using MINITAB® software and regression tree analysis was performed using CART® software. CART proved to have the advantage of providing relative importance of all the independent variables considered. The important factors affecting the monitored vehicular contaminants were found to be different for each month and season on performing the regression and regression tree analyses.

INTRODUCTION

Indoor air quality (IAQ) is one of the major environmental concerns as people spend about 90% of their time indoors and about 7% of their daily time commuting, mostly between their workplace and their residence [Klepeis et al., 2001]. Over the years, vehicular usage has been increasing rapidly. This when combined with the booming population growth, makes air pollution caused due to vehicles a potential health hazard. People are exposed to higher levels of traffic contaminants when they drive in heavy traffic, stand near idling vehicles, spend time at places near roads having high traffic, especially if the location is downwind of the road (OEHHA Report, 2004). The degree of exposure levels to contaminants is more for people commuting in a bus as compared to the levels of exposure occurring at bus stops or during loading and unloading times (CARB Factsheet, 2003). Since the levels of exposure to contaminants are on the higher side in a transport microenvironment and people spend considerable amount of time in travel, it is important to identify all the factors that could influence the vehicular contaminant levels.

Study of the factors affecting in-vehicle contaminants is more complex as compared to the study in buildings due to the fact that the vehicle is always in a mobile condition. To date, numerous vehicular studies have used regression analysis to determine the influential factors affecting IAQ. Regression analysis helped in identifying the effect of the influencing independent variables on the dependent contaminant concentrations. Fitz et al. (2003) studied children's exposure to different contaminants while commuting in a school bus and identified vehicle exhaust and self intrusion as influential factors affecting in-vehicle contaminant levels when the windows were closed, while ventilation settings were found to be playing a major role when the windows were open. Rodes et al. (1998) identified driving lane, roadway type, congestion level, time of the day, and exhaust from lead vehicles as significant factors affecting the vehicular contaminant levels. Fruin (2003) identified road type, following distance between the lead vehicle and follow vehicle, and exhaust location of the lead vehicle as important factors affecting in-vehicle contaminants. Praml and Schierl (2000) identified outdoor concentrations and traffic having an impact on in-vehicle concentrations. The effect of meteorology was studied by Adams et al. (2001) and the study identified wind speed as the major influential factor. Sabin et al. (2005) identified lead vehicle and type of test bus to be the influential factors affecting in-vehicle contaminant levels when the windows were opened and closed respectively. The relationships for indoor air quality were developed by Vijayan and Kumar (2008) for carbon dioxide (CO_2), carbon monoxide (CO), sulfur dioxide (SO_2), and particulate matter (PM) using regression analysis to study the factors influencing the vehicular concentration levels.

Marion and Brent (2003) identified interior temperature, vehicle make, age and deodorizer used playing an important role in influencing the vehicular indoor volatile organic compound (VOC) levels, while road type, driving time, and air conditioning were found to be the influential factors in a study by Chan et al. (1991). Diapouli et al. (2008) observed in-vehicle contaminant levels to be higher in areas of heavy traffic, during the peak hours, and were mainly influenced by the stop and go traffic predominantly found at signals. Chan et al. (2003) studied the exposure of passengers to VOCs in four different public commuting modes – taxi, subway, air conditioned bus and non-air conditioned bus and found passenger exposure to be influenced by commuting mode selected by the passenger. Similar observations were made by Chan et al. (2002a). The study also reported lane of travel, air conditioning system, internal sources, and ventilation settings to impact the in-vehicle levels while driving time was not found to be important. Duci et al. (2003) also identified mode of commuting and route taken to be influencing indoor contaminant levels in vehicles.

Chan and Liu (2001) observed that the vehicular contaminant levels apart from being influenced by heavy traffic and street configuration was increased by 2-3 times in tunnel microenvironment as compared to urban and sub-urban roads. The study also reported vehicle height, size of the vehicle, leakage and intake positions of ventilation systems to affect the contaminant levels while much variation was not observed between levels in air conditioned and non-air conditioned vehicles. Duffy and Nelson (1997) found in-vehicle contaminant levels to be 50% higher inside the bus as compared to new cars and 25% higher than old cars. Gomez-Perales et al. (2004) observed low wind speed contributing to higher contaminant levels inside vehicles. Fernandez-Bremauntz and Ashmore (1995) identified road lane used and vehicle size to mainly affect vehicular contaminant levels. Jo et al. (1996, 1998, 1999, and 2001) observed the contaminant concentration of VOCs to be higher in taxis as compared to buses.

Clifford et al. (1997) observed traffic and wind speed to some extent having an influence on monitored in-vehicle levels while Chan et al. (2002b) observed air conditioning to play a major role in influencing the indoor contaminants. Lau and Chan (2003) identified the mode of transportation to be mainly influencing indoor VOC. Chan et al. (1999) identified vehicle body position, intake point of ventilation, ventilation effect, transportation mode, road type, driving conditions, relative distance from emission source to possibly affect the in-vehicle contaminant levels. Kuo et al. (2000) observed that the VOCs were not correlated to traffic density which was in contradiction to the findings of other studies. Chan (2003) observed that in-vehicle levels were in general affected by ventilation settings and CO2 levels were found to be mainly dependent on the passengers commuting and not the driving environment. Alm et al. (1999) identified higher in-vehicle contaminant levels to be a result of higher vehicular traffic and exposure to contaminants influenced by time of day, average speed, wind speed, and relative humidity.

The Classification and Regression Trees (CART) developed by Breiman et al. (1984) helps in accurately predicting and classifying the data based on a set of if-then logical split conditions developed by the tree building algorithms. Tree methods are more useful when there is very little or no knowledge on any theories or predictions that relate the variables. Since the bus is in motion for most of the time and the factors influencing vehicular indoor contaminant levels keep changing randomly, tree methods are more useful in determining the relationships between vehicular contaminant levels and independent variables that could have gone unnoticed using other analytic techniques like linear models or multiple regression. So far, CART has been used extensively in the fields of marketing, financial services, banking, health care, manufacturing, transportation, environmental, telecommunications, insurance, and education.

In the environmental field, CART has been used for correcting air pollution time series for meteorological variability [Visser and Noordijk (2002)], exploring factors controlling the variability of pesticide concentrations in the willamette river basin using tree-based models [Anderson and Qian (1999)], obtaining atmospheric teleconnection patterns and severity of winters in the laurentian great lakes basin [Assel and Rodionov (2000)], statistical forecasting to describe the surface winds in Sydney harbour [Dunsmuir et al. (2003)], predicting the onset of Australian winter rainfall by nonlinear classification [Firth et al. (2005)], tree-structured modeling of the relationship between great lakes ice cover and atmospheric circulation patterns [Herche et al. (2001)], forecasting daily high ozone concentrations by classification trees [Trivisano et al. (1996)], understanding the relationship between PM and meteorological variables in California [Richard and Jeff (2003)], forecasting ozone levels from meteorological interactions [Burrows et al. (1995), Huang and Smith (1999)] and adjusting ozone trends for meteorological variations [Stoeckenius and Hudischewskyj (1990)]. The United States Environmental Protection Agency (USEPA) had proposed the use of CART in the guideline for developing an ozone forecasting program [USEPA (1999)].

From the literature review, it was observed that there are many variables such as meteorological conditions, traffic intensity, passenger count, ventilation settings, type of lead vehicle, distance from the lead vehicle, position in the bus (front or back), exhaust pipe position of the lead vehicle (up or down), startup conditions (hot start or cold start), etc. that would affect the contaminant concentrations in a moving vehicle. The contaminant level buildup within a vehicle is due to combination of different factors and not a result of variation due to a single variable. However, most of the studies have used only limited number of variables that could possibly influence the in-vehicle contaminant concentrations. The present review found only few studies [Vijayan and Kumar (2008), Kadiyala (2008), Kadiyala and Kumar (2008a), Kadiyala and Kumar (2009), Kadiyala et al. (2010a, 2010b), and Kadiyala and Kumar (2011)] in the literature that have monitored real time on-road variables continuously over a longer period of time. All the studies reported in the literature used regression analysis for identification of the influential variables except for a few studies by Kadiyala (2008), Kadiyala and Kumar (2008a, 2008b, 2009) that have used regression trees. Kadiyala and Kumar (2008b) have demonstrated the advantages of using regression tree analysis over regression analysis with data collected in one month. Since CART helps in predicting the relations between different variables that are also non-parametric, its application to understanding IAQ can be valuable as there are many complexities involved in relating the indoor contaminant levels with different variables. Despite the advantages of applicability of regression tree analysis over regression analysis, none of the vehicular IAQ studies reported in the literature have used regression tree analysis that presents the relationships and patterns in a pictorial form which is easy to understand and remember as compared to a regression equation. This study extends the work done by Kadiyala and Kumar (2008b) to examine two different mathematical techniques: regression and regression trees to identify the important factors affecting multivariate environmental data. Regression analysis was performed using MINITAB® software and regression tree analysis was performed using CART® software.

METHODOLOGY

A 20% biodiesel (BD20) bus and an ultra low sulfur diesel (ULSD) bus were selected from the Toledo Area Regional Transit Authority (TARTA) 500 series fleet which had all the cameras located inside the buses in working condition and run on a single preassigned route daily with a time lag of 10-20 minutes. The fleet selected for the study was the 500 series Thomas built buses (acquired by Detroit Diesel) of the TARTA line up, with a Mercedes Benz MBE 900 engine. The route selected for the study was Route #20 which runs between TARTA garage and Meijer on the Central Avenue strip [TARTA Route, 2011]. The locations of the bus, when on the run, were identified by the GPS unit located inside the bus. Continuous monitoring of the gaseous contaminants of CO_2, CO, nitric oxide (NO), nitrogen dioxide (NO_2), and SO_2 were done simultaneously with two important indoor comfort parameters: indoor temperature (temp.) and indoor relative humidity (RH), in each bus using two YES Plus® instruments [Critical Environment Technologies (2011)]; while PM data were monitored continuously using Grimm 1.108 aerosol spectrometer [Grimm Technologies (2011)]. The instruments drew power continuously from the adapters connected to the bus and a wired mesh box is provided to safeguard the instruments. The instruments are held in position within the wired mesh box using velcro attachments. In order to obtain quality data from the instruments, Yes Plus instruments were calibrated at the end of each week since May 2007 for CO_2, CO, NO, NO_2, and SO_2 sensors using calibration gases supplied by CALGAZ. The Grimm 1.108 instruments were regularly cleaned with canned air and the particulate filters were frequently replaced to prevent clogging of inlets. It was made sure that the Grimm aerosol spectrometers were still under factory calibration and this ensured quality data from it. More details on the experimental setup, test protocol, and instrument maintenance procedures adopted by the researchers have been documented elsewhere [Kadiyala (2008), Kadiyala et al. (2010a), Kadiyala and Kumar (2011)].

Figure 1 provides an overview of the project that helped in determining the factors influencing vehicular contaminant levels on a monthly and seasonal basis. Data collection included downloading the data from instruments, obtaining meteorological data, and monitoring the real-time on-road variables. The data downloaded from all the instruments have been set for 1-minute interval that is averaged to 1-hour for analysis. The variables considered in this study to determine the influential factors affecting vehicular gaseous contaminant levels are ambient temp., ambient RH, wind speed, wind direction, precipitation, visibility, indoor temp., indoor RH, passenger count, bus status (bus position/door position - Idle/Open, Idle/Close, Run/Close), number of cars and buses/trucks ahead; while ambient $PM_{2.5}$ data were considered additionally for PM. Meteorological data were downloaded from the National Climatic Data Center website [NCDC NOAA (2008)] and ambient $PM_{2.5}$ concentrations were obtained from the USEPA on request. Other real-time variables such as passenger count, vehicular traffic in the front, bus operating conditions (run/idle), and door status (open/close) were monitored by analyzing the hard drives taken from the buses at the end of each week that recorded the video from closed circuit cameras. All the collected data were processed to obtain 1-hour averages for analysis. The above procedure helped us in putting together a data base of contaminant concentration and the associated variables as function of time for further analysis. More details on monitoring the variables and database development are discussed elsewhere by Kadiyala (2008), Kadiyala et al. (2010a), and Kadiyala and Kumar (2011).

This chapter considered only the qualitative data collected from both the buses that were representative of the selected route, selected bus, and having all the cameras in proper working condition for analysis. The different seasons in this study are defined as winter (January-March), spring (April-June), summer (July-September) and fall (October-December). The data were divided into these seasons to identify the factors influencing in-vehicle contaminant levels seasonally.

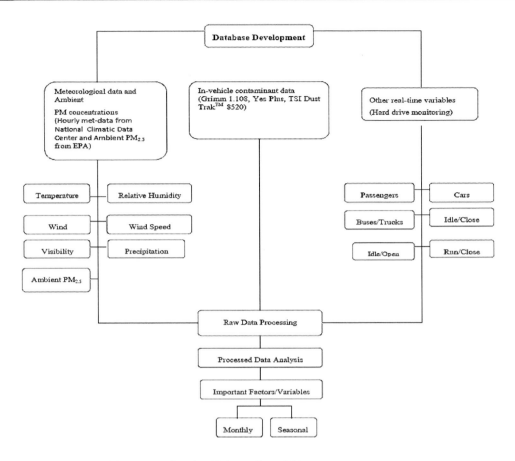

Figure 1. Overview of the Project for Identifying Influential Factors.

RESULTS AND DISCUSSION

Kadiyala and Kumar (2008b) have provided a comprehensive step-by-step procedure to run MINITAB and CART software to determine the factors affecting in-vehicle air quality. This chapter adopts the same methodology and extends the work to determine the performance of two advanced methodologies of regression and regression tree analyses in determining the factors affecting vehicular IAQ on a monthly, seasonal, and yearly basis. Regression and regression tree models were developed for each monitored in-vehicle contaminant in BD20 and ULSD buses for each month, season, and year. Tables 1-4 present the monthly important variables obtained by performing regression and regression tree analyses in BD20 and ULSD buses; while seasonal and yearly results are presented in Tables 5-6 and Tables 7-8 respectively. "Best subset regression" method was used to determine the important factors affecting IAQ on using MINITAB to develop regression tree models. CART identifies the relative importance (0% - 100%) of each variable in the development of corresponding regression tree model. As CART provides the relative importance of all the variables, the important factors affecting IAQ that are summarized in Tables 2, 4, 6, and 8 are those obtained from CART output having a relative importance of greater than 40% (refer to Appendix–A of Kadiyala (2008) for a detailed list of factors and their percentage importance in development of a regression tree model for that month, season or year). To check the working capability of CART software in developing regression trees and identifying important variables for the indoor contaminants, four different runs were performed using each month's data as discussed below:

Run 1: Complete month's contaminant concentration data with missing variables (considering CART can account for missing values). Missing variables are a result of camera error, hard disk problems, and the amount of time required to record the observations on a one-min interval basis.

Run 2: 90% of the data used in Run 1.

Run 3: Only the contaminant concentration data with no missing variables.

Run 4: 90% of the data used in Run 3.

On performing the above four runs, it was observed that the main factors (relative importance > 40%) affecting contaminant concentrations were same. However, the order of relative importance for these variables was found to be changing with each run. The results presented in this paper are those obtained from Run 3.

Table 1. Monthly MINITAB Results in BD20 Bus

Month' Year	In-vehicle Contaminants					
	CO_2	CO	NO	NO_2	SO_2	PM
Apr'07	Idle/Open, Passengers	Indoor RH, Wind speed	Wind speed, Wind direction	NA	NA	Wind direction, Indoor RH, Ambient $PM_{2.5}$
May'07	Passengers, Wind direction, Idle/Close	Indoor RH, Ambient RH, Wind speed	Trucks, Cars, Wind direction	Run/Close	Cars, Ambient temp., Indoor RH	Visibility, Wind direction, Run/Close, Indoor temp.
Jun'07	Passengers, Indoor RH, Ambient temp.	Ambient RH, Wind direction	Indoor RH, Ambient RH	Indoor RH	Wind direction, Ambient RH	Ambient RH, Ambient $PM_{2.5}$, Visibility
Jul'07	Wind direction, Trucks, Ambient RH, Ambient temp.	Indoor RH, Indoor temp., Wind direction	Indoor temp., Ambient temp.	NA	Indoor RH, Indoor temp.	NA
Aug'07	Indoor RH, Trucks	Indoor RH, Run/Close	Wind direction, Idle/Open, Idle/Close	NA	Indoor RH	NA
Sep'07	Passengers	Wind direction, Trucks	Wind direction, Indoor temp.	NA	Ambient RH, Wind direction	Wind direction, Passengers, Wind speed, Cars
Oct'07	Wind direction, Indoor temp, Passengers, Trucks	Wind direction, Indoor temp.	Ambient temp., Indoor temp., Wind direction	Indoor RH	Trucks, Cars, Indoor Temp., Wind speed	Wind direction, Visibility
Nov'07	Wind speed, Passengers, Idle/Close	Wind speed	Indoor temp., Wind speed	Wind speed, Ambient Temp.	Wind speed, Wind direction	Ambient $PM_{2.5}$, Wind speed, Trucks

Table 1. (Continued)

Month'Year	\multicolumn{6}{c}{In-vehicle Contaminants}
	CO$_2$	CO	NO	NO$_2$	SO$_2$	PM
Dec'07	Passengers, Wind direction, Indoor temp., Wind speed	Indoor RH	Wind direction, Indoor temp.	Trucks, Run/Close	Ambient RH	NA
Jan'08	Passengers, Indoor RH	Wind direction, Indoor RH, Indoor temp., Ambient temp.	Indoor RH, Ambient RH	Wind direction	Indoor RH, Ambient temp.	NA
Feb'08	Passengers, Ambient RH, Indoor RH, Wind speed	Indoor RH, Ambient temp.	Trucks, Ambient RH	Passengers, Indoor RH	Indoor temp., Ambient temp., Indoor RH	NA
Mar'08	Wind direction, Passengers, Wind speed, Ambient RH	Indoor temp., Indoor RH	Indoor RH, Indoor Temp.	Indoor temp., Indoor RH, Trucks	Cars, Trucks, Indoor temp.	NA

Table 2. Monthly CART Results in BD20 Bus

Month'Year	\multicolumn{6}{c}{In-vehicle Contaminants}
	CO$_2$	CO	NO	NO$_2$	SO$_2$	PM
Apr'07	Idle/Open, Wind direction, Passengers	Indoor RH, Wind direction, Wind speed	Wind speed, Wind direction, Indoor RH	NA	NA	Wind direction, Indoor RH, Ambient PM$_{2.5}$
May'07	Passengers, Wind direction, Idle/Close, Run/Close	Indoor RH, Wind direction, Ambient RH, Wind speed	Indoor RH, Wind direction, Cars	Run/Close	Wind direction, Ambient temp., Cars, Indoor RH, Ambient RH	Visibility, Wind direction, Run/Close, Indoor temp., Cars, Ambient RH
Jun'07	Passengers, Indoor RH, Ambient RH, Ambient temp.	Ambient RH, Wind direction, Ambient temp.	Indoor RH, Ambient RH	Indoor RH, Wind Speed, Visibility, Wind direction	Wind direction, Ambient RH	Ambient PM$_{2.5}$, Visibility, Wind direction, Ambient RH, Indoor RH, Ambient temp.
Jul'07	Wind direction, Trucks, Indoor RH, Ambient RH, Indoor temp., Ambient temp.	Indoor RH, Indoor temp., Wind direction	Indoor temp., Wind direction, Ambient temp.	NA	Indoor RH, Indoor temp.	NA
Aug'07	Indoor RH	Indoor RH, Indoor temp.	Wind direction, Idle/Open, Idle/Close, Indoor RH, Indoor temp.	NA	Indoor RH	NA
Sep'07	Passengers	Wind direction, Trucks, Indoor temp.	Wind direction, Indoor temp., Passengers	NA	Ambient RH, Wind direction, Ambient temp., Indoor RH, Visibility	Wind direction, Passengers, Wind speed

Month'Year	In-vehicle Contaminants					
	CO$_2$	CO	NO	NO$_2$	SO$_2$	PM
Oct'07	Wind direction, Indoor temp., Wind speed, Passengers, Ambient RH, Trucks	Wind direction, Indoor temp., Passengers	Indoor temp., Wind direction, Ambient temp.	Indoor RH	Trucks, Wind direction, Cars, Indoor temp., Wind speed	Wind direction
Nov'07	Wind speed, Indoor RH Passengers, Idle/Close	Indoor RH, Wind speed	Indoor temp., Wind speed, Ambient temp.	Indoor temp., Wind speed, Ambient temp.	Wind speed, Indoor RH, Wind direction	Ambient PM$_{2.5}$, Wind speed
Dec'07	Passengers, Wind direction, Ambient RH, Indoor temp., Indoor RH, Wind speed	Indoor RH, Wind direction	Indoor temp., Indoor RH, Wind direction	Trucks	Wind direction, Ambient RH	NA
Jan'08	Passengers, Indoor RH	Wind direction, Indoor RH, Indoor temp., Ambient temp.	Indoor RH, Ambient RH	Wind direction	Indoor RH, Ambient temp.	NA
Feb'08	Passengers, Ambient RH, Indoor RH, Wind speed, Ambient temp.	Indoor RH, Ambient temp.	Ambient RH	Passengers, Indoor RH	Ambient temp., Indoor RH, Indoor temp.	NA
Mar'08	Wind direction, Indoor RH, Passengers, Wind speed, Ambient RH	Indoor temp., Indoor RH	Indoor temp., Indoor RH	Indoor temp., Indoor RH	Indoor temp., Cars	NA

Table 3. Monthly MINITAB Results in ULSD bus

Month'Year	In-vehicle Contaminants					
	CO$_2$	CO	NO	NO$_2$	SO$_2$	PM
Jan'07	Passengers	Indoor temp.	Trucks, Run/Close, Indoor temp.	NA	Cars, Indoor temp.	Ambient PM$_{2.5}$, Indoor temp.
Feb'07	Indoor temp., Passengers	Wind direction, Indoor RH	Indoor temp., Wind direction, Ambient temp.	Idle/Open, Cars, Wind direction	Ambient temp., Wind direction	Ambient temp., Ambient PM$_{2.5}$
Mar'07	Wind speed, Indoor RH	Indoor temp., Wind speed, Indoor RH, Idle/Close	Ambient temp., Wind direction, Indoor RH, Trucks	NA	Wind speed, Run/Close, Indoor temp., Wind direction	Indoor RH, Ambient PM$_{2.5}$, Visibility

Table 3. (Continued)

Month' Year	\multicolumn{6}{c}{In-vehicle Contaminants}
	CO$_2$	CO	NO	NO$_2$	SO$_2$	PM
Apr'07	Idle/Open, Trucks, Passengers	Idle/Open, Indoor RH, Ambient temp., Wind direction	NA	Wind direction, Trucks, Ambient temp.	Indoor RH, Ambient RH, Visibility, Idle/Open	NA
May'07	Visibility, Ambient RH	Wind direction, Indoor temp., Wind speed	Indoor RH, Ambient RH, Visibility, Ambient temp.	NA	Ambient RH, Ambient temp.	NA
Jun'07	Ambient RH, Passengers, Wind speed	Wind direction, Ambient RH, Ambient temp., Wind speed	Wind speed, Wind direction, Ambient RH, Ambient temp.	Ambient temp., Wind speed, Ambient RH, Cars	Wind speed, Ambient RH, Ambient temp., Trucks	NA
Jul'07	NA	NA	NA	NA	NA	NA
Aug'07	Passengers	Indoor RH, Wind direction, Run/Close	Cars, Wind direction, Wind speed	NA	Indoor temp., Indoor RH, Ambient RH	NA
Sep'07	Passengers, Indoor RH	Ambient temp., Indoor temp.	Trucks, Wind direction	NA	Indoor RH, Wind direction, Indoor temp.	NA
Oct'07	Passengers, Ambient temp.	Cars, Wind direction	Indoor RH, Ambient RH	NA	Indoor RH, Wind speed	NA
Nov'07	Passengers, Indoor temp., Ambient temp.	Wind speed, Indoor temp., Trucks	Indoor RH, Visibility, Ambient temp.	NA	Cars, Indoor RH, Visibility	Ambient PM$_{2.5}$
Dec'07	Passengers, Indoor temp., Wind direction	Indoor RH, Passengers, Ambient temp.	Trucks, Run/Close, Passengers	Ambient temp., Idle/Open, Wind direction	Idle/Open, Wind direction	NA

Table 4. Monthly CART Results in ULSD Bus

Month' Year	\multicolumn{6}{c}{In-vehicle Contaminants}
	CO$_2$	CO	NO	NO$_2$	SO$_2$	PM
Jan'07	Passengers, Indoor RH	Indoor temp.	Indoor temp.	NA	Indoor temp.	Indoor temp.
Feb'07	Wind direction, Indoor temp., Wind speed, Passengers, Cars	Wind direction, Ambient temp., Indoor RH	Indoor temp., Wind direction, Ambient temp.	Wind direction	Ambient temp., Wind direction, Indoor RH, Ambient RH, Indoor temp, Visibility	Ambient temp., Ambient PM$_{2.5}$
Mar'07	Wind speed, Visibility, Indoor RH	Indoor temp., Wind speed, Wind direction, Indoor RH, Idle/Close	Passengers, Indoor RH, Trucks, Ambient temp., Wind direction	NA	Wind speed, Run/Close, Ambient temp., Ambient RH, Indoor temp., Wind direction	Ambient RH, Ambient temp., Indoor RH, Ambient PM$_{2.5}$, Visibility,

Month' Year	In-vehicle Contaminants					
	CO_2	CO	NO	NO_2	SO_2	PM
Apr'07	Idle/Open, Run/Close, Wind direction, Cars, Trucks, Passengers	Ambient temp., Wind direction, Idle/Open, Run/Close, Indoor RH	NA	Wind direction	Indoor RH, Ambient RH, Visibility	NA
May'07	Wind direction, Visibility, Ambient temp., Ambient RH	Indoor temp., Wind speed, Wind direction	Indoor RH, Ambient RH, Visibility, Ambient temp.	NA	Ambient RH, Indoor RH, Visibility, Ambient temp.	NA
Jun'07	Ambient RH, Passengers, Wind direction, Wind speed, Ambient temp.	Wind direction, Ambient RH, Ambient temp., Wind speed	Ambient RH, Ambient temp., Wind speed, Wind direction	Ambient RH, Ambient temp., Wind speed	Ambient RH, Wind direction, Ambient temp., Wind speed	NA
Jul'07	NA	NA	NA	NA	NA	NA
Aug'07	Passengers	Indoor RH, Wind direction, Run/Close, Passengers	Wind direction, Wind speed	NA	Indoor RH, Ambient RH, Indoor temp.	NA
Sep'07	Passengers, Indoor RH	Indoor RH, Passengers, Wind direction, Ambient temp., Indoor temp.	Wind direction	NA	Indoor RH, Wind direction, Indoor temp.	NA
Oct'07	Passengers, Ambient temp., Indoor RH	Cars, Wind direction, Indoor temp.	Indoor RH, Indoor temp., Ambient RH	NA	Ambient RH, Indoor RH, Wind speed	NA
Nov'07	Indoor temp., Passengers, Ambient temp., Cars	Wind speed, Indoor temp.	Indoor RH, Visibility, Ambient temp., Ambient RH	NA	Indoor temp., Indoor RH, Ambient RH, Visibility, Passengers	Ambient $PM_{2.5}$
Dec'07	Run/Close, Passengers, Indoor temp., Indoor RH, Wind direction, Ambient temp.	Indoor temp., Indoor RH, Ambient temp., Passengers	Passengers	Ambient temp., Wind direction	Wind direction	NA

Table 5. Seasonal MINITAB Results in BD20 and ULSD Buses

Season/Year	BD20						ULSD					
	CO2	CO	NO	NO2	SO2	PM	CO2	CO	NO	NO2	SO2	PM
Winter' 07	NA	NA	NA	NA	NA	NA	Passengers, Indoor RH	Idle/Close, Run/Close, Wind direction, Cars, Indoor RH	Wind direction, Visibility, Indoor temp.	Wind direction, Wind speed	Indoor RH, Ambient RH, Idle/Close	Ambient PM2.5, Indoor RH, Ambient RH, Visibility
Spring' 07	Indoor RH, Passengers	Cars, Ambient temp.	Indoor temp., Cars, Trucks, Indoor RH	Indoor RH	Run/Close, Idle/Close, Trucks	Visibility, Wind direction, Ambient PM2.5, Ambient Temp.	Passengers	Idle/Open, Indoor RH, Ambient RH	Ambient RH, Ambient temp., Indoor RH, Visibility	Ambient RH, Wind direction	Ambient RH, Ambient temp	NA
Summer' 07	Indoor RH, Passengers, Indoor temp.	Indoor temp., Wind speed, Wind direction, Idle/Open	Wind speed, Indoor temp.	NA	Indoor RH, Wind direction	NA	Passengers	Wind direction, Ambient temp., Indoor temp., Ambient RH	Indoor temp., Wind direction, Ambient temp.	NA	Wind direction, Ambient temp., Ambient RH, Indoor temp., Indoor RH	NA
Fall' 07	Passengers	Wind speed, Wind direction, Visibility	Wind speed, Wind direction	Wind speed, Wind direction	Wind direction, Ambient RH, Wind speed	NA	Passengers, Indoor temp.	Indoor temp., Indoor RH, Ambient temp.	Wind speed, Passengers	Ambient temp.	Ambient temp.	NA
Winter' 08	Passengers, Wind direction	Wind direction, Indoor RH	Cars, Indoor Temp.	Cars, Indoor RH	Cars, Indoor Temp.	NA	NA	NA	NA	NA	NA	NA

Table 6. Seasonal CART Results in BD20 and ULSD Buses

Season/ Year	BD20						ULSD					
	CO2	CO	NO	NO2	SO2	PM	CO2	CO	NO	NO2	SO2	PM
Winter' 07	NA	NA	NA	NA	NA	NA	Wind direction, Passengers, Indoor RH	Idle/Close, Ambient temp., Run/Close, Indoor temp, Wind direction, Indoor RH	Indoor temp, Ambient temp, Wind direction, Visibility	Wind direction	Wind direction, Idle/Close, Indoor RH, Ambient RH, Ambient temp.	Ambient RH, Ambient PM2.5, Indoor RH, Visibility, Wind direction
Spring' 07	Indoor RH, Passengers	Ambient temp., Ambient RH	Ambient RH, Cars, Trucks, Indoor RH, Indoor temp., Wind direction	Indoor RH	Passengers, Run/Close, Idle/Close	Visibility, Wind direction, Ambient PM2.5, Ambient RH, Ambient temp.	Passengers	Ambient RH	Indoor RH, Visibility, Ambient RH, Ambient temp.	Ambient RH, Wind direction	Ambient RH, Ambient temp.	NA
Summer' 07	Indoor RH, Ambient RH, Wind direction, Passengers, Indoor temp., Ambient temp.	Indoor temp., Wind direction, Idle/Open	Indoor temp.	NA	Indoor RH, Indoor temp.	NA	Passengers	Wind direction, Ambient temp., Indoor RH, Indoor temp., Ambient RH	Indoor temp., Ambient temp., Wind direction, Ambient RH	NA	Wind direction, Ambient temp., Indoor temp., Ambient RH, Indoor RH	NA

Table 6. (Continued)

Season/Year	In-vehicle Contaminants										
	BD20						ULSD				
	CO2	CO	NO	NO2	SO2	PM	CO2	CO	NO	NO2	SO2	PM
Fall' 07	Passengers	Ambient temp., Wind speed, Wind direction, Visibility, Ambient RH	Wind speed, Wind direction	Wind direction	Wind direction, Wind speed, Ambient RH	NA	Passengers, Indoor temp., Indoor RH	Indoor temp., Indoor RH, Ambient temp.	Passengers	Ambient temp.	Ambient temp.	NA
Winter' 08	Passengers, Wind direction	Wind direction, Indoor RH	Indoor temp.	Indoor temp., Indoor RH, Cars	Cars, Indoor temp.	NA	NA	NA	NA	NA	NA	NA

Table 7. Yearly MINITAB Results in BD20 and ULSD Buses

BD20 (Apr. 07 – Mar. 08)							ULSD (Jan. 07- Dec. 07)					
CO2	CO	NO	NO2	SO2	PM	CO2	CO	NO	NO2	SO2	PM
Passengers	Ambient Temp., Ambient RH	Indoor Temp.	Visibility, Wind speed, Wind direction	Ambient RH	NA	Passengers	Ambient Temp.	Passengers, Ambient Temp.	Ambient Temp.	Ambient Temp., Ambient RH, Indoor RH	NA

Table 8. Yearly CART Results in BD20 and ULSD Buses

BD20 (Apr. 07 – Mar. 08)							ULSD (Jan. 07- Dec. 07)					
CO2	CO	NO	NO2	SO2	PM	CO2	CO	NO	NO2	SO2	PM
Passengers	Ambient Temp., Ambient RH	Indoor Temp.	Visibility, Wind speed, Wind direction, Ambient RH	Ambient RH	NA	Passengers	Ambient Temp.	Passengers, Ambient Temp.	Ambient Temp.	Ambient Temp., Ambient RH, Indoor Temp., Indoor RH	NA

On examining the performance of regression (MINITAB results) and regression tree (CART results) analyses, it can be observed that the important factors affecting contaminant levels are found to be more or less similar on a monthly, seasonal, and yearly basis (refer to Tables 1-8). Regression tree analysis provided a more detailed list of influential variables in association with the variable relative importance when developing regression tree models. The study of contaminant concentrations using regression and regression tree analyses helped find that different factors affect in-vehicle contaminant levels on a monthly, seasonal, and yearly basis. This finding was different to the findings from previous vehicular studies as most of the earlier studies either treated the entire database as single when performing statistical techniques or did not have a good representative sample for each month to get a comprehensive outlook of the important variables that could affect IAQ levels. Some of the general observations made during the study of regression trees are discussed below.

EFFECT OF PASSENGERS

CO_2 levels were predominantly influenced by passenger ridership and peak hours. An increase in passenger ridership during peak hours yielded in-vehicle CO_2 levels to reach a maximum of 1200 ppm irrespective of the buses. Higher levels of PM are observed in both the buses when the driver was smoking within the vehicle cabin (even though the bus is a non-smoking compartment). This observation was made from the video analysis. Similar observations were made by Kadiyala et al. (2010a) and Kadiyala and Kumar (2011).

EFFECT OF CARS AND TRUCKS/BUSES

All the monitored in-vehicle contaminants showed a positive correlation with vehicular traffic ahead of the bus when there is sufficient ventilation. NO_2 and CO_2 levels were found to be mainly influenced by trucks; while the contaminants of NO, CO, PM and SO_2 concentrations were predominantly influenced by cars. Higher in-vehicle contaminant concentrations are a result of penetration of lead vehicular exhaust contaminants inside the vehicle compartment.

EFFECT OF VENTILATION SETTINGS

In general, it was observed that the CO_2 concentrations are on the higher side when the doors are closed and the bus is moving with passengers inside. Relatively lower CO_2 levels are observed when the bus had passengers inside and is in idling position with doors opened (better air exchange rate) as compared to a running bus with doors closed and passengers inside. When there are no passengers inside, higher CO_2 levels are observed during idling condition with doors opened as compared to a case where the bus is running with doors closed and no passengers. When the bus is in idling position with doors closed and no passengers, CO_2 levels are found to be minimum and the concentrations increased with an increase in number of passengers. Higher CO levels are observed when the bus is in idling position with doors opened as compared to a running bus with doors closed and an idling bus with doors closed. NO, NO$_2$ and SO$_2$ have shown similar patterns with regard to the contaminant levels. PM levels are found to be the highest when there is smoking within the bus with the bus status being idle with doors closed. The contaminant concentrations of particulates are relatively higher in an idling bus with door opened or closed as compared to a running bus with doors closed.

EFFECT OF AMBIENT TEMP AND AMBIENT RH

With sufficient ventilation, PM levels increased positively with increase in ambient temperature and negatively with increase in ambient RH. NO, NO$_2$, and SO$_2$ contaminants varied positively with ambient RH and negatively with ambient temperature when there is sufficient ventilation. In-vehicle CO$_2$ and CO levels were found to increase with an increase in ambient temperature and decrease in ambient RH.

EFFECT OF INDOOR TEMP. AND INDOOR RH

Vehicular CO$_2$ levels are found to be dependent on indoor RH to some extent and are not influenced by indoor temperature. Other monitored in-vehicle contaminants did not show a fixed pattern with indoor temperature and indoor RH.

EFFECT OF WIND DIRECTION

The wind direction is found to be a categorical variable from the regression tree analysis, i.e. the best split condition is in the form of wind direction coming at certain angles grouped as one set rather than having wind directions less than a certain value as a single set. CART has identified wind direction as an important factor affecting all the contaminant levels. Relatively higher contaminant levels might be observed when the wind is blowing from the north due to the geographical position of Detroit, an industrialized city to the north of Toledo as compared to the surrounding areas. Since the bus position was not considered with respect to wind direction, more detailed analysis of bus position with respect to wind direction could help find additional facts on contaminant concentration build up.

EFFECT OF WIND SPEED

Higher wind speeds resulted in lower contaminant concentration buildup inside the buses while lower wind speeds helped buildup in-vehicle contaminant levels when sufficient ventilation is provided.

EFFECT OF VISIBILITY AND AMBIENT PM2.5

PM levels inside the buses were found to be positively correlated to ambient $PM_{2.5}$ levels and visibility. All the monitored gaseous contaminants with the exception of NO_2 were found to increase with a decrease in visibility.

CONCLUSION

A field study to continuously collect IAQ data in two public transport buses for a period of over one year was successfully carried out. Two mathematical techniques: regression and regression trees were examined to determine the important variables affecting vehicular IAQ. Both the techniques yielded similar results in terms of identifying the important variables. Regression tree analysis turns out to be a slightly better predictive tool than regression analysis as CART provided the relative importance of all the variables considered. A comprehensive study of the different factors affecting the monitored in-vehicle contaminant levels of CO_2, CO, NO, NO_2, SO_2, and PM on a monthly, seasonal, and yearly basis helped in better understanding of the in-bus air quality. Relationships between the monitored vehicular contaminants and independent variables have been established using regression trees. The factors affecting the monitored vehicular contaminants varied on a monthly, seasonal, and yearly basis. This observation is significantly different from earlier studies and indicates that environmental data needs to be analyzed more on a time-specific basis.

ACKNOWLEDGMENTS

The authors would like to thank the United States Department of Transportation (US DOT) and Toledo Area Regional Transit Authority (TARTA) for the alternative fuel grant awarded to the Intermodal Transportation Institute (ITI) of the University of Toledo. The authors would also like to express their sincere gratitude to the TARTA management and the employees for their continued interest and involvement in this work. The views expressed in this paper are those of the authors alone and do not represent the views of the funding organizations.

REFERENCES

Adams, H.S.; Nieuwenhuijsen, M.J.; Colvile, R.N. Determinants of fine particle ($PM_{2.5}$) personal exposure levels in transport microenvironments, London, UK. *J. Atmos. Environ.* 2001, 35, 4557-4566.

Alm, S.; Jantunen, M.J.; Vartiainen, M. Urban commuter exposure to particulate matter and carbon monoxide inside an automobile. *J. Expo. Anal. Env. Epid.* 1999, 9, 237-244.

Anderson, C.W.; Qian, S.S. Exploring factors controlling the variability of pesticide concentrations in the Willamette River Basin using tree-based models. *J. Environ. Sci. Technol.* 1999, 33, 3332-3340.

Assel, R.; Rodionov, S. Atmospheric teleconnection patterns and severity of winters in the Laurentian Great Lakes basin. *J. Atmos. Ocean.* 2000, 38, 601-635.

Breiman, L.; Friedman, J.H.; Olshen, R.A.; Stone, C.J. (1984). Classification and regression trees. Pacific Grove: Wadsworth.

Burrows, W.R.; Benjamin, M.; Beauchamp, S.; Lord, E.R.; McCollor, D.; Thomson, B. CART decision tree statistical analysis and prediction of summer season maximum surface ozone for the Vancouver, Montreal, and Atlantic Regions of Canada. *J. Appl. Meteorol.* 1995, 34, 1848-1862.

California Air Resources Board (CARB) Factsheet, (2003). Children's school bus exposure study. Available at http://www.arb.ca.gov/research/schoolbus/sbfact.pdf.

Chan, A. Commuter exposure and indoor–outdoor relationships of carbon oxides in buses in Hong Kong. *J. Atmos. Environ.* 2003, 37, 3809–3815.

Chan, L.Y.; Chan, C.Y.; Qin, Y. The effect of commuting microenvironment on commuter exposures to vehicular emission in Hong Kong. *J. Atmos. Environ.* 1999, 33, 1777-1787.

Chan, L.Y.; Lau, W.L.; Lee, S.C.; Chan, C.Y. Commuter exposure to particulate matter in public transportation modes in Hong Kong. *J. Atmos. Environ.* 2002b, 36, 3363–3373.

Chan, L.Y.; Lau, W.L.; Wang, X.M.; Tang, J.H. Preliminary measurements of aromatic VOCs in public transportation modes in Guangzhou, China. *J. Environ. Int.* 2003, 29, 429–435.

Chan, L.Y.; Lau, W.L.; Zou, S.C.; Cao, Z.X.; Lai, S.C. Exposure level of carbon monoxide and respirable suspended particulate in public transportation modes while commuting in urban area of Guangzhou, China. *J. Atmos. Environ.* 2002a, 36, 5831–5840.

Chan, L.Y.; Liu, Y.M. Carbon monoxide levels in popular passenger commuting modes traversing major commuting routes in Hong Kong. *J. Atmos. Environ.* 2001, 35, 2637-2646.

Chan, C.C.; Ozkaynak, H.; Spengler, J.D. Driver exposure to volatile organic compounds, CO, ozone, and NO2 under different driving conditions. *J. Environ. Sci. Technol*, 1991, 25, 964–972.

Clifford, M.J.; Clarke, R.; Riffat, S.B. Driver's Exposure to carbon monoxide in Nottingham, UK. *J. Atmos. Environ.* 1997, 31, 1003–1009.

Critical Environment Technologies, 2011. Available at http://www.critical-environment.

Diapouli, E.; Grivas, G.; Chaloulakou, A.; Spyrellis, N. PM10 and ultrafine particles counts in-vehicle and on-road in the Athens Area. *J. Water, Air, and Soil Pollut:Focus.* 2008, 8, 89–97.

Duci, A.; Chaloulakou, A.; Spyrellis, N. Exposure to carbon monoxide in the Athens urban area during commuting. *J. Sci. Total Environ.* 2003, 309, 47–58.

Duffy, B.L.; Nelson, P.F. Exposure to emissions of 1,3-butadiene and benzene in the cabins of moving motor vehicles and buses in Sydney, Australia. *J. Atmos. Environ.* 1997, 31, 3877–3885.

Dunsmuir, W.T.M.; Spark, E.; Connor, G.J. Statistical forecasting techniques to describe the surface winds in Sydney Harbour. *J. Aust. Meteorol. Mag.* 2003, (In press).

Fernandez-Bremauntz, A.A.; Ashmore, M. Exposure of commuters to carbon monoxide in Mexico City – 1. Measurement of in-vehicle concentrations. *J. Atmos. Environ.* 1995, 29, 525-532.

Firth, L.; Hazelton, M.L.; Campbell, E.P. Predicting the onset of Australian winter rainfall by nonlinear classification. *J. Climate.* 2005, 18, 772-781.

Fitz, R. et al., 2003. Characterizing the range of children's pollutant exposure during school bus commutes. Final report, Contract NO. 00-322. California Air Resources Board, Research Division, Sacramento, CA.

Fruin, S., 2003. Black carbon concentrations inside vehicles: Implications for refined diesel particulate matter exposures, Doctoral thesis, University of California, Los Angeles.

Gomez-Perales, J.E.; Colvile, R.N.; Nieuwenhuijsen, M.J.; Fernandez-Bremauntz, A.A.; Gutierrez-Avedoy, V.J.; Paramo-Figueroa, V.H.; Blanco-Jimenez, S.; Bueno- Lopez, E.; Mandujano, F.; Bernabe-Cabanillas, R.; Ortiz-Segovia, E. Commuters' exposure to PM2.5, CO, and benzene in public transport in the metropolitan area of Mexico City. *J. Atmos. Environ.* 2004, 38, 1219–1229.

Grimm Technologies, Inc., 2011. Available at http://www.dustmonitor.com/index.html.

Herche, L.; Assel, R.; Rodionov, S. Tree-structured modeling of the relationship between Great Lakes ice cover and atmospheric circulation patterns. *J. Great. Lakes. Res.* 2001, 27, 486-502.

Huang, L.; Smith, R.L. Meteorologically-dependent trends in urban ozone. *J. Environmetrics.* 1999, 10, 103-118.

Jo, W.K.; Choi, S.J. Vehicle occupants exposure to aromatic VOCs while commuting on an urban–suburban route in Korea. *J. AandWMA.* 1996, 46, 749–54.

Jo, W.K.; Park, K.H. Concentrations of volatile organic compounds in automobiles' cabins while commuting along a Korean urban area. *J. Environ. Int.* 1998, 24, 259–65.

Jo, W.K.; Park, K.H. Commuter exposure to volatile organic compounds under different driving conditions. *J. Atmos. Environ.* 1999, 33, 409–17.

Jo, W.K.; Park, K.H. Public bus and taxicab drivers' exposure to aromatic work-time volatile organic compounds. *J. Environ Res.* 2001, 86, 66–72.

Kadiyala, A., 2008. Identification of factors affecting contaminant levels and determination of infiltration of ambient contaminants in public transport buses operating on biodiesel and ULSD fuels, Masters Thesis, University of Toledo, Toledo.

Kadiyala, A., Kumar, A. 2008a. Study of factors affecting in-vehicle pollutant levels in the city of Toledo public transport buses running on biodiesel. Paper 508, Proceedings of the 101st AandWMA Annual Conference, Portland, OR.

Kadiyala, A.; Kumar, A. Application of CART and Minitab software to identify variables affecting indoor concentration levels. *Environ. Prog.* 2008b, 27, 160-168.

Kadiyala, A., Kumar, A. 2009. Study of in-vehicle pollutant behavior in public transport buses running on alternative fuels. Paper 571, Proceedings of the 102nd AandWMA Annual Conference, Detroit, MI.

Kadiyala, A.; Kumar, A.; Vijayan, A. Study of occupant exposure of drivers and commuters with temporal variation of in-vehicle pollutant concentrations in public transport buses operating on alternative diesel fuels. *The Open. Environ. Eng. J.* 2010a, 3, 55-70.

Kadiyala, A.; Kaur, D.; Kumar A. Application of MATLAB to select an optimum performing genetic algorithm for predicting in-vehicle pollutant concentrations. *Environ. Prog. and Sust. Energ.* 2010b, 29(4), 398-405.

Kadiyala, A.; Kumar, A. Study of in-vehicle pollutant variation in public transport buses operating on alternative fuels in the city of Toledo, Ohio. *The Open. Environ. Biol. Monit. J.* 2011, 4, 1-20.

Klepeis, N.E.; Nelson, W.C.; Ott, W.R.; Robinson, J.P.; Tsang, A.M.; Switzer, P.; Behar, J.V.; Hern, S.C.; Engelmann, W.H. The national human activity pattern survey (NHAPS): a resource for assessing exposure to environmental pollutants. *J. Expo. Anal. Env. Epid.* 2001, 11, 231-252.

Kuo, H.W.; Wei, H.C.; Liu, C.S.; Lo, Y.Y.; Wang, W.C.; Lai, J.S.; Chan, C.C. Exposure to volatile organic compounds while commuting in Taichung, Taiwan. *J. Atmos. Environ.* 2000, 34, 3331-3336.

Lau, W.L.; Chan, L.Y. Commuter exposure to aromatic VOCs in public transportation modes in Hong Kong. *J. Sci. Total Environ.* 2003, 308, 143–155.

Marion, J.F., Brent, D.K. Measurement of volatile organic compounds inside automobiles. *J. Expo. Anal. Env. Epid.* 2003, 13, 31–34.

Office of Environmental Health Hazard Assessment (OEHHA) Report, 2004. Air pollution from nearby traffic and children's health: information for schools. Available at http://www.oehha.ca.gov/public_info/facts/pdf/Factsheetschools.pdf.

Praml, G.; Schierl, R. Dust exposure in Munich public transportation: a comprehensive 4-year survey in buses and trams. *J. Int. Arch. Occ. Environ. Hea.* 2000, 73, 209-214.

Richard J.H., Jeff A. 2003. Application of the categorical and regression tree (CART) model (trademark Salford Systems, San Diego, Ca) to understand the relationship between PM and meteorological variables in the San Joaquin Valley. Poster #83, 22nd Annual Conference of American Association for Aerosol Research (AAAR), Anaheim, CA.

Rodes, C. et al., 1998. Measuring concentrations of selected air pollutants inside California vehicles. Final Report, Contract NO. 95-339. California Air Resources Board, Research Division, Sacramento, CA.

Sabin, L.; Kozawa, K.; Behrentz, E.; Winer, A.; Fitz, D.; Pankratz, D.; Colome, S.; Fruin, S. Analysis of real-time variables affecting children's exposure to diesel-related pollutants during school bus commutes in Los Angeles. *J. Atmos. Environ.* 2005, 39, 5243–5254.

Stoeckenius, T.E.; Hudischewskyj, A.B. 1990. Adjustment of ozone trends for meteorological variation. Paper # SYSAPP-90/008, Systems Applications Inc., San Rafael, CA.

TARTA Routes and Timings, 2011. Available at http://www.tarta.com/wp content/uploads/routes/20.pdf.

Trivisano, C.; Cocchi, D.; Bruno, F. Forecasting daily high ozone concentrations by classification trees. *J. Environmetrics.* 1996, 15, 141-153.

United States Environmental protection Agency, 1999. Guideline for developing an ozone forecasting program, EPA-454/R-99-009.

Vijayan, A., Kumar, A. 2008. Characterization of indoor air quality inside public transport buses using alternative diesel fuels. Proceedings of TRB Conference, Transportation Research Board, Washington, D.C.

Visser, H., Noordijk, H., 2002. Correcting air pollution time series for meteorological variability. With an application to regional PM10 concentrations. National Institute for Public Health and the Environment (RIVM) Report 722601007, Bilthoven, Netherlands.

In: Handbook of Genetic Algorithms: New Research
Editors: A. Ramirez Muñoz and I. Garza Rodriguez
ISBN: 978-1-62081-158-0
© 2012 Nova Science Publishers, Inc.

Chapter 5

OPTIMIZATION WITH GENETIC ALGORITHMS OF A HYBRID DISTILLATION/MELT CRYSTALLIZATION PROCESS

Cristofer Bravo-Bravo,[a] Juan Gabriel Segovia-Hernández,[a,*] Salvador Hernández,[a] Fernando Israel Gómez-Castro,[a] Claudia Gutiérrez-Antonio[b] and Abel Briones-Ramírez[c]

[a]Universidad de Guanajuato, Campus Guanajuato, División de Ciencias Naturales y Exactas, Departamento de Ingeniería Química, Noria Alta s/n, 36050, Guanajuato, Gto., México
[b]CIATEQ, A.C., Ingeniería de plantas, Av. del Retablo 150, Col. Fovissste, 76150, Querétaro, Querétaro, México
[c]Exxerpro Solutions, Av. del Sol 1B Interior 4B, Plaza Comercial El Sol, El Sol Querétaro, Querétaro, 76113, México

ABSTRACT

Large scale purification of reaction products in a chemical plant is typically accomplished through distillation. Innovative hybrid processes offer significant cost savings, particularly for azeotropic or close-boiling mixtures, and allow the cost-efficient synthesis of new products. Hybrid separation processes are characterized by the combination of two or more different unit operations, which contribute to the separation task by different physical separation principles such that separation boundaries or inefficiencies of a single unit operation can be overcome. Despite of the inherent advantages of hybrid separation processes, they are not systematically exploited in industrial applications. A major reason relies on the complexity of the design and optimization of these highly integrated processes. In this work we study the design and optimization of the hybrid distillation/melt crystallization process, using conventional and thermally coupled distillation sequences. The design and optimization was carried out using a multiobjective genetic algorithm with restrictions coupled with the process

---

[*] Corresponding author, e-mail: gsegovia@ugto.mx, Phone: +52 (473) 732-0006 ext 8142.

simulator Aspen Plus^{TM}, for the evaluation of the objective function. The results show that this hybrid configuration with thermally coupled arrangements is a feasible option in terms of energy savings and capital investment.

Keywords: hybrid process, melt crystallization, energy savings, multiobjective optimization, control properties

1. INTRODUCTION

Due to the instability on the worldwide energetic reserves, the proper use of energetic sources has taken special importance on the last years. Energy is an important resource for the most of the human activities. Among those activities, the manufacturing industry, particularly chemical industry, has the largest contribution to the energy consumption, representing about 6% of all domestic energy use, and 24% of the total U.S. manufacturing energy use. The second largest consumer of energy is the petroleum refining industry, with a contribution of about 10% to the total U.S. manufacturing energy use. For both, chemical and refining industries, the purification stages accounts for approximately the 60% of the energy requirements for the different production plants; with distillation systems as the main contributors for the energy use (about 95%) of the total energy required for separations, and is one of the more used unit operations on the chemical and petrochemical industries (Eldridge et al., 2005). This is due to the flexibility of the distillation systems, its relatively low capital investment and low operational risk. Furthermore, there is much information available for the design and controllability of the distillation columns. Nevertheless, one of the main disadvantages of the distillation systems is its low thermodynamic efficiency, with typical values of 10% or lower. This is reflected on high external energy requirements to achieve the desired separation.

Considering the high percent of industries on which distillation represents an important proportion of the purification processes, it results mandatory the development of purification systems with significant energy savings. Particularly, the reduction on energy requirements of distillation sequences where the separation of mixtures of components with low relative volatilities, and for distillation systems operating at cryogenic or very high temperatures, represents a major research opportunity. Solvent extraction and absorption are important alternatives to compliment distillation and generate processes with lower energy requirements. By applying process intensification and equipment with reduced solvent needs, the use of more selective mass-separating agents may be achieved. There are also applications for membrane hybrid systems, and significant investment in membrane technology has already occurred with limited industrial implementation. Only few opportunities have been identified for crystallization hybrid systems.

Among the proposed alternatives for improving equipment in existing distillation systems there are divided wall columns, enhanced packing designs, heat integrated distillation, and improving mass transfer efficiencies. In the case of divided wall columns, they allow reducing considerably energy requirements and investment costs, thus they are widely used in the industry, in important enterprises such as BASF, UOP and others (Kaibel et al., 2006, Dejanovic et al., 2010). By the other hand, in the case of hybrid systems, improved mass separating agents and process equipment there is only little application on industry, mainly because of the lack of tools to evaluate their performance for specific applications (Eldridge et al., 2005).

Many studies focused on the use of hybrid processes for purification have been developed on the last years (Berry and Ng, 1997; Lipnizki et al., 1999; Myasnikov et al., 2003; Franke et al., 2008; Lima and Grossmann, 2009, among others). Among the advantages on the use of such hybrid systems there is the increasing in the separation efficiency, which results on reductions in the energy consumption and, with a proper design, may also result in a low investment cost. One example of a hybrid process consists on the coupling of crystallization and distillation. They are applicable to systems where three-phase equilibrium occurs, i.e. solid–liquid–vapor equilibrium. In such systems a part of impurities is separated in form of crystals, thus, the liquid vaporizes at a low pressure, near the triple-point pressure of the main component. Therefore, combining all these processes in one installation is expected to raise the separation efficiency without requiring any considerable additional investment (Myasnikov et al., 2003). A related hybrid separation process consists on the combination of distillation and melt crystallization for separation of close-boiling isomer mixtures. This hybrid system combines advantages of both individual separation processes, in which high separation factors per stage can be reached. Furthermore, the distillation/melt crystallization hybrid systems have the capacity to avoid the main limitations of the individual unit operations: high energy-consumption for distillation, and limitations of yield due to the eutectics for melt crystallization. Many studies for hybrid processes combining distillation and crystallization have been reported for the separation of terphenyl, xylene, dichlorobenzene and diphenylmethane diisocyanate isomers (Franke et al., 2008). Other related studies include that reported by Berry and Ng (1997), who used models based on constant separation factors for the synthesis of hybrid distillation/melt crystallization processes, also proposing guidelines for flowsheet selection. Mathematical programming, especially mixed-integer nonlinear programming (MINLP), is presented as other alternative for the study of hybrid systems, offering the possibility to solve the design problem simultaneously. Wallert et al. (2005) reported the use of shortcut methods for the synthesis and evaluation of a hybrid distillation/layer crystallization processes by a general disjunctive approach. Franke et al. (2008) proposed a three-step design method for hybrid distillation/melt crystallization processes. In a first step different sequences are generated by heuristic rules. These sequences are evaluated in a second step by shortcut methods on the basis of energy consumption, in order to identify the most promising alternatives. In the third and last step a reduced number of promising sequences is rigorously optimized by MINLP methods, and the best sequence on the basis of total annualized costs is chosen. Marquardt et al. (2010) reviewed an optimization-based framework composed of shortcut and rigorous design steps for the robust and efficient synthesis of hybrid distillation/melt hybrid processes. A multitude of hybrid processes composed of distillation and melt crystallization units are evaluated with shortcut models. Then, some promising designs are subsequently rigorously optimized by an economic objective function and discrete-continuous optimization techniques. It is shown that the design of the cost-optimal hybrid process within the systematic synthesis framework can be accomplished with robustness and efficiency. Nevertheless, given the nonconvex nature of the equations modeling the hybrid systems, the use mathematical programming methods does not guarantee finding the global optimum.

Recently, stochastic methods have been successfully applied to the solution of the optimization problem for different process engineering situations (Gross and Rossen, 1998; Leboreiro and Acevedo, 2004; Gómez-Castro et al., 2008, Gutiérrez-Antonio et al., 2009; Vázquez-Castillo et al., 2009; Bravo-Bravo et al., 2010). Stochastic optimization methods are robust numerical tools, with only a reasonable computational effort required for the optimization of multivariable functions; they are also applicable to unknown structure problems, requiring only calculations of the objective function, and can be used with all models without problem reformulation. Among the stochastic methods, genetic algorithms have been studied with special attention, with several available multiobjective techniques such as VEGA, MOGA, NSGA, Niche Pareto GA, and NSGA-II (Gutiérrez-Antonio and Briones-Ramírez, 2009). These stochastic methods are very useful for the reliable design and optimization of chemical processes, where several decision variables are involved and models are usually nonlineal and nonconvex. To the best of our knowledge, multiobjective stochastic methods have not been reported for design and optimization of hybrid process with rigorous models. In this work, the design and optimization of the hybrid distillation/melt crystallization process is presented, using as optimization tool a multiobjective genetic algorithm with restrictions. The use of thermally coupled distillation sequences is considered for further reduction of the external energetic demand of the distillation section. An industrial process has been taken as our case of study, consisting on the separation of a mixture of xylene isomers (Jenkins, 1949). For the evaluation of the objective function, the multi-objective genetic algorithm is coupled with the process simulator Aspen Plus^{TM}; thus, all the results obtained are rigorous. Since the optimization strategy considers multiple objectives, not only one optimal design but a set of optimal designs, namely the Pareto front, is obtained. The results show that the proposed hybrid configuration with thermally coupled arrangements is a reliable alternative in terms of energy savings and capital investment, with reductions in emissions of greenhouse gases.

2. OPTIMIZATION STRATEGY

Given the high number of degrees of freedom involved in the design and optimization problem of a hybrid separation process, its solution is not trivial (Franke et al., 2008). To this problem for a hybrid distillation/melt crystallization process with both, conventional and thermally coupled distillation sequences, we used the multiobjective genetic algorithm with constraints developed by Gutiérrez-Antonio and Briones-Ramírez (2009). This algorithm is coupled to the Aspen Plus process simulator, and manages the multiobjective optimization using the NSGA-II, and the constraints are handled with based on the concept of non-dominance proposed by Coello-Coello (2000). The evolutionary algorithm used to solve the multiobjective optimization problem formulated in this study has a population-based nature, and has the capacity to find multiple optimums simultaneously.

In terms of multiobjective optimization, when a minimization takes place and the algorithm reaches a point where there is no feasible vector that can decrease the value of one objective without simultaneously increasing the value of another objective, it is said that point in the search space is the Pareto optimum.

By definition we can say that one point $\vec{z}^* \in \mathfrak{I}$ is a Pareto optimum if for each $\vec{z} \in \mathfrak{I}$:

$$\bigwedge_{i \in I}\left(f_i(\vec{z}) \geq f_i(\vec{z}^*)\right) \tag{1}$$

or at least there is some $i \in I$, where I represents the set of objective functions to optimize, that:

$$f_i(\vec{z}) > f_i(\vec{z}^*) \tag{2}$$

Then, we define that $\vec{z}$ dominates $\vec{w}$ when $f(\vec{z}) < f(\vec{w})$, if $W \subseteq \mathfrak{I}$ and $\vec{w} \in W$ if none $\vec{z} \in W$ dominates $\vec{w}$, we say that $\vec{w}$ is not dominated with respect to W. As established by Mezura-Montes (2001), the set of solutions which are not dominated and optimums of Pareto integrates the Pareto front. For unit operations, the Pareto front represents all optimal designs, from minimum energy consumption to minimal size of equipment. The adequate design shall be chosen by selection of a point along the Pareto front.

In the case of the optimization problem of the hybrid distillation/melt crystallization process, our objective functions consider the simultaneous minimization of the total energy requirements and the size of the equipments involved in the sequence. Next, the minimization formulation problem for each sequence analyzed in this work is presented, considering constraints, objectives and variables involved.

Hybrid Distillation Melt/Crystallization Process with Conventional Distillation Arrangement, C-DSI-C

The hybrid distillation melt/crystallization process with conventional distillation arrangement, C-DSI-C, is shown in Figure 1. In this sequence, there are two crystallizers and two conventional distillation columns. For the crystallizers, we ensure the use of the cheapest refrigerant. For the distillation columns, we minimize the heat duty, Q_i, and the number of stages, in columns B1 and B2, N_i, subject to achieving the required recoveries or purities in each product stream. This minimization can be formulated as:

$$\text{Min}(Q_i, N_i) = f(Q_i, R, N_i, N_F)$$
$$\text{st}$$
$$\vec{y}_k \geq \vec{x}_k \tag{3}$$

where R is the reflux ratio, N_i is the total number of stages of the column i, N_F is the feed stage number in column B2, $\vec{x}_k$ and $\vec{y}_k$ are vectors of required and obtained purities or recoveries, respectively. As can be noticed, in the optimization problem four variables in competition are considered for optimization: heat duty of and number of stages in each column of the sequence, B1 and B2.

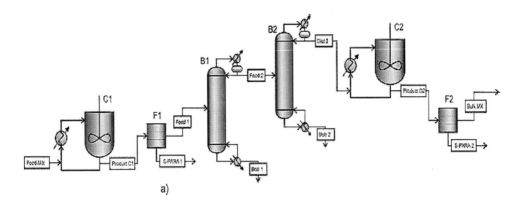

Figure 1. Hybrid distillation/melt crystallization processes using conventional distillation arrangement (C-DSI-C).

Hybrid Distillation Melt/Crystallization Process with Conventional Distillation Arrangement, C-TCIS-C

The hybrid distillation melt/crystallization process with an indirect thermally coupled distillation sequence, C-TCIS-C, is shown in Figure 2. In this sequence, there are two crystallizers and two thermally coupled distillation columns. For the crystallizers, we ensure the use of the cheapest refrigerant. For the distillation columns, we minimize the heat duty, Q_i, and the number of stages, in columns B1 and B2, N_i, subject to meet the required recoveries or purities in each product stream. This minimization can be formulated as:

$$Min(Q_i, N_i) = f(Q_i, R, N_i, N_j, N_F, F_j)$$
st
$$\vec{y}_k \geq \vec{x}_k \tag{4}$$

where R is the reflux ratio, N_i is the total number of stages of the column i, N_j is the stage number of the interconnection flow j, N_F is the feed stage number in the prefractionator, F_j is the interconnection flow j, $\vec{x}_k$ and $\vec{y}_k$ are vectors of required and obtained purities or recoveries. For this case, the optimization problem has four variables in competition: heat duty and number of stages in each column of the sequence.

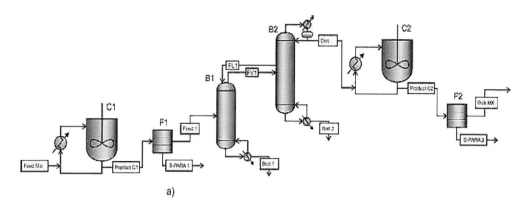

Figure 2. Hybrid distillation/melt crystallization processes using an indirect thermally coupled distillation sequence (C-TCIS-C).

Hybrid Distillation Melt/Crystallization Process with Modified Distillation Arrangement, C-MTCIS-C

The hybrid distillation melt/crystallization process with a modified arrangement of the indirect thermally coupled distillation sequence, C-MTCIS-C, is shown in Figure 3. In this sequence, there are three crystallizers and two conventional distillation columns, one of them with a sidestream. For the crystallizers, we ensure the use of the cheaper refrigerant. For the distillation columns, we minimize the heat duty, Q_i, and the number of stages, in columns B1 and B2, N_i, subject to meet the required recoveries or purities in each product stream. This minimization can be formulated as:

$$\text{Min}(Q_i, N_i) = f(Q_i, R, N_i, N_s, N_F)$$
$$\text{st}$$
$$\vec{y}_k \geq \vec{x}_k \qquad (5)$$

where R is the reflux ratio, N_i is the total number of stages of the column i, N_s is the side stream stage, N_F is the feed stage number in the prefractionator, $\vec{x}_k$ and $\vec{y}_k$ are vectors of required and obtained purities or recoveries. For this case, four variables in competition are considered for optimization: heat duty and number of stages in each column of the sequence.

Hybrid Distillation Melt/Crystallization Process with a Petlyuk Distillation Arrangement, C-PC-C

The hybrid distillation melt/crystallization process with a Petlyuk sequence, C-PC-C, is shown in Figure 4. In this sequence, there are two crystallizers along with a prefractionator and a main column. For the crystallizers, we ensure the use of the cheapest refrigerant.

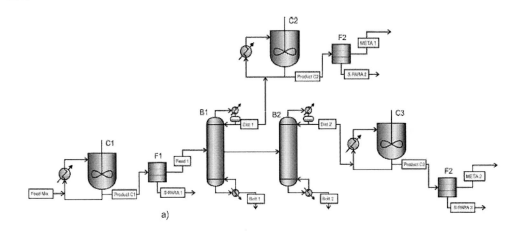

Figure 3. Hybrid distillation/melt crystallization processes using a modified arrangement from the indirect thermally coupled distillation sequence (C-MTCIS-C).

For the distillation columns, we minimize the heat duty of the sequence, Q, and the number of stages, in prefractionator B1 and main column B2, N_i, subject to meet the required recoveries or purities in each product stream. This minimization can be formulated as:

$$\text{Min}(Q, N_i) = f(Q, R, N_i, N_j, N_s, N_F, F_j)$$
st
$$\vec{y}_k \geq \vec{x}_k \tag{6}$$

where R is the reflux ratio, N_i is the total number of stages of the column i, N_j is the stage number of the interconnection flow j, N_s is the side stream stage, N_F is the feed stage number in the prefractionator, F_j is the interconnection flow j, $\vec{x}_k$ and $\vec{y}_k$ are vectors of required and obtained purities or recoveries. For this case, the optimization problem considers three variables in competition: heat duty of the sequence and number of stages in prefractionator and main column.

In this work the implemented multiobjective algorithm is based on the NSGA-II (Deb et al., 2000) and constraints are handling using a modification of the work of Coello-Coello (2000). A brief description of the process is presented below. The entire population is divided into sub-populations using as criterion the total number of satisfied constraints. Thus, the best individuals of the generation are those that satisfied the n constraints, and they are followed by the individuals which satisfies only n-1 constraints, and so on. Within each sub-population, individuals are ranked using the NSGA-II, but considering now as other objective function to minimize the difference between the obtained and required constraints. Next, dominance calculation of each subgroup is as follows:

$$\text{dominance}\{Q_i, N_i, \min[0, (\vec{x}_k - \vec{y}_k)]\} \tag{7}$$

The main objective of this strategy is minimizing both, the objective functions and the number of violated constraints, for each subgroup. The link to Aspen Plus allows obtaining optimal designs using rigorous simulations; however, the 95% of the total time of the optimization procedure is employed by the evaluation of the objective functions and constraints on the process simulator. Thus, the use artificial neuronal networks (ANN) is considered to speed up a multiobjective genetic algorithm with constraints, as proposed by Gutiérrez-Antonio and Briones-Ramírez (2010). The artificial neuronal networks create approximated functions for objective and constraints, which are used to evaluated the individuals of the population. In this way, the original objectives and constraints functions are used just every *m* generations, and the approximated functions in the rest of them, decreasing the total computational time. This allows reaching the Pareto front very quickly. Figure 5 shows a block diagram for the evolutionary strategy coupled to the artificial neuronal networks.

For the optimization hybrid system, we used 800 individuals and 40 generations as parameters of the multiobjective genetic algorithm with a frequency parameter was 5. These parameters were obtained through a tuning process, where several runs of the algorithm were performed with different numbers of individuals and generations.

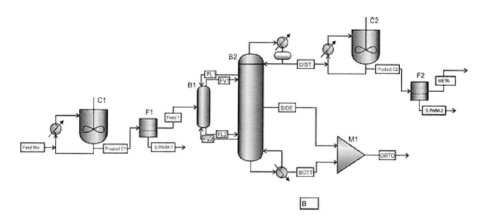

Figure 4. Hybrid distillation/melt crystallization processes using a Petlyuk column (C-PC-C).

3. CASE OF STUDY

In this work, the separation of a mixture of ortho, meta, and para xylene isomers is analyzed. The mixture contains 1% of the low boiling meta component (M), 66% of the intermediate boiling para component (P), and 33% of the high boiling ortho component (O). For each component, a molar purity of 99% is required. The separation of this mixture is feasible by using only distillation, but it results economically unattractive because of the low relative volatilities between the adjacent components (Table 1). It can be noticed that the separation of components P and O is especially difficult using conventional distillation. Nevertheless, the combination of distillation with melt crystallization offers the advantage of obtaining almost pure product. Physical properties for the case of study are displayed in Table 2, and compositions and temperatures for the eutectic point for the xylene isomers are showed in Table 3. We analyze four basic configurations obtained according to the rules of synthesis for hybrid systems proposed by Berry and Ng (1997): *i)* hybrid distillation/melt crystallization processes using conventional distillation arrangement (Figure 1); *ii)* hybrid distillation/melt crystallization processes using an indirect thermally coupled distillation sequence (Figure 2); *iii)* hybrid distillation/melt crystallization processes using a modified arrangement from the indirect thermally coupled distillation sequence (Tamayo-Galván et al., 2008, Figure 3); *iv)* hybrid distillation/melt crystallization processes using a Petlyuk column (Figure 4).

For this class of systems, thermodynamic models such as NRTL can be used to calculate equilibrium vapor-liquid data and, using the temperatures and compositions correspondent to eutectic points, the activity parameters are calculated using the Wilson model in order to calculate the liquid-solid equilibrium. The design pressure for each distillation arrangement was chosen to ensure the use of cooling water in the condensers. For the crystallizers, we assume that the separation of a pure isomer from the remaining melt at eutectic composition can be accomplished in one crystallization stage. In each crystallization region, one pure isomer crystallizes as product when the temperature is lowered in the crystallizer (Franke et al., 2008). The volume and residence time of the crystallizers were calculated rigorously, based on operating conditions and solubility data of the mixture of isomers.

Table 1. Composition in the feed and separation factors $\alpha_{i,j}$ for the isomer mixture

Component	Formula	x_f	α_{ij}
M-Xylene		0.01	1.25
P-Xylene		0.66	1.09
O-Xylene		0.33	1.00

Table 2. Physical properties of the components for the case of study

Component	Formula	Molecular Weight	T_m [°C]	ΔH_m [kJ/mol]	T_b [°C]	ΔH_v [kJ/mol]
p-Xylene	C8H10	106.160	13.260	17.110	138.370	36.070
m-Xylene	C8H10	106.160	-47.870	11.569	139.120	36.400
o-Xylene	C8H10	106.160	-25.180	13.598	144.410	36.820

4. RESULTS

In this section, the resulting Pareto fronts for the four hybrid configurations studied is presented.

Hybrid Configuration C-DSI-C

Figure 6 shows the Pareto front for case C-DSI-C, which includes the objectives to minimize: heat duty and the number of stages of the distillation sequence and the total annual cost of the system. All the optimal designs satisfy the specified purities and recoveries in the distillation column and crystallizers. Figure 7 and 8 show the profiles of temperature and composition for the distillation columns. The composition profiles in the o-xylene isomer meet the purity constraint specification of 99% for the bottom of the first column (B1). In the top of the column B1 a ternary mixture is obtained, containing m-xylene and p-xylene, together with the o-xylene that could not be separated in the first column.

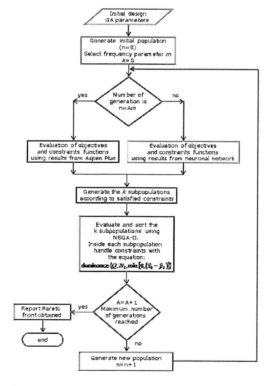

Figure 5. Optimization strategy.

In the second column (B2) the remaining o-xylene is separated, obtaining a binary mixture of m-xylene and p-xylene in the top of the column, with purities of 49.32% and 50.32%, respectively. Both xylene isomers are then obtained with purities of 99% in the crystallizers. The Pareto front shows two extreme feasible designs: a design with minimum number of stages in the distillation columns and a design with minimum total annual cost, calculated using the method of Guthrie (Turton et al., 2004), and minimum reboiler duty in the distillation arrangement. Those designs contrast the objectives of minimizing the number of stages in the distillation columns (1,724,715 $/yr) with minimizing the total annual cost and energy consumption in reboilers (1,410,871 $/yr). Thus, the engineer must decide which one is the appropriate design for his particular needs. Each design in the Pareto front is an optimal design, and this set includes designs from minimum number of stages to minimum total annual cost, along with all designs between these extremes. Also, from Figure 6 we can observe a good diversity in the designs that made up the Pareto front: number of stages and heat duties in the distillation system and total annual cost of the system covers a wide range of values. Tables 3 and 4 display parameters for the distillation columns and crystallizers for both optimal cases selected. As seen in Table 3, design with the lowest total annual cost required 20 stages more than the design with minimum number of stages in the distillation columns.

On the other hand, the design with minimum number of stages in the distillation configurations used 27.87% more energy, and showed an increment of 18.19% in the total annual cost when compared to that of the most economical design.

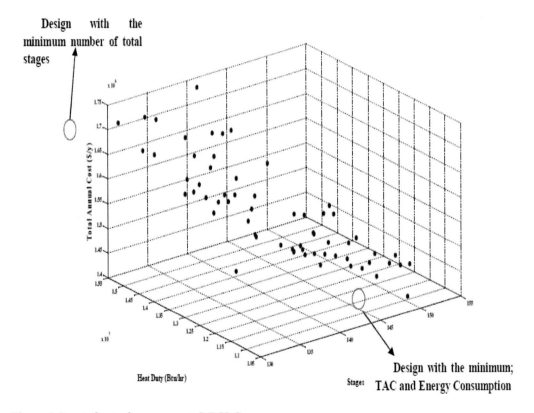

Figure 6. Pareto front of arrangement C-DSI-C.

Table 3. Optimal design of distillation columns for arrangement C-DSI-C with lowest total annual cost and lowest number of stages in the distillation columns

Distillation Sequences [DSI]	Design with the minimum number of stages	Design with the minimum; TAC, Green House Gas Emissions, Energy Consumption, Environmental Impact and the most thermodynamic efficient
Number of Stages Column B1	43	58
Number of Stages Column B2	87	92
Reflux Ratio Column B1	22.0255	16.3892
Reflux Ratio Column B2	105.2136	78.8435
Feed Stage Column B1	16	15
Feed Stage Column B2	49	53
Heat Duty Column B1 (Btu/hr)	7895879.2100	5465899.5700
Heat Duty Column B2 (Btu/hr)	7260385.3300	5465899.5700
Pressure Design Column B1 (bar)	1.0135	1.0135
Pressure Design Column B2 (bar)	1.0135	1.0135
Bottom Pressure Column B1 (bar)	1.7030	1.7030
Bottom Pressure Column B2 (bar)	1.7030	1.7030
Temperature One Stage before the Condenser of Column B1(°F)	290.2033	289.9865
Condenser Temperature Column B1 (°F)	288.8343	288.9045
Temperature One Stage before the Reboiler of Column B1(°F)	328.2485	328.4468
Reboiler Temperature Column B1 (°F)	328.9997	329.0024
Temperature One Stage before the Condenser of Column B2(°F)	282.2702	282.2478
Condenser Temperature Column B2 (°F)	281.7281	281.7351
Temperature One Stage before the Reboiler of Column B2(°F)	328.6253	328.6465
Reboiler Temperature Column B2 (°F)	329.0000	329.0009
Diameter of Column B1 (ft)	4.7548	4.2989
Diameter of Column B2 (ft)	4.7618	4.1032
FEED1 (lbmol/hr)	77.7196	77.7196
DIST2 (lbmol/hr)	4.3828	4.3828
BOTT1 (lbmol/hr)	58.7035	58.7035
BOTT2(lbmol/hr)	14.6333	14.6333
FEED2 (lbmol/hr)	19.0161	19.0161
Total Annual Cost ($/y)	1724715.92	1410871.56

Hybrid Configuration C-TCIS-C

Figure 9 presents the Pareto front of the arrangement C-TCIS-C. This figure highlights three designs that minimize the design objectives. One is the design with fewer stages in the distillation columns, other design that minimizes the total annual cost of the system, and the last one represents the design that minimizes the energy consumption in the reboilers of the distillation columns. The design with minimum number of stages presents a total annual cost of 1,341,946 $/yr; the configuration with minimum energy consumption in the reboilers shows a total annual cost of 1,212,290 $/yr and the arrangement with minimum total annual cost presents a value of 1,206,375 $/yr. It is also possible to observe that the design with minimum number of stages used 11 stages less than the design with minimum total annual cost, and 21 stages less than the design with minimum energy consumption in the reboilers.

On the other hand, the first design consumes 23.40% and 16.03% more energy than the second and third design respectively, and the total annual cost is 11.23% and 10.7% higher than those of the second and third design respectively. As for the application of process synthesis, although the first configuration minimizes the number of stages used in the separation is not a good design choice, since its cost and energy consumption are higher compared to those of the other two designs.

a)
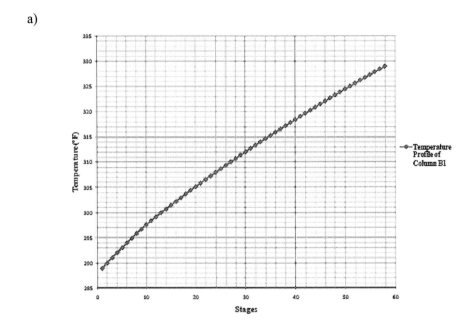

b)
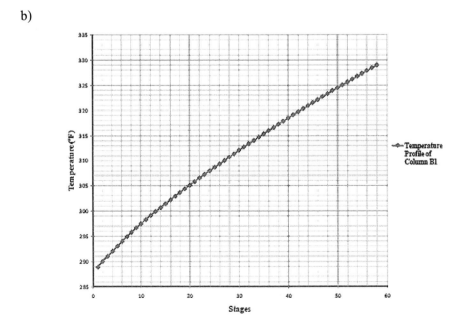

Figure 7. Temperature profiles for the distillation columns in arrangement C-DSI-C with the minor total annual cost: a) column B1; b) column B2.

The second design uses more energy but has a lower total annual cost in contrast to the third design, so for this particular case, the benefit in the selection of a separation scheme should be focused on minimizing the total annual cost. Therefore the design selected for comparative purposes, is the one with the lowest total annual cost. Composition profiles in hybrid design with a thermally coupled distillation column show that o-xylene accomplishes with the purity restrictions, obtaining a purity of 99% in the bottom of the first column (B1). The second column (B2) separates the rest of o-xylene, with a composition of 99% in the bottom, and a binary mixture of m-xylene and p-xylene in the top with purities of approximately 49.32%, 50.32%, respectively, which meet the purity constraints. It can be seen that the purities imposed in the crystallizers were achieved. At this point it is possible to highlight that the best design of hybrid system with a thermally coupled distillation column shows a decrease of 14.5% in the total annual cost in comparison with the best design of the hybrid system with a system of conventional distillation columns.

Hybrid Configuration C-MTCIS-C

The third hybrid system shows a distillation system which is a modification of a thermally coupled distillation column. The presence of recycle streams in distillation columns with thermal coupling has influenced the notion of control problems during the operation of these systems, with respect to the rather well-known behavior of conventional distillation sequences. For this reason, it has been proposed several alternate configurations to distillation schemes with thermal coupling, as the configuration discussed in this hybrid system, eliminating the recycle streams that appear to have some operational advantages over expected dynamic properties of the thermally coupled distillation sequences (Tamayo-Galván et al., 2008). Figure 10 shows the Pareto front with two designs that minimize an objective of the optimization, one is the design with minimum number of stages (2,038,549 $/yr), the other turns out to be the design that minimizes the total annual cost (1,405,896 $/yr). In this case, the configuration with minimum number of steps shows that the total annual cost and energy consumption are 45.5% and 53% higher than those of the design with the lowest total annual cost. The design that has a minimum number of stages used 22 stages less than the design with the lowest total annual cost.

For use in process synthesis, although the first configuration minimizes the number of stages used in the separation, it is not showing to be a good choice in economic terms, and therefore discarded for further analysis. Also in this hybrid system purities as restrictions imposed on the genetic algorithm (99%) for the recoveries of the three isomers of xylene are achieved in the distillation system and crystallizers. When comparing the best design selected for the three hybrid systems that have so far been analyzed, it can be noticed that the best option is the setting that uses a thermally coupled distillation system, since it is the one showing the lowest total annual cost.

Hybrid Configuration C-PC-C

Figure 11 displays the Pareto front for the hybrid system with a Petlyuk distillation column. The Pareto front highlights two designs that minimize an objective, one of them is the design with minimum number of stages, and the other one is the design that minimizes the energy consumption in the reboiler of the Petlyuk column and the total annual cost of the system. The selected designs comply with the restrictions of the purities in the distillation column and the crystallizer. The first design has 6 stages less than the second design, but consumes 7.3% more energy and its total annual cost is 4% higher than the second.

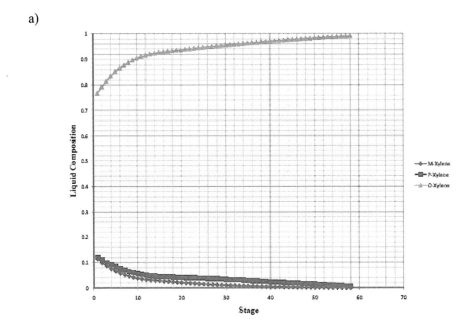

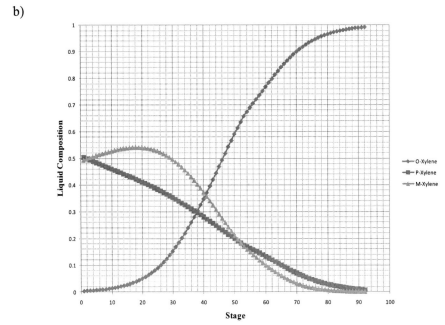

Figure 8. Composition profiles for the distillation columns in arrangement C-DSI-C with the minor total annual cost: a) column B1; b) column B2.

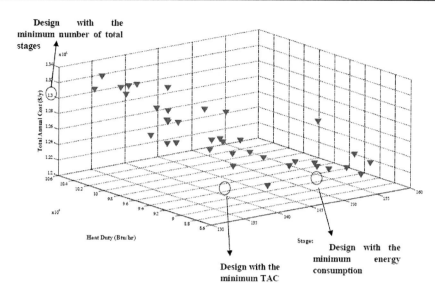

Figure 9. Pareto front of arrangement C-TCIS-C.

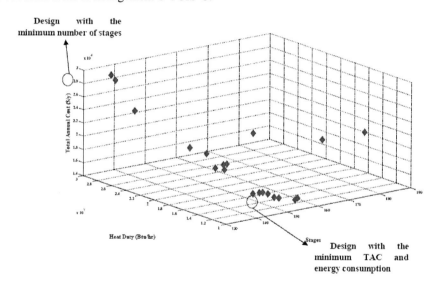

Figure 10. Pareto front of arrangement C-MTCIS-C.

So the design selected for comparison is one that shows the lowest total annual cost and minimum energy consumption. The total annual cost of the design that turns out to be the cheapest is 1,226,443 $/yr. At this point emerges an important point to highlight, the best hybrid design with a thermally coupled distillation column (C-TCIS-C) has a total annual cost of 1,206,375 $/yr and a total number of stages in the distillation column of 170 while the best hybrid design with a Petlyuk column (C-PC-C) shows a total annual cost of 1,226,443 $/yr with a total number of stages 132. The difference between the total annual cost between the system more affordable compared to the more expensive is 1.6%. This result implies that for practical purposes the energy consumption between the two systems is almost similar. However, the difference between the number of stages of the hybrid configuration with the lowest value compared with that shows the highest number is 28.8%. Given these results, it is possible to conclude that the best design among the four case studies reviewed is the hybrid system that includes a Petlyuk distillation column.

According to the previously presented results, the best design option is the hybrid system that uses a thermally coupled distillation column, since it shows the lowest total annual cost. Although the thermally coupled column has a larger number of stages in comparison with the hybrid system that uses a Petlyuk column, the total annual costs are very similar. Thus making a comparative analysis the best design hybrid options in terms of total annual cost are C-TCIS-C and C-PC-C.

Table 4. Optimal design of crystallizers for arrangement C-DSI-C with lowest total annual cost and lowest number of stages in the distillation columns

Crystallizers	Design with the minimum number of stages	Design with the minimum; TAC, Green House Gas Emissions, Energy Consumption, Environmental Impact and the most thermodynamic efficient
Operation Temperature Crystallizer C1 (°C)	-4.9	-4.9
Heat Duty Crystallizer C1 [Btu/hr]	-112248.46	-112248.46
Operation Pressure Crystallizer C1 [bar]	1.0000	1.0000
Volume Crystallizer C1 (lt)	10586.9487	10586.9487
Residence Time (hr)	1.0000	1.0000
Crystals production Masic Flow Crystallizer C1 (lb/hr)	15158.6320	15158.6320
Magma Density at the outlet from Crystallizer C1(lb/cuft)	40.5447	40.5447
Operation Temperature Crystallizer C2 (°C)	-2.5	-2.5
Heat Duty Crystallizer C2 [Btu/hr]	-53018.7730	-53018.7730
Operation Pressure Crystallizer C2 [bar]	1.0000	1.0000
Volume Crystallizer C2 (lt)	219.1019	219.1019
Residence Time (hr)	1.0000	1.0000
Crystals production Masic Flow Crystallizer C2 (lb/hr)	233.8554	233.8554
Magma Density at the outlet from Crystallizer C2(lb/cuft)	30.2236	30.2236
Total Annual Cost of Crystallizers ($/y)	184.2373	184.2373

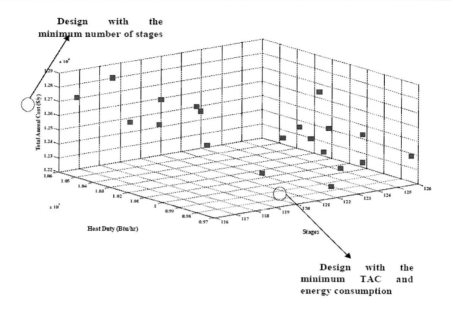

Figure 11. Pareto front of arrangement C-PC-C.

CONCLUSION

We have presented a design and optimization strategy for hybrid distillation/melt crystallization systems. The proposed strategy is based on the use of genetic algorithms. The methodology is illustrated by an industrial case study, where a ternary mixture of close-boiling ortho, meta and para isomers of xylene is separated into pure products by hybrid process of melt crystallization and distillation. The stochastic procedure allows manipulation of continuous and integer variables simultaneously. All resulting optimal designs are rigorous, since the optimization procedure is coupled to Aspen Plus. The Pareto fronts obtained for hybrid systems present good diversity, in terms of the different structures of the columns, and also with respect to energy consumption. Moreover, it was found that the optimum energy consumption design can be related to the minimum total annual operating cost. According to the results obtained it can be concluded that the best design options in terms of total annual cost are the hybrid system that uses a thermally coupled distillation column (C-TCIS-C) and the system with a Petlyuk column (C-PC-C). The benefit of our approach lies in the ability to deal with complex hybrid flowsheets and design of the optimal process with paramount robustness and efficiency, considering several objectives and constraints simultaneously.

ACKNOWLEDGMENTS

We acknowledge the financial support provided by Universidad de Guanajuato and CONACyT (Mexico) through project 84552.

REFERENCES

Berry, D. A., Ng, K. M., 1997, Synthesis of Reactive Crystallization Processes, *AIChE J.*, 43, 1737.

Bravo-Bravo, C., Segovia - Hernández, J. G., Gutiérrez - Antonio, C., Duran, A. L., Bonilla-Petriciolet, A., Briones – Ramírez, A., 2010, Extractive Dividing Wall Column: Design and Optimization, *Ind. Eng. Chem. Res.*, 49, 3672.

Coello-Coello, C. A., 2000, Constraint-Handling Using and Evolutionary Multiobjective Optimization Technique, *Civ. Eng. Environ. Syst.*, 17, 319.

Deb, K., Agrawal, S., Pratap, A., Meyarivan, T., A fast elitist non-dominated sorting genetic algorithm for multiobjective-optimization: NSGA-II, *KanGAL report 200001*, Indian Institute of Technology, Kanpur, India, 2000.

Dejanovic, I., Matijasevic, Lj., Olujic, Z., 2010, Dividing wall column – A breakthrough towards sustainable distilling, *Chem. Eng. Process.*, 49, 559.

Eldridge, R. B., Seibert, A. F., Robinson, S., Hybrid Separations/Distillation Technology Research Opportunities for Energy and Emissions Reduction, *Report for Industrial Technologies Program*, U.S. Department of Energy, Energy Efficiency and Renewable Energy, 2005.

Franke, M. B., Nowotny, N., Ndocko, E. N., Gorak, A., Strube, J., 2008, Design and Optimization of a Hybrid Distillation/Melt Crystallization Process, *AIChE J.*, 54, 2925.

Gabor, M., Mizsey, P., 2008, A Methodology to Determine Controllability Indices in the Frequency Domain. *Ind. Eng. Chem. Res.*, 47, 4807.

Gómez – Castro, F. I., Segovia - Hernández, J. G., Hernández, S., Gutiérrez – Antonio, C., Briones - Ramírez, A., 2008, Dividing Wall Distillation Columns: Optimization and Control Properties, *Chem. Eng. Technol.*, 31, 1246.

Gross, B., Roosen, P., 1998, Total Process Optimization in Chemical Engineering with Evolutionary Algorithms, *Comput. Chem. Eng.* 1998, 22, S229.

Gutiérrez-Antonio, C., Briones-Ramírez, A., 2009, Pareto front of Ideal Petlyuk Sequences Using a Multiobjective Genetic Algorithm with Constraints, *Comput. Chem. Eng.*, 33, 454.

Gutiérrez-Antonio, C., Briones-Ramírez, A., 2010, Speeding multiobjective genetic algorithm with constraints through artificial neuronal networks, *Computer Aided Process Engineering*, 28, 391.

Jenkins, R. L., 1949, Separation of Diphenylbenzene Isomers by Distillation and Crystallization, U.S. Patent 2,489,215.

Kaibel, B., Jansen, H., Zich, E., 2006, Unfixed dividing wall technology for packed and tray distillation columns, In: Distillation and Absorption Conference Proceedings, E. Sorensen (Ed.), London, UK, 252.

Leboreiro, J., Acevedo, J., 2004, Processes Synthesis and Design of Distillation Sequences Using Modular Simulators: A Genetic Algorithm Framework, *Comput. Chem. Eng.*, 28, 1223.

Lima, R. M., Grossmann, I. E., 2009, Optimal Synthesis of p-Xylene Separation Processes Based on Crystallization Technology, *AIChE J.*, 55, 354.

Lipnizki, F., Field, R. W., Ten, P.-K., 1999, Pervaporation- Based Hybrid Process: A Review of Process Design, Applications and Economics, *J. Membr. Sci.*, 153, 183.

Marquardt, W., Kraemer, K., Harwardt, A., 2010, Systematic Optimization-Based Synthesis of Hybrid Separation Processes, in: Distillation and Absorption Conference Proceedings, A. B. de Haan, H. Kooijman and A. Gorak (Ed.), Eindhoven, The Netherlands, 29.

Mezura-Montes, E., Uso de la Técnica Multiobjetivo NPGA para el Manejo de Restricciones en Algoritmos Genéticos, Maestría en Inteligencia Artificial, Thesis, Universidad Veracruzana, 2001.

Myasnikov, S. K., Uteshinsky, A. D., Kulov, N. N., 2003, Hybrid of Pervaporation and Condensation–Distillation Crystallization: A New Combined Separation Technology, *Theor. Found. Chem. Eng.*, 37, 527.

Tamayo-Galván, V. E., Segovia-Hernández, J. G., Hernández, S., Cabrera-Ruiz, J., Alcántara-Ávila, J. R., 2008, Controllability Analysis of Alternate Schemes to Complex Column Arrangements with Thermal Coupling for the Separation of Ternary Mixtures, *Comput. Chem. Eng.*, 32, 3057.

Turton, R., Bailie, R. C., Whiting, W. B., Shaeiwitz, J. A., *Analysis, Synthesis and Design of Chemical Process*. Second Edition, Prentice Hall, USA, 2004.

Vázquez-Castillo, J. A., Venegas-Sánchez, J. A., Segovia-Hernández, J. G., Hernández-Escoto, H., Hernández, S., Gutiérrez-Antonio, C., Briones-Ramírez, A., 2009, Design and Optimization, using Genetic Algorithms, of Intensified Distillation Systems for a Class of Quaternary Mixtures, *Comput. Chem. Eng.*, 33, 1841.

Wallert, C., Marquardt, W., Leu, J. T., Strube, J., 2005, Design and Optimization of Layer Crystallization Processes. *Computer Aided Chemical Engineering*, 20, 871.

In: Handbook of Genetic Algorithms: New Research
Editors: A. Ramirez Muñoz and I. Garza Rodriguez
ISBN: 978-1-62081-158-0
© 2012 Nova Science Publishers, Inc.

Chapter 6

HYBRID GENETIC ALGORITHMS: MODELING AND APPLICATION TO THE SCHEDULING PROBLEMS

*Jorge J. Magalhães Mendes**
Civil Engineering Department
School of Engineering, Polytechnic of Porto
Rua Dr. António Bernardino de Almeida, Porto, Portugal

ABSTRACT

The key idea of the Hybrid Genetic Algorithms (HGA) is to use traditional genetic algorithms (GAs) to explore in several regions of the search space and simultaneously incorporates a good mechanism to intensify the search around some selected regions.

Genetic algorithms are search algorithms based on the mechanics of natural selection and natural genetics. They combine survival of the fittest among string structures with a structured yet randomized information exchange to form a search algorithm with some of the innovative flair of human search [1].

One fundamental advantaged of GAs from traditional methods is described by Goldberg [1]: in many optimization methods, we move gingerly from a single solution in the decision space to the next using some transition rule to determine the next solution. This solution-to-solution method is dangerous because it is a perfect prescription for locating false peaks in multimodal search spaces. By contrast, GAs work from a rich database of solutions simultaneously (a population of chromosomes), climbing many peaks in parallel; thus the probability of finding a false peak is reduced over methods that go solution to solution.

Problems which appear to be particularly appropriate for solution by genetic algorithms include timetabling and scheduling problems [2].

The scheduling problems consist of determining the starting and finishing times of the activities. These activities are linked by precedence relations and their processing requires one or more resources. The resources are renewable, that is, the availability of each resource is renewed at each period of the planning horizon.

---

* Email: jjm@isep.ipp.pt

Usually, the objective of the well-known job shop scheduling problem (JSSP), resource constrained project scheduling problem (RCPSP) and the resource-constrained project scheduling problem with multiple modes (MRCPSP) is minimizing the *makespan*.

This paper presents some approaches for the JSSP, RCPSP and MRCPSP. These approaches are compared with some other powerful heuristics and show that our results are competitive.

Keywords: Evolutionary algorithms, Genetic algorithms, JSSP, RCPSP, MRCPSP.

1. INTRODUCTION

Hybrid genetic algorithms (HGA), also called 'memetic algorithms' in some other literature, are combination between genetic algorithms (GAs) and local search methods.

Genetic algorithms (GAs) are search algorithms based on the mechanics of natural selection and natural genetics. They combine survival of the fittest among string structures with a structured yet randomized information exchange to form a search algorithm with some of the innovative flair of human search [1].

The GAs follows the principles of The Origin of Species proposed by Charles Darwin [6], see Figure 1.

One fundamental advantaged of GAs from traditional methods is described by Goldberg [1]: in many optimization methods, we move gingerly from a single solution in the decision space to the next using some transition rule to determine the next solution. This solution-to-solution method is dangerous because it is a perfect prescription for locating false peaks in multimodal search spaces. By contrast, GAs work from a rich database of solutions simultaneously (a population of chromosomes), climbing many peaks in parallel; thus the probability of finding a false peak is reduced over methods that go solution to solution.

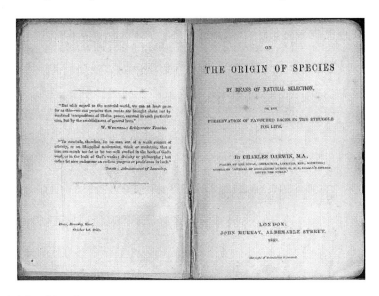

Figure 1. The Origin of Species.

First of all, an initial population of potential solutions (individuals) is generated randomly. A selection procedure based on a fitness function enables to choose the individuals candidate for reproduction. The reproduction consists in recombining two individuals by the crossover operator, possibly followed by a mutation of the offspring. Therefore, from the initial population a new generation is obtained. From this new generation, a second new generation is produced by the same process and so on. The stop criterion is normally based on the number of generations.

The general schema of GAs may be illustrated as follows, see Figure 2.

There are many variations of genetic algorithms obtained by altering the reproduction, crossover, and mutation operators. Reproduction is a process in which individual (chromosome) is copied according to their fitness values.

In this paper reproduction is accomplished by first copying some of the best individuals from one generation to the next, in what is called an elitist strategy. The fitness proportionate selection, also known as roulette-wheel selection, is the genetic operator for selecting potentially useful solutions for reproduction. The characteristic of the roulette wheel selection is stochastic sampling.

---

procedure GENETIC-ALGORITHM

Generate initial population P_0;
Evaluate population P_0;
Initialize generation counter g ← 0;
While stopping criteria not satisfied repeat
Select some elements from Pg to copy into $Pg+1$;
   Crossover some elements of Pg and put into $Pg+1$;
   Mutate some elements of Pg and put into $Pg+1$;
   Evaluate some elements of Pg and put into $Pg+1$;
   Increment generation counter: g ← g+1;
End while

End GENETIC-ALGORITHM;

Figure 2. Pseudo-code of a genetic algorithm.

The fitness value is used to associate a probability of selection with each individual chromosome. If f_i is the fitness of individual i in the population, its probability of being selected is,

$$p_i = \frac{f_i}{\sum_{i=1}^{N} f_i} , \quad i = 1,...,n$$

An example is presented in Table 1.

A roulette wheel model is established to represent the survival probabilities for all the individuals in the population. Then the roulette wheel is rotated for several times [1], see

Figure 3. After selection the mating population consists of the chromosomes (individuals): 1, 2, 3, 4, 5 and 6.

Table 1. Selection probability and fitness value

Number of chromosome	Fitness value	Selection probability
1	14	0,20
2	12	0,17
3	10	0,14
4	9	0,13
5	8	0,11
6	7	0,10
7	4	0,06
8	3	0,04
9	2	0,03
10	1	0,01

After selecting, crossover may proceed in two steps. First, members of the newly selected (reproduced) chromosomes in the mating pool are mated at random. Second, each pair of chromosomes undergoes crossover as follows: an integer position k along the chromosome is selected uniformly at random between 1 and the chromosome length l. Two new chromosomes are created swapping all the genes between $k+1$ and l [1], see Figure 4.

The mutation operator preserves diversification in the search. This operator is applied to each offspring in the population with a predetermined probability. We assume that the probability of the mutation in this paper is 0.001. With 200 genes positions we should expect 200 x 0.001 = 0.2 genes to undergo mutation for this probability value.

Figure 5 shows the evolutionary strategy with these operators of selection, recombination (or crossover) and mutation.

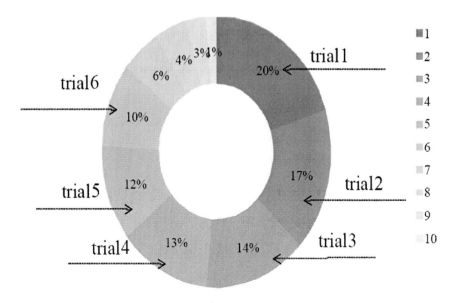

Figure 3. Roulette-wheel selection.

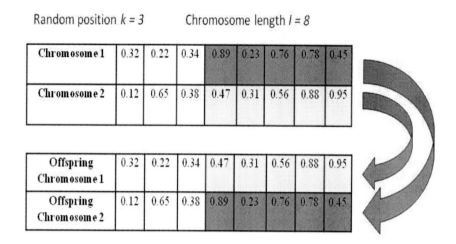

Figure 4. Crossover operator example.

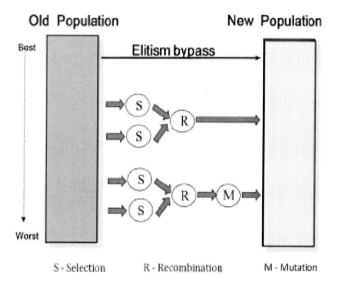

Figure 5. Evolutionary strategy.

2. SCHEDULE GENERATION SCHEMES

Schedule generation schemes (SGS) are the core of most heuristic solution procedures for the JSSP, RCPSP and MRCPSP. SGS start from scratch and build a feasible schedule by stepwise extension of a partial schedule. A partial schedule is a schedule where only a subset of the $n+2$ operations have been scheduled.

There are two different classics methods SGS available. They can be distinguished into operation and time incrementation. The so called serial SGS performs operation-incrementation and the so called parallel SGS performs time-incrementation [8].

The constructive heuristic used to construct active schedules is based on a scheduling generation scheme that does time incrementing, called parallel modified.

This heuristic makes use of the priorities and the delay times defined by the genetic algorithm and constructs active schedules. This heuristic is described by Mendes [9, 10], Gonçalves et al. [4] and Mendes et al. [3].

Figure 6 illustrates where the set of parameterized active schedules is located relative to the class of semi-active, active, and non-delay schedules.

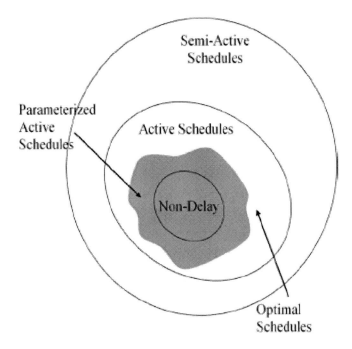

Figure 6. Diagram of various schedules with regard to optimal schedules.

3. TYPES OF SCHEDULES

Classifying schedules is the basic work to be done before attacking scheduling problems [7].

Schedules can be classified into one of the following three types of schedules:

i. Feasible schedules. A schedule is said to be feasible if it is non-preemptive and if the precedence and resource constraints are satisfied.
ii. Semi-active schedules. These are feasible schedules obtained by scheduling operations as early as possible. In a semi-active schedule the start time of a particular operation is constrained by the processing of a different operation on the same resource or by the processing of the directly preceding operation on a different resource.
iii. Active schedules. These are feasible schedules in which no operation could be started earlier without delaying some other operation or breaking a precedence constraint. Active schedules are also semi-active schedules. An optimal schedule is always active.

iv. Non-delay schedules. These are feasible schedules in which no resource is kept idle at a time when it could begin processing some operation. Non-delay schedules are active and hence are also semi-active.

In this work are generated active schedules. The constructive heuristic used to construct these active schedules is based on a parameterized active generation scheme (PAS).

4. AN OVERALL FRAMEWORK

The overall framework for the HGA combines a genetic algorithm, a schedule generation scheme that generates parameterized active schedules and a local search procedure. The genetic algorithm is responsible for evolving the chromosomes which represent the priorities of the operations.

For each chromosome the following three phases are applied:

1. Schedule parameters - this phase is responsible for transforming the chromosome supplied by the genetic algorithm into the priorities of the operations and delay time;
2. Schedule generation - this phase makes use of the priorities and the delay time and constructs active schedules;
3. Schedule improvement - this phase makes use of a local search procedure to improve the solution obtained in the schedule generation phase.

After a schedule is obtained, the quality is feedback to the genetic algorithm. Figure 7 illustrates the sequence of steps applied to each chromosome. Details about each of these phases will be presented in the next sections.

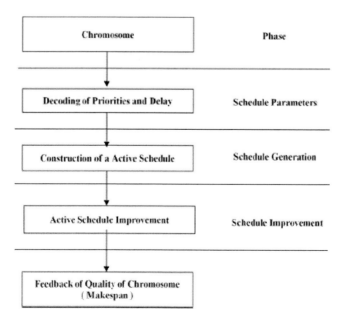

Figure 7. Phases of the framework.

In the next sections of this paper we describe applications for solving scheduling problems in JSSP, RCPSP and MRCPSP.

5. GAs IN JSSP

A job shop scheduling problem (JSSP) consists of a set of different machines (e.g., lathes, milling machines, drills, etc.) that perform operations on jobs. Each job consists of a sequence of operations, each of which uses one of the machines for a fixed duration. Once started, the operation cannot be interrupted. Each machine can process at most one operation at a time. A schedule is an assignment of operations to time intervals on the machines. The problem is to find a schedule of minimal time to complete all jobs, French [11].

In the JSSP there are a set of jobs $J = \{1,..., n\}$, a set of machines $M = \{1,..., m\}$, and a set of operations $O = \{o_0, o_1, ..., o_{ji}, ... o_{nm}, o_{nm+1}\}$. Set O contains all the operations of each job. Each job has m operations. Each machine can process at most one operation at time.

Let $O = \{o_0, o_1, ..., o_{ji}, ... o_{nm}, o_{nm+1}\}$ denote the set of all operations ($n \times m$) to be scheduled and $M = \{1,..., m\}$ the set of machines. The operations o_0 and o_{nm+1} are dummy, have no duration and represent the initial and final operations. The operations are interrelated by two kinds of constraints:

- First, the precedence constraints, which force each operation o_{ji} to be scheduled after all predecessor operations are completed P_{ji};
- Second, operation o_{ji} can only be scheduled if the machine it requires is idle. Further, let d_{ji} denote the (fixed) duration (processing time) of operation o_{ji}.

Let F_{ji} represent the finish time of operation o_{ji}. A schedule can be represented by a vector of finish times $(F_{11}, F_{ji}, ..., F_{nm+1})$.

Let $A(t)$ be the set of operations being processed at time t, and let $r_{ji} = 1$ if operation o_{ji} requires machine m to be processed and $r_{ji} = 0$ otherwise.

The conceptual model of the JSSP can be described the following way:

$$\text{Min } F_{nm+1} \tag{1}$$

subject to:

$$F_{kl} \leq F_{ji} - d_{ji} \qquad j=1,...,n \; ; \; i=1,...,m \; ; \; kl \in P_{ji} \tag{2}$$

$$\sum_{ji \in A(t)} r_{ji} \leq 1 \qquad i \in M \; ; \; t \geq 0 \tag{3}$$

$$F_{ji} \geq 0 \qquad j=1,...,n \; ; \; i=1,...,m \tag{4}$$

The objective function (1) minimizes the finish time of operation o_{nm+1} (the last operation), and therefore minimizes the *makespan*. Constraints (2) impose the precedence relations between operations and constraints (3) state that one machine can only process one operation at a time. Finally (4) forces the finish times to be non-negative.

5.1. HGA Implementation

The genetic algorithm uses a random key alphabet which is comprised of real random numbers between 0 and 1.

A chromosome represents a solution to the problem and is encoded as a vector of random keys (random numbers). Each solution chromosome is made of *2nm* genes where *nm* is the number of operations:

Chromosome = (gene₁, .., geneₙₘ, gene ₙₘ₊₁, ... , gene ₂ₙₘ)

The priority decoding expression used the following expression:

$$PRIORITY_j = gene_j \qquad j = 1,...,nm$$
$$Delay_g = gene_{nm+g} \times 1.5 \times mdur \qquad g = 1,...,nm.$$

where *mdur* is the maximum duration of all operations. The factor 1.5 was obtained after some experimental tuning.

Initial Population

The initial populations are generated randomly. The quality of this population is poor and one way to improve it is to incorporate some chromosomes generated by priority rules.

In this paper are selected the priority rules GRPW (greatest rank positional weight) and SPT (shortest processing time) to improve some chromosomes of the initial population.

GA Configuration

Though there is no straightforward way to configure the parameters of a genetic algorithm, we obtained good results with values: population size of 2 × number of operations in the problem; mutation probability of 0.001; top (best) 1% from the previous population chromosomes are copied to the next generation; stopping criterion maximum number of generations; initial population: 1% chromosomes calculated by priority rules GRPW and SPT.

Local Search

Since there is no guarantee that the schedule obtained in the construction phase is locally optimal with respect to the local neighborhood being adopted, local search may be applied to improve the solution quality.

The local search procedure begins by identifying the critical path in the solution obtained by the schedule generation procedure. Any operation on the critical path is called a critical operation. It is possible to decompose the critical path into a number of blocks where a block is a maximal sequence of adjacent critical operations that require the same machine. The critical path thus gives the neighborhood of moves.

Table 2. Experimental results for instances FT06, FT10 and FT20

Jobs	Operations /job	Instance	BKS	GA-MC	Seeds	Maximum number of generations
6	6	FT06	55	55	20	400
10	10	FT10	930	930	20	400
20	5	FT20	1165	1165	20	400

Table 3. Experimental results for instances LA01-LA40

Jobs	Operations /job	Instance	BKS	GA-MC	Seeds	Maximum number of generations
10	5	LA01	666	666	20	400
10	5	LA02	655	655	20	400
10	5	LA03	597	597	20	400
10	5	LA04	590	590	20	400
10	5	LA05	593	593	20	400
15	5	LA06	926	926	20	400
15	5	LA07	890	890	20	400
15	5	LA08	863	863	20	400
15	5	LA09	951	951	20	400
15	5	LA10	958	958	20	400
20	5	LA11	1222	1222	20	400
20	5	LA12	1039	1039	20	400
20	5	LA13	1150	1150	20	400
20	5	LA14	1292	1292	20	400
20	5	LA15	1207	1207	20	400
10	10	LA16	945	946	20	400
10	10	LA17	784	784	20	400
10	10	LA18	848	848	20	400
10	10	LA19	842	842	20	400
10	10	LA20	902	902	20	400
15	10	LA21	1046	1054	20	400
15	10	LA22	927	932	20	400
15	10	LA23	1032	1032	20	400
15	10	LA24	935	944	20	1472
15	10	LA25	977	985	20	439
20	10	LA26	1218	1218	20	400
20	10	LA27	1235	1252	20	400
20	10	LA28	1216	1231	20	400
20	10	LA29	1157	1180	20	400
20	10	LA30	1355	1355	20	400
30	10	LA31	1784	1784	20	400
30	10	LA32	1850	1850	20	400
30	10	LA33	1719	1719	20	400
30	10	LA34	1721	1721	20	400
30	10	LA35	1888	1888	20	400
15	15	LA36	1268	1278	20	400
15	15	LA37	1397	1408	20	400
15	15	LA38	1196	1213	20	400
15	15	LA39	1233	1244	20	400
15	15	LA40	1222	1233	20	400

After the critical path will be known, the Monte Carlo Method randomly selects the block where we make the swapping operations. In fact, this process selects the most critical resource to make the swapping between two consecutive operations, see Mendes [12].

The performance of GA-MC (Genetic Algorithm – Monte Carlo) was evaluated on a standard set of 43 benchmark instances belonging to two classical sets known from FT from Fisher and Thompson [13] and LA from Lawrence [14]. The problem size varies between 6 and 30 jobs and between 5 and 15 machines.

Tables 2 and 3 summarize the experimental results. It lists number of jobs, number of operations, instance, best known solution (BKS), GA-MC, number of seeds and number of maximum generations for each instance.

In most of the problems, the GA-MC performs better than others algorithms. In terms of average relative deviation (ARD), GA-MC outperforms all others genetic algorithms reported in this paper.

As different authors used different number of problems, the comparison is based only on those problems that authors considered. The result is showed in Table 4.

Table 4. Comparison of the % deviations for the different number of problems authors considered

NIS	Test problems	Authors	Algorithm	ARD(%)
43	FT06 - FT10 FT20 and LA01-LA40	Mendes [12]	GA-MC	0.29
		Gonçalves and Beirão [15]	-	0.90
		Gonçalves et al. [4]	P. Active	0.39
			Non-Delay	1.22
			Active	1.10
		Mendes [16]	GA-RKV-JSP	0.38
		Nowicki and Smutnicki [17]	-	0.05
		F. Pezella and E. Merelli [18]	-	0.10
		Zhang et al. [19]	-	0.00
		Sha and Hsu [20]	HPSO	0.02
		Sha and Hsu [20]	PSO	0.37
		Binato et al. [21]	-	1.77
		Aiex et al. [22]	-	0.43
40	LA01-LA40	Mendes [12]	GA-MC	0.31
		Mendes [16]	GA-RKV-JSP	0.41
		Rego and Duarte [23]	F&F-PRD	0.29
		Hasan et al. [24]	GR-SA-RA	0.97
		Gonçalves et al. [4]	P. Active	0.41
			Non-Delay	1.20
			Active	1.10
		Aarts et al. [25]	GLS1	2.05
			GLS2	1.75
		Dorndorf et al. [26]	PGA	4.00
			SBGA	1.25
		Binato et al. [21]	-	1.87
		Adams et al. [27]	SBI	3.67

5.2. Partial Conclusions

This HGA implementation is based on a genetic algorithm. The chromosome representation of the problem is based on random keys. Reproduction, crossover and mutation are applied to successive chromosome populations to create new chromosome populations.

The schedules are constructed using a heuristic rule in which the priorities are defined by the genetic algorithm.

After a schedule is obtained, a local search heuristic is applied to improve the solution using the Monte Carlo Method.

The HGA was tested on a set of 43 standard instances taken from the literature and compared with the best state-of-the-art approaches. The algorithm produced good results when compared with other approaches.

Further work could be conducted to explore the possibility of genetically correct the chromosomes supplied by the genetic algorithm to reflect the solutions obtained by the local search heuristic.

6. GAs IN RCPSP

The resource constrained project scheduling problem (RCPSP) can be stated as follows. A project consists of $n+2$ activities where each activity has to be processed in order to complete the project. Let $J = \{0, 1, ..., n, n+1\}$ denote the set of activities to be scheduled and $K = \{1, ..., k\}$ the set of resources. The activities 0 and $n+1$ are dummy, have no duration and represent the initial and final activities.

The activities are interrelated by two kinds of constraints:

1. The *precedence constraints*, which force each activity j to be scheduled after all predecessor activities, P_j, are completed;
2. Performing the activities requires resources with *limited capacities*.

While being processed, activity j requires $r_{j,k}$ units of resource type $k \in K$ during every time instant of its non-preemptable duration d_j. Resource type k has a limited capacity of R_k at any point in time. The parameters d_j, $r_{j,k}$ and R_k are assumed to be non-negative and deterministic. For the project start and end activities we have $d_0 = d_{n+1} = 0$ and $r_{0,k} = r_{n+1,k} = 0$ for all $k \in K$.

The problem consists in finding a schedule of the activities, taking into account the resources and the precedence constraints, that minimizes the *makespan* (C_{max}).

Let F_j represent the finish time of activity j. A schedule can be represented by a vector of finish times $(F_1, ..., F_m, ..., F_{n+1})$. The *makespan* of the solution is given by the maximum of all predecessors activities of activity $n+1$, i.e. $F_{n+1} = Max_{l \in P_{n+1}} \{F_l\}$.

The conceptual model of the RCPSP was described by Christofides et al. [28] in the following way:

Min F_{n+1} \hfill (1)

subject to:

$$F_l \leq F_j - d_j \qquad j=1,...,N+1 \ ; \ l \in P_j \qquad (2)$$

$$\sum_{j \in A(t)} r_{j,k} \leq R_k \qquad k \in K \ ; \ t \geq 0 \qquad (3)$$

$$F_j \geq 0 \qquad j=1,...,N+1 \qquad (4)$$

The objective function (1) minimizes the finish time of activity $n+1$, and therefore minimizes the makespan. Constraints (2) impose the precedence relations between activities and constraints (3) limit the resource demand imposed by the activities being processed at time t to the capacity available. Finally (4) forces the finish times to be non-negative.

6.1. HGA Implementation

The genetic algorithm uses a random key alphabet which is comprised of real random numbers between 0 and 1.

A chromosome represents a solution to the problem and is encoded as a vector of random keys (random numbers). Each solution chromosome is made of $2n$ genes where n is the number of activities:

Chromosome = $(gene_1, .., gene_n, gene_{n+1}, ..., gene_{2n})$

The priority decoding expression used the following expression

$$PRIORITY_j = \frac{LLP_j}{LCP} \times \left[\frac{1+gene_j}{2} \right] \qquad j=1,...,n$$

where LLP_j is the longest length path from the beginning of the activity j to the end of the project and LCP is the length along the critical path of the project, see Mendes [9].

The genes between $n+1$ and $2n$ are used to determine the delay times used when scheduling an activity. The delay time used by each scheduling iteration g, $Delay_g$, is given by the following expression:

$Delay_g = gene_{n+g} \times 1.5 \times MaxDur$

where MaxDur is the maximum duration of all activities. The factor 1.5 was obtained after some experimental tuning.

GA Configuration

Though there is no straightforward way to configure the parameters of a genetic algorithm, we obtained good results with values: population size of 5 × number of activities in the problem; mutation probability of 0.001; top (best) 1% from the previous population chromosomes are copied to the next generation; stopping criterion of 250 generations.

Local Search

Local search algorithms move from solution to solution in the space of candidate solutions (the search space) until a solution optimal or a stopping criterion is found. In this paper was applying backward and forward improvement based on Klein [29].

Initially is constructed a schedule by planning in a forward direction starting from the project's beginning. After is applying backward and forward improvement based on Klein [29].

The backward planning consists in reversing the project network and applying the scheduling generator scheme.

An example is illustrated in Mendes [9].

The performance of GA-RKV (Genetic Algorithm – Random Key Variant) was tested on the instance sets:

- J30 (480 instances each with 30 activities)
- J60 (480 instances each with 60 activities)
- J120 (600 instances each with 120 activities)

available in PSPLIB. All problem instances require four resource types. Instances details are described in Kolisch et al. [30].

The results obtained are given in Tables 5, 6 and 7.

Table 5, column 3, summarizes the average deviation percentage from the optimal *makespan* (D_{OPT}), for the instance set J30. The GA-RKV obtained $D_{OPT} = 0.01$. The number of instances for which the algorithm obtains the optimal solution is 476. For the set J30, GA-RKV ranks five.

Table 5. Top-ten computational results for J30 instances

Algorithm	Reference	J30 D_{OPT}
MAPS	Mendes and Gonçalves [31]	0.00
GA-TS path relinking	Kochetov and Stolyar [32]	0.00
Decomp. & local opt.	Palpant et al. [33]	0.00
F&F(5)	Ranjbar [34]	0.00
GA - RKV	**Mendes [9]**	**0.01**
GA - RKV -AS	Mendes [41]	0.01
GAPS	Mendes et al. [3]	0.01
Scatter Search - FBI	Debels et al. [35]	0.01
VNS-activity list	Fleszar and Hindi [36]	0.01
GA - DBH	Debels and Vanhoucke [37]	0.02
GA – hybrid, FBI	Valls et al. [38]	0.02
GA - FBI	Valls et al. [39]	0.02

Table 6. Top-ten computational results for J60 instances

Algorithm	Reference	J60 D_{LB}
F&F(5)	Ranjbar [34]	10.56
MAPS	Mendes and Gonçalves [31]	10.64
GAPS	Mendes et al. [3]	10.67
GA - DBH	Debels and Vanhoucke [37]	10.68
Scatter Search - FBI	Debels et al. [35]	10.71
GA – hybrid, FBI	Valls et al. [38]	10.73
GA, TS – path relinking	Kochetov and Stolyar [32]	10.74
GA - FBI	Valls et al. [39]	10.74
GA - RKV -AS	Mendes [41]	10.81
Decomp. & local opt.	Palpant et al. [33]	10.81
GA - RKV	Mendes [9]	**10.88**
VNS-activity list	Fleszar and Hindi [8]	10.94

Tables 6 and 7, columns 3, summarize the average deviation percentage from the well-known critical path-based lower bound (D_{LB}) for the instance set J60 and J120, respectively. For the instances set J60 and J120, GA-RKV ranks eleven. The lower bound values (D_{LB}) are reported by Stinson et al. [42].

Table 7. Top-ten computational results for J120 instances

Algorithm	Reference	J120 D_{LB}
GA-DBH	Debels and Vanhoucke [37]	30.82
MAPS	Mendes and Gonçalves [31]	31.19
GAPS	Mendes et al. [3]	31.20
GA – hybrid, FBI	Valls et al. [38]	31.24
F&F(5)	Ranjbar [34]	31.42
Scatter Search - FBI	Debels et al. [35]	31.57
GA - FBI	Valls et al. [39]	31.58
GA, TS – path relinking	Kochetov and Stolyar [32]	32.06
Decomp. & local opt.	Palpant et al. [33]	32.41
GA - RKV -AS	Mendes [41]	32.47
GA - RKV	Mendes [9]	**32.50**
GA - Self adapting	Hartmann [40]	33.21

6.2. Partial Conclusions

This HGA is based on a genetic algorithm (a variant of the genetic algorithm proposed by Goldberg [1] with binary code) for the resource constrained project scheduling problem. The chromosome representation of the problem is based on random keys.

The schedules are constructed using a heuristic in which the priorities are defined by the genetic algorithm with a constructive heuristic. The constructive heuristic for constructing feasible schedules is extended by the flexible use of different planning directions including the backward and forward planning.

The HGA was tested on a set of 1560 standard instances taken from the literature and compared with the best state-of-the-art approaches. The algorithm produced good results when compared with other approaches therefore validating the effectiveness of the proposed algorithm.

7. GAs IN MRCPSP

The MRCPSP problem can be stated as follows. A project consists of $n+2$ activities where each activity has to be processed in order to complete the project. Let $J = \{0, 1, ..., n, n+1\}$ denote the set of activities to be scheduled and $K = \{1, ..., k\}$ the set of renewable resources. Each resource type k has a limited capacity of R_k at any point in time.

The activities 0 and $n+1$ are dummy, have no duration and represent the initial and final activities. The activities are interrelated by two kinds of constraints:

1. The precedence constraints, which force each activity j to be scheduled after all predecessor activities, P_j, are completed;
2. Performing the activities requires resources with limited capacities.

Each activity can be performed in one of several different modes. A mode represents a combination of different resources and/or levels of resources with an associated duration. Once an activity is started in one mode, it may not be changed. One activity j can be executed in m modes given by the set $M_j = \{1, ..., m_j\}$. The duration of activity j being performed in mode m_j is given by d_{jm}. The activity j executed in mode m_j uses r_{jmk} units of renewable resource k, where $r_{jmk} \leq R_k$ for each renewable resource k.

While being processed, activity j requires r_{jmk} units of resource type $k \in K$ during every time instant of its non-preemptable duration d_{jm}. The parameters d_{jm}, r_{jmk} and R_k are assumed to be non-negative and deterministic.

The problem depends on finding a schedule of the activities, taking into account the resources and the precedence constraints, which minimize the *makespan* (C_{max}).

Let F_j represent the finish time of activity j. A schedule can be represented by a vector of finish times $(F_1, ..., F_m, ..., F_{n+1})$. The *makespan* of the solution is given by the maximum of all predecessors activities of activity $n+1$, i.e., $F_{n+1} = Max_{l \in P_{n+1}} \{F_l\}$.

7.1. HGA Implementation

The genetic algorithm uses a random key alphabet U (0, 1) and an evolutionary strategy identical to the one proposed by Goldberg [1].

A chromosome represents a solution to the problem and it is encoded as a vector of random keys (random numbers). Each solution encoded as *initial chromosome* is made of $mn+1$ genes where n is the number of activities (first level):

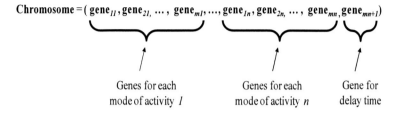

For each activity j we have a set of execution modes:

$$\text{Activity}_1 = (gene_{11}, gene_{21}, \ldots, gene_{m1})$$
$$\text{Activity}_2 = (gene_{12}, gene_{22}, \ldots, gene_{m2})$$
$$\ldots\ldots\ldots$$
$$\text{Activity}_n = (gene_{1n}, gene_{2n}, \ldots, gene_{mn})$$

After the choice of the execution mode m_j for each activity j, the solution chromosome is composed by $n+1$ genes (second level).

The priority decoding expression uses the following expression:

$$PRIORITY_j = \frac{LLP_j}{LCP} \times \left[\frac{1+gene_{m_j}}{2}\right] \qquad j=1,\ldots,n$$

where,

LLP_j is the longest length path from the beginning of the activity j to the end of the project

LCP is length along the critical path of the project, see Mendes [9].

m_j is the gene of the selected mode for activity j.

The gene $mn+1$ is used to determine the delay time used when scheduling the activities. The delay time used by each chromosome is given by the following expression:

$$\textit{Delay time} = gene_{mn+1} \times 1.5 \times \textit{MaxDur}$$

where MaxDur is the maximum duration of all activities. The factor 1.5 is obtained after some experimental tuning.

This genetic *algorithm* is based on a two-level mechanism, see Mendes [5]. The first level is composed by the chromosome with all genes to solve the initial problem – MRCPSP.

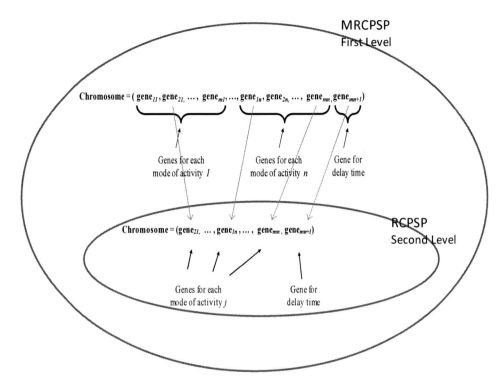

Figure 8. A Two-Level Genetic Algorithm.

For each *initial chromosome*, we must select only one gene by activity. With a gene by activity we must solve the *solution chromosome*, that is, the second level. In the second level we must solve a problem *RCPSP*, see Figure 8.

GA Configuration

Though there is no straightforward way to configure the parameters of a genetic algorithm, we obtained good results with values: population size of 5 × number of activities in the problem; mutation probability of 0.001; top (best) 1% from the previous population chromosomes are copied to the next generation; stopping criterion of 50 generations.

Local Search

Initially is constructed a schedule by planning in a forward direction starting from the project's beginning. After it is applied backward and forward improvement trying to get a better solution. The backward planning consists in reversing the project network and applying the scheduling generator scheme. An example is described by Mendes [9].

The performance of RKV-PAS-MM (Random Key Variant Parameterized Active Schedules for Multi-Mode) it is compared with the approaches proposed by Vaca [43] and Mendes [44] for the problem instances (multi-mode). To illustrate its effectiveness, we consider a total of 26 instances (with a number of activities between 7 and 80) proposed by Vaca [43] which are adapted to the field of construction. The problem instances require between two and ten renewable resource types. All resources are given as renewable resources and the availabilities associated with all resources are assumed to be constant over time. Instance details are described by Vaca [43].

Table 8, Column 5, summarizes the obtained results by RKV-PAS-MM. The results are the best when compared with the values presented by Vaca [43] and Mendes [44]. The maximal computational time dispended is 60 seconds for each instance.

Table 9 shows the *number of instances solved* (NIS) and the *average relative deviation* (ARD) with respect to the *best known solution* (BKS).

Table 8. Experimental results for multi-mode instances

Probl. No.	Duration without limited resources	BKS	Best duration for multi-mode [43]	RKV-NDS-MM [44]	RKV-PAS-MM
1	71	96	107	107	96
2	74	77	77	77	78
3	238	306	309	309	306
4	23	28	28	28	28
5	125	132	132	132	132
6	46	75	75	75	75
7	26	42	42	42	42
8	75	126	141	141	126
9	37	66	66	66	66
10	85	103	103	103	103
11	93	140	140	140	140
12	9	11	11	11	11
13	8	9	9	9	9
14	48	88	88	88	88
15	18	19	19	19	19
16	11	12	12	12	12
17	16	22	23	23	22
18	16	17	17	17	17
19	36	37	37	37	37
20	8	13	14	13	13
21	80	80	80	80	80
22	28	38	38	38	38
23	33	34	40	34	34
24	33	35	37	35	35
25	31	56	61	58	56
26	60	84	86	85	84

For comparison between the methods, we have used one measure, namely the *average relative deviation* (ARD):

$$RE = \sum_{i=1}^{NIS} \frac{C_{max_i} - BKS_i}{BKS_i}$$

$$ARD = \frac{RE}{NIS}$$

Table 9. Average relative deviation

Algorithm	NIS	ARD
Oscar Lopez Vaca [43]	26	0.027
J. Magalhaes-Mendes [44]	26	0.013
RKV-PAS-MM	26	0.000

Overall, we solved 26 instances with RKV-PAS-MM and obtained an ARD of 0.000%. The RKV-PAS-MM obtained the best-known solution for 25 instances, i.e. in 96% of problem instances. RKV-PAS-MM presented an improvement with respect to almost all others algorithms.

7.2. Partial Conclusions

This HGA is based on a genetic algorithm (a variant of the genetic algorithm proposed by Goldberg [1] with binary code) for the resource-constrained project scheduling problem with multiple modes. The chromosome representation of the problem is based on random keys.

The schedules are constructed using priorities defined by the genetic algorithm with a constructive heuristic. The constructive heuristic for constructing feasible schedules is extended by the flexible use of different planning directions including the backward and forward improvement (FBI) planning. For some instance, a combination of a heuristic constructive and the genetic algorithm may yield a good result, but in another instance the FBI can improve the initial schedule.

The approach was tested on a set of 26 standard instances with multi-modes in the field of construction, taken from the literature and compared with the others approach. The algorithm produced good results when compared with other approaches therefore validating the effectiveness of the proposed algorithm.

CONCLUSION

In this work were presented HGAs applied to JSSP, RCPSP and MRCPSP. The ingredients of the HGAs, evolutionary operators and local neighborhood search, were described or referenced. The performance of each HGA was investigated on a set of JSSP, RCPSP and MRCPSP instances and compared to the best-know results know from the literature. In the particular case of MRCPSP instances are used scheduling instances in the field of construction.

The results seem to confirm the suitability of genetic algorithms to solve difficult problems in the field of scheduling.

The results also demonstrate the utility of using hybrid algorithms, which allow to overcome the apparent drawbacks of genetic algorithms.

Further research can be extended to problems where the resource availability varies within time and in contexts where several projects must be concurrently executed.

REFERENCES

[1] D.E. Goldberg, *Genetic Algorithms in Search Optimization and Machine Learning*, Addison-Wesley, 1989.

[2] E. Burke and J. Landa Silva, *The Design of Memetic Algorithms for Scheduling and Timetabling Problems*, Recent Advances in Memetic Algorithms, Studies in Fuzziness and Soft Computing (Eds. Krasnogor, N., Hart, W. & Smith, J.), pp. 289-312, Springer, 2004.

[3] J.J.M. Mendes, J.F.Gonçalves, and M.C.G. Resende, A Random Key Based Genetic Algorithm for the Resource Constrained Project Scheduling Problem, *Computers and Operations Research*, 36, pp. 92 – 109, 2009.

[4] J.F. Gonçalves, J.J.M. Mendes, and M.C.G. Resende, A hybrid genetic algorithm for the job shop scheduling problem, *European Journal of Operational Research*, 167, 77-95, 2005.

[5] J. Magalhães-Mendes, A Two-level Genetic Algorithm for the Multi-Mode Resource-Constrained Project Scheduling Problem, *International Journal of Systems Applications, Engineering and Development*, Issue 3, Volume 5, 2011, pp. 271-278, ISSN: 2074-1308.

[6] C. Darwin, *The Origin of Species by Means of Natural Selection*, London, John Murray, Albemarle Street, 1859.

[7] R. Kolisch, *Project Scheduling under Resource Constraints*, Physica-Verlag, Germany, 1995.

[8] R.Kolisch and S.Hartmann, *Heuristic Algorithms for Solving the Resource-Constrained Project Scheduling Problem: Classification and Computational Analysis*, J. Weglarz (editor), Kluwer, Amsterdam, the Netherlands, 1999, pp. 147–178.

[9] J. Magalhães-Mendes, Project scheduling under multiple resources constraints using a genetic algorithm, *WSEAS TRANSACTIONS on BUSINESS and ECONOMICS*, Issue 11, Volume 5, November 2008, pp. 487-496.

[10] J. J. Magalhães-Mendes, *Sistema de Apoio à Decisão para Planeamento de Sistemas de Produção do Tipo Projecto*. Ph. D. Thesis. Departamento de Engenharia Mecânica e Gestão Industrial, Faculdade de Engenharia da Universidade do Porto, Portugal, 2003.

[11] S. French, *Sequencing and Scheduling: An Introduction to the Mathematics of the Job Shop*, Horwood, Chichester, UK, 1982.

[12] J. Magalhães-Mendes, *A genetic algorithm for the job shop scheduling with a new local search using Monte Carlo method*, Proceedings of the 10th WSEAS international conference on Artificial intelligence, knowledge engineering and data bases, pp.26-31, February 20-22, 2011, Cambridge, UK.

[13] H. Fisher and G.L. Thompson, *Probabilistic Learning Combinations of Local Job-Shop Scheduling Rules, in: Industrial Scheduling*, J.F. Muth and G.L. Thompson (eds.), Prentice-Hall, Englewood Cliffs, NJ, 1963, pp. 225-251.

[14] S. Lawrence, *Resource Constrained Project Scheduling: An Experimental Investigation of Heuristic Scheduling Techniques*, GSIA, Carnegie Mellon University, Pittsburgh, PA, 1984.

[15] J.F. Gonçalves and N.C. Beirão, Um Algoritmo Genético Baseado em Chaves Aleatórias para Sequenciamento de Operações, *Revista Associação Portuguesa de Desenvolvimento e Investigação Operacional*, Vol. 19, 1999, pp. 123-137, (in Portuguese).

[16] J. Magalhães-Mendes, An Optimization Approach for the Job Shop Scheduling Problem, *In Proceedings of the 14th WSEAS International Conference on Applied Mathematics (MATH'09)*, Tenerife, Canary Islands, Spain, 2009, pp. 120-125.

[17] E. Nowicki and C. Smutnicki, A Fast Taboo Search Algorithm for the Job-Shop Problem, *Management Science*, Vol. 42, No. 6, 1996, pp. 797-813.

[18] F. Pezzela and E. Merelli, A tabu search method guided by shifting bottleneck for the job shop scheduling problem, *European Journal of Operational Research*, Vol. 120, 2000, pp. 297-310.

[19] C. Y. Zhang, P. Li. and Z. Guan, A very fast TS/SA algorithm for the job shop scheduling problem, *Computers and Operations Research*, Vol. 35, 2008, pp. 282-294.

[20] D. Y. Sha and C. Hsu, A hybrid particle swarm optimization for job shop scheduling problem. *Computers and Industrial Engineering*, 51, 4, 2006, pp. 791-808.

[21] S. Binato, W.J.Hery, D.M. Loewenstern and M.G.C.Resende, *A GRASP for Job Shop Scheduling*. In: Essays and Surveys in Metaheuristics, Ribeiro, Celso C., Hansen, Pierre (Eds.), Kluwer Academic Publishers, 2002.

[22] R.M.Aiex, S.Binato and M.G.C. Resende, Parallel GRASP with Path-Relinking for Job Shop Scheduling, *Parallel Computing*, Vol. 29, Issue 4, 2003, pp. 393 - 430.

[23] C. Rego and R. Duarte, A filter-and-fan approach to the job shop scheduling problem, *European Journal of Operational Research*, Vol. 194, 2009, pp. 650–662.

[24] S.M.K. Hasan, R. Sarker and D. Cornforth, GA with Priority Rules for Solving Job-Shop Scheduling Problems. In *Proceedings of the IEEE Congress on Evolutionary Computation CEC(2008)*, 2008, pp. 1913-1920.

[25] E.H.L.Aarts, P.J.M.Van Laarhoven, J.K. Lenstra and N.L.J.Ulder, A computational study of local search algorithms for job shop scheduling, *ORSA Journal on Computing*, 6, 1994, pp. 118-125.

[26] U. Dorndorf, and E. Pesch, Evolution Based Learning in a Job Shop Environment, *Computers and Operations Research*, Vol. 22, 1995, pp. 25-40.

[27] J. Adams, E. Balas and Z. Zawack, The shifting bottleneck procedure for job shop scheduling, *Management Science*, Vol. 34, 1988, pp. 391-401.
[28] N. Christofides, R.Alvarez-Valdés and J. Tamarit, Problem scheduling with resource constraints: A branch and bound approach, *European Journal of Operational Research*, Vol. 29, 1987, pp. 262-273.
[29] R. Klein, Bidirectional planning: improving priority rule-based heuristics for scheduling resource-constrained projects, *European Journal of Operational Research*, Vol. 127, 2000, pp. 619-638.
[30] R. Kolisch, Schwindt, A.Sprecher, *Benchmark instances for scheduling problems*. In J.Weglarz, (ed.) Handbook on recent advances in project scheduling, Kluwer, Amsterdam, 1998, pp. 197-212.
[31] J.J.M. Mendes and J.F. Gonçalves, A Memetic Algorithm-Based Heuristics for the Resource Constrained Project Scheduling Problem, Proceedings of *II International Conference on Computational Methods for Coupled Problems in Science and Engineering*, Spain, 2007, pp. 644-648.
[32] Y. Kochetov and A. Stolyar. Evolutionary local search with variable neighborhood for the resource constrained project scheduling problem. In *Proceedings of the 3rd International Workshop of Computer Science and Information Technologies*, Russia, 2003.
[33] M. Palpant, C. Artigues and P. Michelon, LSSPER: Solving the resource-constrained project scheduling problem with large neighbourhood search, *Annals of Operations Research*, Vol.131, 2004, pp. 237-257.
[34] M. Ranjbar, Solving the resource-constrained project scheduling problem using filter-and-fan approach, *Applied Mathematics and Computation*, Vol. 201, 2008, pp. 313–318.
[35] D. Debels, B. De Reyck, R.Leus and M.Vanhoucke, A Hybrid Scatter Search/Electromagnetism Meta-Heuristic for Project Sheduling, *European Journal of Operational Research*, Vol. 169, 2006, pp. 638-653.
[36] K. Fleszar and K.S. Hindi, Solving the resource-constrained project scheduling problem by a variable neighbourhood search, *European Journal of Operational Research*, Vol. 155, 2004, pp. 402-413.
[37] D. Debels and M. Vanhoucke. *A Decomposition-Based Heuristic for the Resource-Constrained Project Scheduling Problem*. Working Paper 2005/293, Faculty of Economics and Business Administration, University of Ghent, Ghent, Belgium, 2005.
[38] V. Valls, F. Ballestin and M.S. Quintanilla. *A hybrid genetic algorithm for the RCPSP*. Technical report, Department of Statistics and Operations Research, University of Valencia, 2003.
[39] V. Valls, F. Ballestin and M.S. Quintanilla, Justification and RCPSP: A technique that pays, *European Journal of Operational Research*, Vol.165, 2005, pp. 375-386.
[40] R. Kolisch and S. Hartmann, Experimental investigation of heuristics for resource-constrained project scheduling: an update, *European Journal of Operational Research*, Vol.174 (1), 2006, pp. 23-37.
[41] J. Magalhães-Mendes, *Project Scheduling* in Progress in Management Engineering, Editors: Lucas P. Gragg and Jan M. Cassell, Nova Publishers, New York, USA, pp. 117-134, 2009.

[42] J.P. Stinson, E.W. Davis and B.M. Khumawala, Multiple Resource-Constrained Scheduling Using Branch and Bound, *AIIE Transactions*, Vol. 10, 1978, pp. 252-259.
[43] O. C. L. Vaca, *Um Algoritmo Evolutivo para a Programação de Projectos Multi-modos com Nivelamento de Recursos Limitados*, Ph.D. Thesis, Universidade Federal de Santa Catarina, Florianópolis, Brasil, 1995. (In portuguese)
[44] J. Magalhães-Mendes, Project Scheduling with Multiple Modes: A Genetic Algorithm based Approach, in *Proceedings of the First International Conference on Soft Computing Technology in Civil, Structural and Environmental Engineering*, B.H.V. Topping, Y. Tsompanakis, (Editors), Civil-Comp Press, Stirlingshire, United Kingdom, paper 12, 2009. doi:10.4203/ccp.92.12.

In: Handbook of Genetic Algorithms: New Research
Editors: A. Ramirez Muñoz and I. Garza Rodriguez

ISBN: 978-1-62081-158-0
© 2012 Nova Science Publishers, Inc.

Chapter 7

GA OPTIMIZATION FOR A SPACE-CONSTRAINED MACHINE ROOM USING MULTI-LAYER SOUND ABSORBERS AND ONE-LAYER ACOUSTICAL HOODS

Min-Chie Chiu[1,*]

[1]Department of Mechanical and Automation Engineering, Chung Chou University of Science and Technology, Taiwan, ROC

ABSTRACT

Noise control is essential in an enclosed machine room where there is a high level of reverberant and direct sound. Additionally, the noise level can be reduced using a sound absorber and an acoustical enclosure. The traditional method for designing sound absorbers and acoustical enclosures is time-consuming. Therefore, to efficiently control noise levels, shape optimization of multi-layer sound absorbers and one-layer close-fitting acoustical hoods is being considered.

In this paper, a Genetic Algorithm (*GA*) in conjunction with a theoretical model of multi-layer sound absorbers and acoustic hoods is applied in the following numerical optimizations. Before noise abatement is carried out, the reliability of the *GA* method will be checked by optimizing the sound absorption coefficient of three kinds of multi-layer sound absorbers at a pure tone (500Hz). Moreover, the noise abatement of a piece of equipment within a machine room using three kinds of multi-layer sound absorbers and one close-fitting acoustical hood in conjunction with the *GA* method has been exemplified and fully explored. The results reveal that both the acoustical panel and the acoustical enclosure can be precisely designed.

Consequently, this paper provides a quick and effective method for reducing noise levels by optimally designing a shaped multi-layer sound absorber and a one-layer close-fitting acoustical hood using the *GA* method.

Keywords: Close fitting; acoustical hood; multi-layer sound absorber, genetic algorithm

---

[*] Corresponding author. E-mail: minchie.chiu@msa.hinet.net; Mailing address: Department of Mechanical and Automation Engineering, Chung Chou University of Science and Technology, No. 6, Lane 2, Sec.3, Shanchiao Rd., Yuanlin, Changhua 51003, Taiwan, R.O.C.

1. NOMENCLATURE

This paper is constructed on the basis of the following notations:

a: the length of the panel
b: the width of the panel
c_o: sound speed (m s^{-1})
$chrm$:   bit length
d_i: diameter of perforated hole on the i-th front plate (m)
dd: the distance between the equipment and the hood
D: the panel's complex bulk modulus
DD: the total thickness of the acoustic panel
Df_T: the total thickness of the acoustic fiber
D_{fi}:   the thickness of the i-layer acoustic fiber
E: the panel's Young's modulus
elt: elitism (1 for yes, 0 for no)
f: cyclic frequency (Hz)
gen: maximum number of iteration
h: the thickness of the panel
IL_j: the sound insertion loss for the j-th acoustical hood (dB)
j: imaginary unit
k: wave number ($= \dfrac{\omega}{c_o}$)
k_{1i}:   real part of complex $k_{fiber-i}$
k_{2i}:   image part of complex $k_{fiber-i}$
$k_{fiber-i}$:   complex propagation constant of the i-th layer of the acoustic fiber
K_{pi}:   complex propagation constant of the i-th layer of the perforated front plate
K_{1A}:   real part of complex K_{fiber}
K_{2A}:   image part of complex K_{fiber}
L_i: air depth of the i-th layer of the sound absorber (m)
L_x, L_y, L_z: the outline dimension of the plenum (m)
N: hole's number on the perforated front plate per 1m^2
OBJ: objective function
pc: crossover ratio
$p_i\%$:   porosity of the perforated plate ($= \varepsilon$ *100%) (%)
p_i: acoustic pressure at i (Pa)
pm: mutation ratio

pop: number of population
q_i: thickness of the i-th layer of the perforated plate (m)
R_i: acoustic flow resistance of the i-th layer of the acoustic fiber (MKS rayls m^{-1})
$R_{fiber-i}$: real part of the i-th layer of complex Z_{fiber}
S_w: the total area of the wall
SPL_T: the silenced sound power level at RV1
$SWLO$: the unsilenced sound power level for equipment A
u_i: acoustic particle velocity at i (kg s^{-1})
Z_i: specific normal impedance at i.
$Z_{fiber-i}$: characteristic impedance of the i-th layer of the acoustic fiber
Z_p: characteristic impedance of the perforated front plate
$X_{fiber-i}$: image part of the i-th layer of complex $Z_{fiber-i}$
α_i: sound absorption coefficient of the i-layer sound absorber
$\overline{\alpha}_i$: average sound absorption coefficient within a plenum using the i-layer sound absorber
ω: angular frequency (rad s^{-1})
v: kinematic viscosity of air (=15*10^{-6} m^2/s)
ε_i: porosity of the i-th layer of the perforated plate (m)
δ_i: viscous boundary layer thickness of the i-th layer of the perforated plate (m)
ρ_o: air density (kg m^{-3})
$\rho_o c$: the acoustic impedance
η: the panel's internal damping coefficient
vv: the poison ratio of the panel
ρ: the panel's density.

2. INTRODUCTION

Controlling venting noise using an acoustical plenum is vital [1, 2]. Lan and Chiu proposed a strategy to economically reduce the noise level by adjusting the location of noisy equipment [3]. In order to efficiently control the noise level specified by the Environmental Protection Agency, Chiu [4] proposed a noise control strategy that optimally adjusted equipment allocation and searched for a shaped sound barrier around the plant's boundary. Roberts and Murray [5] developed a muffler system for eliminating direct sound waves emitted from a venting device. Sanders [6] introduced a composite sound barrier panel used for depressing a direct sound wave emitted from a piece of equipment. However, in a space-constrained machine room, two primary sound energies (direct sound and reverberant sound) dominate the sound field. Moreover, neither the allocation adjustment nor the shaped sound barrier is useful for a reverberant sound propagating within an enclosed building. To efficiently depress the hybrid noise within a machine room, an optimal assessment of a multi-layer acoustical panel and an one-layer close fitting acoustical enclosure that is often used to reduce reflected and direct sound energy is necessary.

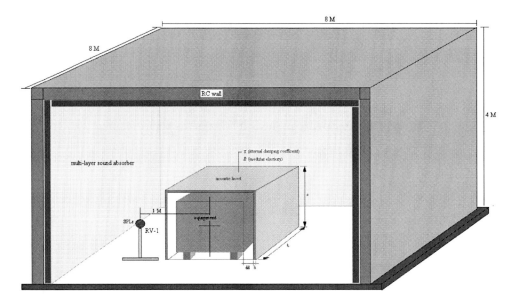

(A) elevation view

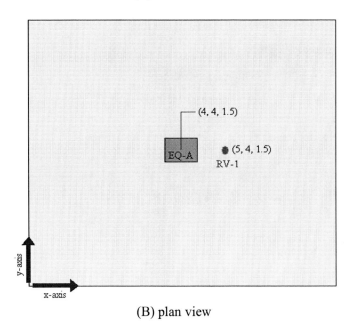

(B) plan view

Figure 1. Noise abatement of a machine room with a multi-layer sound absorber and an acoustical hood.

Addressing the effect of the vibrating mode on a hood, Oldham [7, 8] successfully assessed the *IL* of a close-fitting acoustic hood within simple supported boundary condition and the clamped boundary condition. Results revealed that the theory and the experimental data were in agreement. In addition, in previous work [9], an assessment of a single-layer sound absorber was successfully deduced and applied in space-constrained design. Therefore, the theories of Oldham [7, 8] and Chang *et al.* [9] used to predict the *IL* of an acoustical hood and the sound absorbing coefficient of a multi-layer sound absorber will be adopted.

This paper provides a quick and effective method to reduce noise by optimally designing three kinds of shaped multi-layer sound absorbers (one-layer, two-layer, and three-layer) and a close-fitting acoustical hood using a *GA* method.

3. MATHEMATICAL MODELS

A space-constrained machine room (8 meters in length, 8 meters in width, and 4 meters in height) with one piece of equipment (EQ-A) is shown in Figure 1. As indicated in Figure 1, the wall is lined with three kinds of multi-layer sound absorbers (one-layer, two-layer, and three-layer) and the EQ-A is covered with a close-fitting acoustical hood. The mathematical models for the multi-layer sound absorbers and the one-layer close-fitting acoustic hood are described below.

3.1. The One-layer Close-fitting Acoustic Hood

According to Oldham [7, 8], for a one-layer close-fitting acoustical hood within a clamped boundary condition, the *IL* is

$$IL(\bar{X}) = 10 * \log_{10}\left[\left(\cos(k \cdot dd) + \left(\frac{\pi^2}{4K\omega\rho_o c}\right) * \sin(k \cdot dd)\right)^2\right] \quad (1a)$$

$$K = \frac{(1.35)}{\left[3.86 * D_1\left(\frac{129.6}{a^4} + \frac{78.4}{a^2 b^2} + \frac{129.6}{b^4}\right) - \omega^2 \rho h\right]} \quad (1b)$$

$$D_1 = \left[\frac{Eh^3}{12(1-vv^2)}\right] * (1+i\eta) \quad (1c)$$

$$\bar{X} = (f, h, dd, \eta, a, b, E, \rho, vv) \quad (1d)$$

where *k* is the wave number, *dd* is the distance between the equipment and the hood, $\rho_o c$ is the acoustic impedance, *E* is the panel's Young module, *D* is the panel's complex bulk module, a is the length of the panel, b is the width of the panel, η is the panel's internal damping coefficient, h is the thickness of the panel, vv is the poison ratio of the panel, and ρ is the panel's density.

3.2. The Sound Absorption Coefficients for a Multi-layer Sound Absorber

The derivation of sound absorption coefficients with respect to three kinds of multi-layer sound absorbers (one-layer, two-layer, and three layer) lined inside the machine room's inner wall are described below.

3.2.1. A One-layer Sound Absorber

As indicated in Figure 2(A), the acoustic impedance on the perforated front plate is obtained from the bottom wall where the value of the impedance is infinity. There exists four points representing the absorbing impedance within the absorber. The absorber is composed of a "rigid-backing plate $+L_1$ thickness of air $+D_{f1}$ thickness of the acoustic fiber $+q_1$ thickness of the perforated front plate." As derived in a previous study [9], for a wave propagating normally in a quiescent medium symbolized by "m," the general matrix form between point 1 and point 2 is expressed as

$$\begin{pmatrix} p_2 \\ u_2 \end{pmatrix} = \begin{bmatrix} \cos(k_m L) & jZ_m \sin(k_m L) \\ j\frac{1}{Z_m}\sin(k_m L) & \cos(k_m L) \end{bmatrix} \begin{pmatrix} p_1 \\ u_1 \end{pmatrix} \quad (2)$$

Therefore, the relationship of the acoustic pressure p and the acoustic particle velocity u between point 0 and point 1 is expressed as the transfer matrix and shown below.

$$\begin{pmatrix} p_1 \\ u_1 \end{pmatrix} = \begin{bmatrix} \cos(\omega L_1/c_o) & j\rho_o c_o \sin(\omega L_1/c_o) \\ j\frac{\sin(\omega L_1/c_o)}{\rho_o c_o} & \cos(\omega L_1/c_o) \end{bmatrix} \begin{pmatrix} p_o \\ u_o \end{pmatrix} \quad (3)$$

Development of Eq. (3) yields

$$Z_1 = -j\rho_o c_o \cot(\frac{\omega L_1}{c_o}) \quad (4)$$

The relationship of the acoustic pressure p and the acoustic particle velocity u with respect to point 1 and point 2 is expressed in the transfer matrices below.

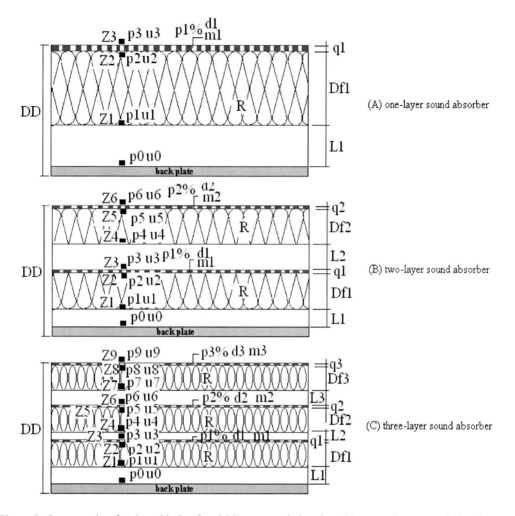

Figure 2. Cross section for three kinds of multi-layer sound absorbers [A: a one-layer sound absorber; B: a two-layer sound absorber; C: a three-layer sound absorber].

$$\begin{pmatrix} p_2 \\ u_2 \end{pmatrix} = \begin{bmatrix} \cos(k_{fiber-1}D_{f1}) & jZ_{fiber-1}\sin(k_{fiber-1}D_{f1}) \\ j\dfrac{1}{Z_{fiber-1}}\sin(k_{fiber-1}D_{f1}) & \cos(k_{fiber-1}D_{f1}) \end{bmatrix} \begin{pmatrix} p_1 \\ u_1 \end{pmatrix} \qquad (5)$$

By developing Eq. (5), an alternative form of Eq.(5) yields

$$Z_2\left(=\frac{p_2}{u_2}\right)$$
$$= Z_{fiber-1}\frac{Z_1\cosh[k_{fiber-1}D_{f1}]+Z_{fiber-1}\sinh[k_{fiber-1}D_{f1}]}{Z_1\sinh[k_{fiber-1}D_{f1}]+Z_{fiber-1}\cosh[k_{fiber-1}D_{f1}]} \qquad (6)$$

By adopting the formula of the specific normal impedance and wave number that is derived by Delany & Bazley [10], Eq. (6) is written as

$$Z_2 = (R_{fiber-1} + jX_{fiber-1})[\frac{\sinh(k_{21}D_{f1})\cos(k_{11}D_{f1}) - j\sin(k_{11}D_{f1})\cosh(k_{21}D_{f1})}{\cos(k_{11}D_{f1})\cosh(k_{21}D_{f1}) - j\sinh(k_{21}D_{f1})\sin(k_{11}D_{f1})}]$$
(7a)

$$k_{11} = \frac{\omega}{c_o}[1 + c_1(\frac{\rho_o f}{R_1})^{c_2}]; k_{21} = \frac{\omega}{c_o}[c_3(\frac{\rho_o f}{R_1})^{c_4}];$$

$$R_{fiber-1} = \rho_o c_o[1 + c_5(\frac{\rho_o f}{R_1})^{c_6}]; X_{fiber-1} = \rho_o c_o[c_7(\frac{\rho_o f}{R_1})^{c_8}]$$
(7b)

where $c_1, c_2, c_3, c_4, c_5, c_6, c_7$ and c_8 are the material constants of the porous acoustic fiber.

For sound flowing into the perforated plate, it is assumed that the incident sound passes through the holes of the perforated plate and is immediately transmitted to the porous material behind the perforated plate. The particle velocity is barely reduced [11, 12]. The continuity of the particle velocity is then applied and expressed as

$$u_2 = u_3$$
(8)

Acoustic impedance yields

$$p_3 = Z_{p_1} u_2 + p_2$$
(9)

Combining Eqs. (8)~(9), the transfer matrix between point 2 and point 3 yields

$$\begin{pmatrix} p_3 \\ u_3 \end{pmatrix} = \begin{bmatrix} 1 & Z_{p1} \\ 0 & 1 \end{bmatrix} \begin{pmatrix} p_2 \\ u_2 \end{pmatrix}$$
(10)

By developing Eq. (10) and substituting Eq. (2), the specific normal impedance at point 3 is

$$Z_3 = Z_2 + Z_{p1}$$
(11)

Adopting the formula of the specific normal impedance and the wave number of the perforated plate from Beranek & Ver [13] yields

$$Z_{p1} = \frac{\rho_o}{\varepsilon_1}\sqrt{8v\omega}\left(1+\frac{q_1}{2d_1}\right) + j\frac{\omega\rho_o}{\varepsilon_1}\left[\sqrt{\frac{8v}{\omega}}\left(1+\frac{q_1}{2d_1}\right)+q_1+\delta_1\right] \quad (12a)$$

$$\delta_1 = 0.85(2d_1)\left(1-1.47\sqrt{\varepsilon_1}+0.47\sqrt{\varepsilon_1^3}\right) \quad (12b)$$

For normal incidence, the sound absorption coefficient [11, 12, 14] is

$$\alpha_1(f,\varepsilon_1,d_1,R_1,q_1,D_{f1},L_1,DD)$$

$$= 1 - \left|\frac{Z_3 - \rho_o c_o}{Z_3 + \rho_o c_o}\right|^2 \quad (13a)$$

$$= \alpha_1(f,p_1\%,d_1,R_1,q_1,D_{f1},L_1,DD)$$

where $p_1\% = \varepsilon_1 * 100$; $q_1 = q_T$; $D_{f1} = Df_T$; $DD = D_{f1}+L_1+q_1$ \quad (13b)

3.2.2. A Two-layer Sound Absorber

Similarly, for the two-layer sound absorber shown in Figure 2(B), the relationship of the acoustic pressure p and the acoustic particle velocity u between point 3 and point 4 is

$$\begin{pmatrix} p_4 \\ u_4 \end{pmatrix} = \begin{bmatrix} \cos(\omega L_2/c_o) & j\rho_o c_o \sin(\omega L_2/c_o) \\ j\dfrac{\sin(\omega L_2/c_o)}{\rho_o c_o} & \cos(\omega L_2/c_o) \end{bmatrix}\begin{pmatrix} p_3 \\ u_3 \end{pmatrix} \quad (14)$$

The relationship of the acoustic pressure p and the acoustic particle velocity u with respect to point 4 and point 5 is expressed in the transfer matrices below.

$$\begin{pmatrix} p_5 \\ u_5 \end{pmatrix} = \begin{bmatrix} \cos(k_{fiber-2}D_{f2}) & jZ_{fiber-2}\sin(k_{fiber-2}D_{f2}) \\ j\dfrac{1}{Z_{fiber-2}}\sin(k_{fiber-2}D_{f2}) & \cos(k_{fiber-2}D_{f2}) \end{bmatrix}\begin{pmatrix} p_4 \\ u_4 \end{pmatrix} \quad (15)$$

By developing Eq. (15), an alternative form of Eq.(15) yields

$$Z_5\left(=\frac{p_5}{u_5}\right)$$

$$= Z_{fiber-2}\frac{Z_4\cosh[k_{fiber-2}D_{f2}]+Z_{fiber-2}\sinh[k_{fiber-2}D_{f2}]}{Z_4\sinh[k_{fiber-2}D_{f2}]+Z_{fiber-2}\cosh[k_{fiber-2}D_{f2}]}$$

$$= (R_{fiber-2} + jX_{fiber-2})[\frac{\sinh(k_{22}D_{f2})\cos(k_{12}D_{f2}) - j\sin(k_{12}D_{f2})\cosh(k_{22}D_{f2})}{\cos(k_{12}D_{f2})\cosh(k_{22}D_{f2}) - j\sinh(k_{22}D_{f2})\sin(k_{12}D_{f2})}]$$
(16a)

$$k_{12} = \frac{\omega}{c_o}[1 + c_1(\frac{\rho_o f}{R_2})^{c_2}]; k_{22} = \frac{\omega}{c_o}[c_3(\frac{\rho_o f}{R_2})^{c_4}];$$

$$R_{fiber-2} = \rho_o c_o[1 + c_5(\frac{\rho_o f}{R_2})^{c_6}]; X_{fiber-2} = \rho_o c_o[c_7(\frac{\rho_o f}{R_2})^{c_8}]$$
(16b)

For sound flowing into the second perforated plate, the transfer matrix between point 5 and point 6 yields

$$\begin{pmatrix} p_6 \\ u_6 \end{pmatrix} = \begin{bmatrix} 1 & Z_{p2} \\ 0 & 1 \end{bmatrix} \begin{pmatrix} p_5 \\ u_5 \end{pmatrix}$$
(17)

By developing Eq. (17) and substituting Eq. (14), the specific normal impedance at point 6 is

$$Z_6 = Z_5 + Z_{p2}$$
(18a)

where

$$Z_{p2} = \frac{\rho_o}{\varepsilon_2}\sqrt{8v\omega}\left(1 + \frac{q_2}{2d_2}\right) + j\frac{\omega\rho_o}{\varepsilon_2}\left[\sqrt{\frac{8v}{\omega}}\left(1 + \frac{q_2}{2d_2}\right) + q_2 + \delta_2\right]$$
(18b)

$$\delta_2 = 0.85(2d_2)\left(1 - 1.47\sqrt{\varepsilon_2} + 0.47\sqrt{\varepsilon_2^3}\right)$$
(18c)

For normal incidence, the sound absorption coefficient is

$$\alpha_2(f, \varepsilon_1, \varepsilon_2, d_1, d_2, R_1, R_2, q_1, q_2, D_{f1}, D_{f2}, L_1, L_2, DD)$$

$$= 1 - \left|\frac{Z_6 - \rho_o c_o}{Z_6 + \rho_o c_o}\right|^2$$
(19a)

$$= \alpha_2(f, p_1\%, p_2\%, d_1, d_2, R_1, R_2, q_1, q_2, D_{f1}, D_{f2}, L_1, L_2, DD)$$

where $p_1\% = \varepsilon_1 * 100; p_2\% = \varepsilon_2 * 100; q_1 + q_2 = q_T; D_{f1} + D_{f2} = Df_T;$

$$DD = D_{f1} + L_1 + q_1 + D_{f2} + L_2 + q_2; \tag{19b}$$

3.2.3. A Three-layer Sound Absorber

Likewise, for a three-layer sound absorber shown in Figure 2(C), the relationship of the acoustic pressure p and the acoustic particle velocity u between point 6 and point 7 is

$$\begin{pmatrix} p_7 \\ u_7 \end{pmatrix} = \begin{bmatrix} \cos(\omega L_3/c_o) & j\rho_o c_o \sin(\omega L_3/c_o) \\ j\dfrac{\sin(\omega L_3/c_o)}{\rho_o c_o} & \cos(\omega L_3/c_o) \end{bmatrix} \begin{pmatrix} p_6 \\ u_6 \end{pmatrix} \tag{20}$$

The relationship of the acoustic pressure p and the acoustic particle velocity u with respect to point 7 and point 8 is expressed in the transfer matrices below.

$$\begin{pmatrix} p_8 \\ u_8 \end{pmatrix} = \begin{bmatrix} \cos(k_{fiber-3} D_{f3}) & jZ_{fiber-3}\sin(k_{fiber-3} D_{f3}) \\ j\dfrac{1}{Z_{fiber-3}}\sin(k_{fiber-3} D_{f3}) & \cos(k_{fiber-3} D_{f3}) \end{bmatrix} \begin{pmatrix} p_7 \\ u_7 \end{pmatrix} \tag{21}$$

By developing Eq. (21), an alternative form of Eq.(21) yields

$$Z_8 = Z_{fiber-3} \dfrac{Z_7 \cosh[k_{fiber-3} D_{f3}] + Z_{fiber-3}\sinh[k_{fiber-3} D_{f3}]}{Z_7 \sinh[k_{fiber-3} D_{f3}] + Z_{fiber-3}\cosh[k_{fiber-3} D_{f3}]}$$

$$= (R_{fiber-3} + jX_{fiber-3})\left[\dfrac{\sinh(k_{23} D_{f3})\cos(k_{13} D_{f3}) - j\sin(k_{13} D_{f3})\cosh(k_{23} D_{f3})}{\cos(k_{13} D_{f3})\cosh(k_{23} D_{f3}) - j\sinh(k_{23} D_{f3})\sin(k_{13} D_{f3})}\right] \tag{22a}$$

$$k_{13} = \dfrac{\omega}{c_o}[1 + c_1(\dfrac{\rho_o f}{R_3})^{c_2}]; \quad k_{23} = \dfrac{\omega}{c_o}[c_3(\dfrac{\rho_o f}{R_3})^{c_4}];$$

$$R_{fiber-3} = \rho_o c_o [1 + c_5(\dfrac{\rho_o f}{R_3})^{c_6}]; \quad X_{fiber-3} = \rho_o c_o [c_7(\dfrac{\rho_o f}{R_3})^{c_8}] \tag{22b}$$

For sound flowing into the third perforated plate, the transfer matrix between point 8 and point 9 yields

$$\begin{pmatrix} p_9 \\ u_9 \end{pmatrix} = \begin{bmatrix} 1 & Z_{p3} \\ 0 & 1 \end{bmatrix} \begin{pmatrix} p_8 \\ u_8 \end{pmatrix} \tag{23}$$

By developing Eq. (23) and substituting Eq. (20), the specific normal impedance at point 9 is in the form

$$Z_9 = Z_8 + Z_{p3} \tag{24a}$$

where

$$Z_{p3} = \frac{\rho_o}{\varepsilon_3}\sqrt{8v\omega}\left(1+\frac{q_3}{2d_3}\right) + j\frac{\omega\rho_o}{\varepsilon_3}\left[\sqrt{\frac{8v}{\omega}}\left(1+\frac{q_3}{2d_3}\right) + q_3 + \delta_3\right] \tag{24b}$$

$$\delta_3 = 0.85(2d_3)\left(1 - 1.47\sqrt{\varepsilon_3} + 0.47\sqrt{\varepsilon_3^3}\right) \tag{24c}$$

For normal incidence, the sound absorption coefficient is

$$\alpha_3(f, \varepsilon_1, \varepsilon_2, \varepsilon_3, d_1, d_2, d_3, R_1, R_2, R_3, q_1, q_2, q_3, D_{f1}, D_{f2}, D_{f3}, L_1, L_2, L_3, DD)$$

$$= 1 - \left|\frac{Z_9 - \rho_o c_o}{Z_9 + \rho_o c_o}\right|^2 \tag{25a}$$

$$= \alpha_3(f, p_1\%, p_2\%, p_3\%, d_1, d_2, d_3, R_1, R_2, R_3, q_1, q_2, q_3, D_{f1}, D_{f2}, D_{f3}, L_1, L_2, L_3, DD)$$

where $\quad p_1\% = \varepsilon_1 * 100; \ p_2\% = \varepsilon_2 * 100; \ p_3\% = \varepsilon_3 * 100; \ q_1 + q_2 + q_3 = q_T;$
$D_{f1} + D_{f2} + D_{f3} = Df_T; DD = D_{f1} + L_1 + q_1 + D_{f2} + L_2 + q_2 + D_{f3} + L_3 + q_3;$ \hfill (25b)

3.3. Acoustical Calculation within a Machine Room

For an enclosed machine room (LL in width, WW in length, and HH in height), the sound pressure level (SPL_{ijk}) of a specific sound receiver (j) emitted from a specific machine (i) that is equipped with an acoustical enclosure (IL_{ik}) at k frequency is

$$SPL_{ijk}\left(SWL(f_{ik}), x_i, y_i, z_i, r_j, yr_j, zr_j, \alpha_k, LL, WW, HH\right)$$

$$= SWL(f_{ik}) - IL_{i,k} - \Psi_{ijk}(r_{ij}, \phi, f_k) + 10\log\left|\frac{Q_i}{4\pi r_{ij}^2} + \frac{4}{RR_k}\right| \tag{26a}$$

$$\Psi_{ijk}(r_{ij}, \phi, f_k) = 7.4\left(\frac{R_{ij}f^2}{\phi}\right)\cdot 10^{-8}; \ RR_k = \frac{\sum_{l=1}^{6} SS_l \alpha_{lk}}{1-\overline{\alpha}_k}; \ \overline{\alpha}_k = \frac{\sum_{l=1}^{6} SS_l \alpha_{lk}}{\sum_{l=1}^{6} SS_l} \tag{26b}$$

where $IL_{i,k}$ is the sound insertion loss of the i-th equipment at the k-th octave band and is shown in Eq. (1), $\Psi_{ijk}(r_{ij},\phi,f_k)$ is the air's sound absorption at $20°C$, f is the sound frequency and ϕ is the humidity in air.

By integrating the sound from all the equipment (n), the sound pressure level (SPL_{jk}) of a specific sound receiver (j) at k frequency is

$$SPL_{jk} = 10*\log\{\sum_{i=1}^{n} 10^{SPL_{ijk}/10}\} \qquad (27)$$

For a j-th sound receiver (j), the overall SPL is calculated by summing the SPL of the individual octave band k ($k=1$ to m).

$$SPL_{T(j)} = 10*\log\{\sum_{k=1}^{m} 10^{SPL_{jk}/10}\} \qquad (28)$$

3.4. Objective Function

As indicated in Figure 1, for a one piece of equipment within the machine room, the objective function at the receiving point (RV-1) is

$$OBJ(IL_{i,k}, D_f, R_f, dH, p\%) = SPL_T(\overline{X}, \overline{Y}) \qquad (29)$$

where $\overline{X}$ is the design parameter group of the multi-layer sound absorber and $\overline{Y}$ is the design parameter group of the close-fitting acoustic hood. The objective functions with respect to various strategies are described below.

(A) Case I: A One-layer Sound Absorber and a One-layer Close-fitting Acoustic Hood

On the basis of Eqs. (1), (13), and (28), the related objective function and the ranges of the parameters are

$$\begin{aligned}&OBJ_1(p\%, dH, D_f, R_f, h, dd) \\ &= SPL_T(RT_1, RT_2, RT_3, RT_4, RT_5, RT_6)\end{aligned} \qquad (30a)$$

$RT_1(=p\%):[5,\ 50];\quad RT_2(=dH):[0.003,\ 0.02];\quad RT_3(=D_f):[0,\ 0.197];$
$RT_4(=R_f):[3000, 22000];\ RT_5(=h):[0.002, 0.002];\ RT_6(=dd):[0.4, 0.8];\quad (30b)$

(B) Case II: A Two-layer Sound Absorber and a One-layer Close-fitting Acoustic Hood

On the basis of Eqs. (1), (19), and (28), the related objective function and the ranges of the parameters are

$$OBJ_2(RT_1, RT_2, RT_3, RT_4, RT_5, RT_6, RT_7, RT_8, RT_9, RT_{10})$$
$$= SPL_T(RT_1, RT_2, RT_3, RT_4, RT_5, RT_6, RT_7, RT_8, RT_9, RT_{10})$$
(31a)

$RT_1(=p_1\%)$: [5, 50]; $RT_2(=dH_1)$: [0.003, 0.02]; $RT_3(=p_2\%)$: [5, 50]; $RT_4(=dH_2)$: [0.003, 0.02]; $RT_5(=(D_{f1}+L_1)/(DD-q_1-q_2))$: [0, 1]; $RT_6(=D_{f1}/(DD-q_1-q_2)/RT_5)$: [0, 1]; $RT_7(=D_{f2}/(DD-q_1-q_2)/(1-RT_5))$: [0, 1]; $RT_8(=R_f)$: [3000, 22000]; $RT_9(=h)$: [0.002, 0.002]; $RT_{10}(=dd)$: [0.4, 0.8]; (31b)

(C) Case III: A Three-layer Sound Absorber and a One-layer Close-fitting Acoustic Hood

On the basis of Eqs. (1), (25), and (28), the related objective function and the ranges of the parameters are

$$OBJ_2(RT_1, RT_2, RT_3, RT_4, RT_5, RT_6, RT_7, RT_8, RT_9, RT_{10}, RT_{11}, RT_{12}, RT_{13}, RT_{14})$$
$$= SPL_T(RT_1, RT_2, RT_3, RT_4, RT_5, RT_6, RT_7, RT_8, RT_9, RT_{10}, RT_{11}, RT_{12}, RT_{13}, RT_{14})$$
(32a)

$RT_1(=p_1\%)$: [5, 50]; $RT_2(=dH_1)$: [0.003, 0.02]; $RT_3(=p_2\%)$: [5, 50]; $RT_4(=dH_2)$: [0.003, 0.02]; $RT_5(=p_3\%)$: [5, 50]; $RT_6(=dH_3)$: [0.003, 0.02]; $RT_7(=(D_{f1}+L_1)/(DD-q_1-q_2-q_3))$: [0, 1]; $RT_8(=D_{f1}/(DD-q_1-q_2-q_3)/RT_7)$: [0, 1]; $RT_9(=(D_{f2}+L_2)/(DD-q_1-q_2-q_3)/(1-RT_7))$: [0, 1]; $RT_{10}(=D_{f2}/(DD-q_1-q_2-q_3)/(1-RT_7)/RT_9)$: [0, 1]; $RT_{11}(=D_{f3}/(DD-q_1-q_2-q_3)/(1-RT_7)/(1-RT_9))$: [0, 1]; $RT_{12}(=R_f)$: [3000, 22000]; $RT_{13}(=h)$: [0.002, 0.002]; $RT_{14}(=dd)$: [0.4, 0.8]; (32b)

4. CASE STUDIES

An aluminum-made acoustical hood (a=b=0.8 (M), E=69*10^9 (Pa), vv =0.33, ρ=2700 kg/m^3) used to depress a noise from a piece of motor-driven equipment is adopted and shown in Figure 1. The sound power level (*SWL*) of that piece of equipment (EQ-A) is shown in Table 1 where the overall *SWL* reaches 118.5 dB(A). Current *SPL* distribution within the machine at 1 meter above the ground is depicted in Figure 3. As indicated in Figure 3, the *SPL* at the receiver (RV-1) is 122.5 dB(A). To depress the direct noise within the machine room, an aluminum-made acoustical hood with a one-layer close-fitting cover is used. Additionally, to reduce the reverberant noise, three kinds of multi-layer sound absorbers (one-layer, two-layer, and three-layer) are adhered to the inner wall and ceiling of the machine room Moreover, to optimally shape the multi-layer sound absorbers and the acoustical hoods within a fixed space, numerical assessments linked to a *GA* optimizer are applied. Before the minimization of a broadband noise is executed, a reliability check of the *GA* method by the maximization of the sound absorption coefficient for three kinds of multi-layer sound absorbers at a targeted tone (500 Hz) has been carried out. Three cases of strategies for eliminating the noise level within the machine room are performed.

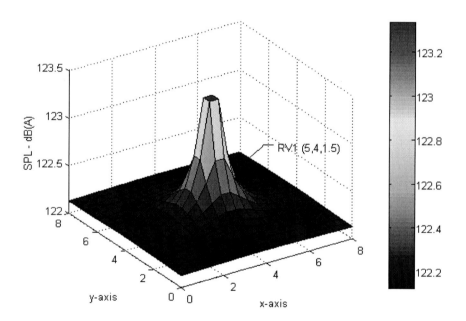

Figure 3. original *SPL* distribution curve within a machine room at 1 meter above the ground without noise abatement.

Table 1. **The spectrum of sound power level (*SWLO*) emitted from a piece of equipment (EQ-A)**

Frequency - Hz	125	250	500	1000	2000	4000	overall
SWLO – dB(A)	106.0	116.0	113.0	107.0	103.0	96.0	118.5

5. GENETIC ALGORITHM

The concept of Genetic Algorithms, first formalized by Holland [15] and then extended to functional optimization by D. Jong [16], involves the use of optimization search strategies patterned after the Darwinian notion of natural selection.

For the optimization of the objective function (*OBJ*), the design parameters of $(X_1, X_2,...,X_k)$ were determined. When the *chrm* (the bit length of the chromosome) was chosen, the interval of the design parameter (X_k) with $[Lb,Ub]_k$ was then mapped to the band of the binary value. The mapping system between the variable interval of $[Lb,Ub]_k$ and the kth binary chromosome of

$$[\underbrace{0\ 0\ 0\ 0\ \bullet\ \bullet\ \bullet\ 0\ 0\ 0}_{chrm} \sim$$

$$\underbrace{1\ 1\ 1\ 1\ \bullet\ \bullet\ \bullet\ 1\ 1\ 1}_{chrm}$$

was then built. The encoding from x to *B2D* (binary to decimal) can be performed as

$$B2D_k = \text{integer}\left\{\frac{x_k - Lb_k}{Ub_k - Lb_k}(2^{chrm} - 1)\right\} \tag{33}$$

The initial population was built up by randomization. The parameter set was encoded to form a string which represented the chromosome. By evaluating the objective function (*OBJ*), the whole set of chromosomes $[B2D_1, B2D_2, ..., B2D_k]$ that changed from binary form to decimal form was then assigned a fitness by decoding the transformation system.

fitness=*OBJ*($X_1, X_2, ..., X_k$); (34a)

where $X_k = B2D_k *(Ub_k - Lb_k)/(2^{chrm}-1) + Lb_k$ (34b)

As indicated in Figure 4, to process the elitism of a gene, the tournament selection, a random comparison of the relative fitness from pairs of chromosomes, was applied.

During the *GA* optimization, one pair of offspring from the selected parent was generated by uniform crossover with a probability of *pc*. By using the masked genes randomly generated, the gene information between parents will be internally exchanged if the mapping gene is 1.

Genetically, mutation occurred with a probability of *pm* where the new and unexpected point was brought into the *GA* optimizer's search domain. Likewise, by using the masked genes randomly generated, the mapped gene will be converted from 1 to 0, or from 0 to 1, if the mapping gene is 1.

To prevent the best gene from disappearing and to improve the accuracy of optimization during reproduction, the elitism scheme of keeping the best gene (one pair) in the parent generation with the tournament strategy was developed.

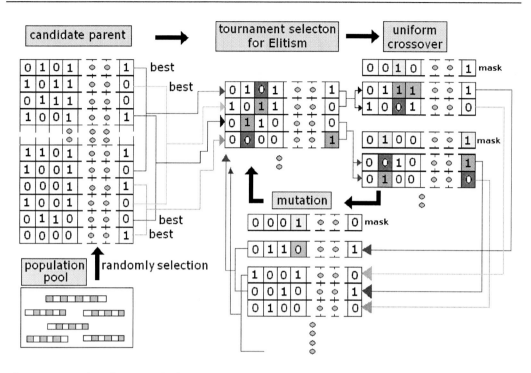

Figure 4. Operations in *GA* method.

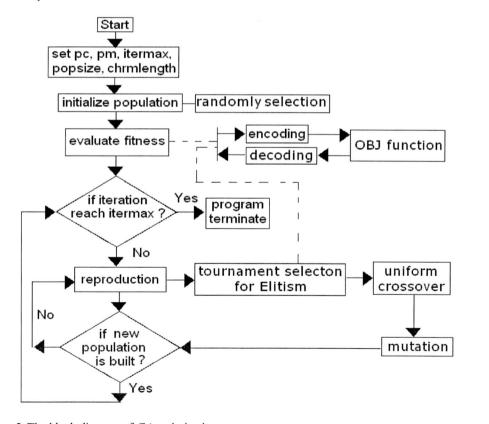

Figure 5. The block diagram of *GA* optimization.

Table 2. Optimal sound absorption coefficient of a three-layer sound absorber at a target tone of 500 Hz using various GA parameters

Item	GA parameters					Results				
	pop	gen	chrm	pc	pm	elt				
1	20	10	10	0.2	0.01	1	RT_1	RT_2	RT_3	RT_4	α_{500}
							16.53	0.01002	44.59	0.01537	0.86197
							RT_5	RT_6	RT_7	RT_8	
							48.42	0.01719	0.14	0.58	
							RT_9	RT_{10}	RT_{11}	RT_{12}	
							0.60	0.19	0.73	3155.6	
2	20	10	10	0.5	0.01	1	RT_1	RT_2	RT_3	RT_4	α_{500}
							21.27	0.00536	20.94	0.00571	0.85279
							RT_5	RT_6	RT_7	RT_8	
							17.76	0.01055	0.51	0.76	
							RT_9	RT_{10}	RT_{11}	RT_{12}	
							0.37	0.85	0.71	3083.6	
3	20	10	10	0.8	0.01	1	RT_1	RT_2	RT_3	RT_4	α_{500}
							42.69	0.00861	25.14	0.00950	0.86317
							RT_5	RT_6	RT_7	RT_8	
							38.97	0.00924	0.42	0.57	
							RT_9	RT_{10}	RT_{11}	RT_{12}	
							0.23	0.30	0.54	3111.6	
4	20	10	10	0.8	0.05	1	RT_1	RT_2	RT_3	RT_4	α_{500}
							47.76	0.01507	14.83	0.01596	0.86480
							RT_5	RT_6	RT_7	RT_8	
							43.43	0.01821	0.11	0.36	
							RT_9	RT_{10}	RT_{11}	RT_{12}	
							0.67	0.34	0.84	3007.5	
5	20	10	10	0.8	0.09	1	RT_1	RT_2	RT_3	RT_4	α_{500}
							31.47	0.01226	45.34	0.01408	0.85277
							RT_5	RT_6	RT_7	RT_8	
							13.66	0.01360	0.22	0.47	
							RT_9	RT_{10}	RT_{11}	RT_{12}	
							0.005	0.86	0.29	3099.3	
6	20	10	20	0.8	0.05	1	RT_1	RT_2	RT_3	RT_4	α_{500}
							10.24	0.01237	46.04	0.00336	0.87655
							RT_5	RT_6	RT_7	RT_8	
							27.36	0.01348	0.20	0.20	
							RT_9	RT_{10}	RT_{11}	RT_{12}	
							0.02	0.66	0.36	3001.3	
7	20	10	30	0.8	0.05	1	RT_1	RT_2	RT_3	RT_4	α_{500}
							21.81	0.00303	12.13	0.00315	0.88393
							RT_5	RT_6	RT_7	RT_8	
							43.89	0.01592	0.01	0.90	

Item	GA parameters					Results				
	pop	gen	chrm	pc	pm	elt				
							RT_9	RT_{10}	RT_{11}	RT_{12}	
							0.05	0.18	0.27	3032.7	
8	40	10	30	0.8	0.05	1	RT_1	RT_2	RT_3	RT_4	α_{500}
							36.23	0.01795	30.44	0.00308	0.88398
							RT_5	RT_6	RT_7	RT_8	
							38.56	0.01595	0.46	0.74	
							RT_9	RT_{10}	RT_{11}	RT_{12}	
							0.45	0.48	0.98	3002.8	
9	60	10	30	0.8	0.05	1	RT_1	RT_2	RT_3	RT_4	α_{500}
							44.33	0.00973	11.84	0.00531	0.87887
							RT_5	RT_6	RT_7	RT_8	
							49.38	0.01687	0.062	0.014	
							RT_9	RT_{10}	RT_{11}	RT_{12}	
							0.40	0.28	0.49	3012.1	
10	40	100	30	0.8	0.05	1	RT_1	RT_2	RT_3	RT_4	α_{500}
							49.11	0.01319	12.48	0.00306	0.88981
							RT_5	RT_6	RT_7	RT_8	
							49.83	0.01923	0.09	0.14	
							RT_9	RT_{10}	RT_{11}	RT_{12}	
							0.29	0.68	0.47	3019.9	
11	40	500	30	0.8	0.05	1	RT_1	RT_2	RT_3	RT_4	α_{500}
							48.74	0.00912	48.67	0.00300	0.89200
							RT_5	RT_6	RT_7	RT_8	
							49.89	0.01999	0.18	0.61	
							RT_9	RT_{10}	RT_{11}	RT_{12}	
							0.52	0.53	0.76	3000.0	
12	40	1000	30	0.8	0.05	1	RT_1	RT_2	RT_3	RT_4	α_{500}
							45.32	**0.01461**	**15.95**	**0.00300**	**0.89216**
							RT_5	RT_6	RT_7	RT_8	
							49.99	**0.01999**	**0.41**	**0.56**	
							RT_9	RT_{10}	RT_{11}	RT_{12}	
							0.29	**0.079**	**0.70**	**3000.0**	

The process was terminated when a number of generations exceeded a pre-selected value of *gen*. The block diagram of *GA* optimization is depicted in Figure 5.

6. RESULTS AND DISCUSSION

6.1. Results

To achieve good optimization, five kinds of *GA* parameters, including population size (*pop*), chromosome length (*chrm*), maximum generation (*gen*), crossover ratio (*pc*), and mutation ratio (pm) are varied step by step during optimization. Before the optimization of multi-layer sound absorbers and acoustical hoods within a machine room is performed, the reliability of the *GA* method will be checked by the maximization of the sound absorption coefficient for three kinds of multi-layer sound absorbers (one-layer, two-layer, and three-layer) at 500 Hz. Subsequently, three kinds of strategies (*Case* I, *Case* II, and *Case* III) used in reducing the noise level within a space-constrained machine room will be assessed.

6.1.1. Reliability Check of the GA Method

Based on Eqs. (13), (19), and (25), the related objective functions with respect to three kinds of multi-layer sound absorbers (one-layer, two-layer, and three-layer) at 500 Hz are

$$OBJ_{11}(\overline{X}_1) = \alpha_1(RT_1, RT_2, RT_3, RT_4) \tag{35}$$

$$OBJ_{12}(\overline{X}_2) = \alpha_2(RT_1, RT_2, RT_3, RT_4, RT_5, RT_6, RT_7, RT_8) \tag{36}$$

$$OBJ_{13}(\overline{X}_3) = \alpha_3(RT_1, RT_2, RT_3, RT_4, RT_5, RT_6, RT_7, RT_8, RT_9, RT_{10}, RT_{11}, RT_{12}) \tag{37}$$

For a three-layer sound absorber, the maximization of the sound absorption coefficient (α) at 500 (Hz) using Eq. (37) in conjunction with the *GA* method was performed first. As indicated in Table 2, twelve sets of *GA* parameters are tried in the sound absorber's optimization. Obviously, the optimal design data can be obtained from the last set of *GA* parameters at (*pop, gen, chrm, pc, pm, elt*) = (40, 1000, 30, 0.8, 0.05, 1). Using the optimal design data in a theoretical calculation, the resulting curves of the sound absorption coefficient (α) with respect to various *GA* parameters (*pop, gen, chrm, pc, pm, elt*) are listed in Table 3 and shown in Figures 6~8.

Table 3. Comparison of sound absorption coefficient for three-kinds of sound absorbers optimally shaped at a target tone of 500 Hz

category of sound absorbers	Design parameters				α_{500}
One-layer sound absorber	RT_1	RT_2	RT_3	RT_4	0.99959
	31.03	0.01614	0.16	21490.8	
Two-layer sound absorber	RT_1	RT_2	RT_3	RT_4	0.89197
	48.23	0.00300	49.99	0.00300	
	RT_5	RT_6	RT_7	RT_8	
	0.25	0.65	0.38	3000.0	
Three-layer sound absorber	RT_1	RT_2	RT_3	RT_4	0.86197
	16.53	0.01002	44.59	0.01537	
	RT_5	RT_6	RT_7	RT_8	
	48.42	0.01719	0.14	0.58	
	RT_9	RT_{10}	RT_{11}	RT_{12}	
	0.60	0.19	0.73	3155.6	

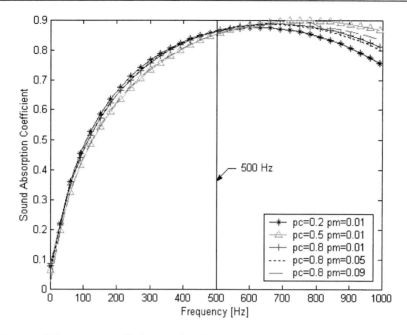

Figure 6. The sound absorption coefficient (α) with respect to various *GA* parameters (*pc, pm*).

Using the *GA* parameters in the maximization of Eq. (36) and (37) and plotting the sound absorption coefficient (α), the related profiles of sound absorption coefficients for three kinds of multi-layer sound absorbers (one-layer, two-layer, and three-layer) are illustrated in Figure 9. As revealed in Figure 9, the sound absorption coefficient (α) is maximized at the desired frequency.

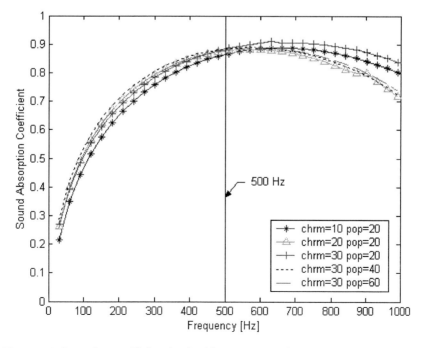

Figure 7. The sound absorption coefficient (α) with respect to various *GA* parameters (*chrm, pop*).

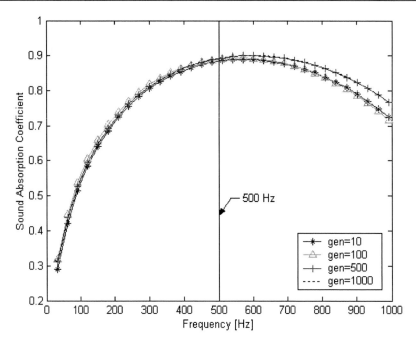

Figure 8. The sound absorption coefficient (α) with respect to various *GA* parameters (*gen*).

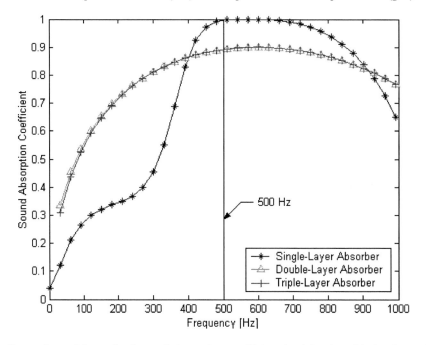

Figure 9. Comparison of the optimal sound absorption coefficient (α) for three kinds of sound absorbers (one-layer, two-layer, and three-layer sound absorbers).

6.1.2. Minimization of Broadband Noise within the Machine Room

To reduce the sound pressure level SPL_T at the receiver (*RV*-1), the optimization of three kinds of strategies (*Case* I: one-layer sound absorber + acoustical hood; *Case* II: two-layer sound absorber + acoustical hood; and *Case* III: three-layer sound absorber + acoustical hood) are performed using Eq. (30)-(32) in conjunction with the *GA* parameters of (*pop*=40, *gen*=1000, *chrm*=30, *pc*=0.8, *pm*=0.05, *elt*=1). The optimal result with respect to *Case* I ~ *Case* III is shown in Table 4. Using the optimal design data (Table 4) in a theoretical calculation, the resulting curves of the *SPL* distribution within the machine at 1 meter above the ground with respect to *Case* I ~ *Case* III are plotted and shown in Figs. 10~12. As illustrated in Table 4, for *Case* I's strategy (one-layer sound absorber + acoustical hood), the *SPL* at RV-1 can be improved from 122.5 dB(A) to 73.6 dB(A). The related *IL* of the optimally shaped acoustical hood is shown in Figure 13. For *Case* II's strategy (two-layer sound absorber + acoustical hood), the *SPL* at RV-1 is reduced from 122.5 dB(A) to 73.3 dB(A). The related *IL* of the optimally shaped acoustical hood is shown in Figure 14. For *Case* III's strategy (three-layer sound absorber + acoustical hood), the *SPL* at RV-1 is reduced from 122.5 dB(A) to 73.0 dB(A). The related *IL* of the optimally shaped acoustical hood is shown in Figure 15.

To appreciate the acoustical efficiency with respect to the sound absorber (used to reduce the reverberant noise) and the acoustical hood, a minimization of the *SPL* using a one-layer sound absorber is performed. The resulting *SPL* distribution within the machine room using a one-layer sound absorber only is plotted in Figure 16. Similarly, the resulting *SPL* distribution within the machine room using an acoustical hood only is assessed and plotted in Figure 17.

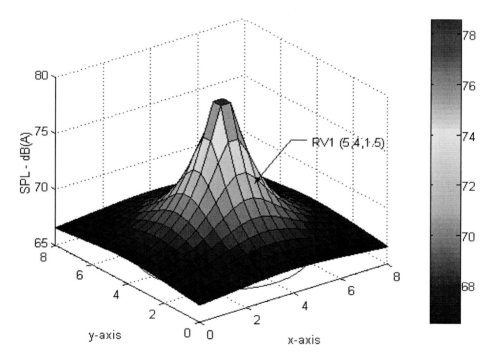

Figure 10. *SPL* distribution curve within a machine room using an optimally shaped one-layer sound absorber and an acoustical hood (*Case* I).

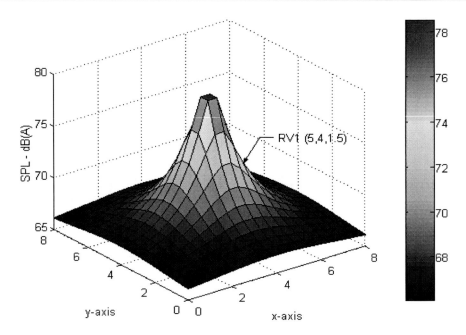

Figure 11. *SPL* distribution curve within a machine room using an optimally shaped two-layer sound absorber and an acoustical hood (*Case* II).

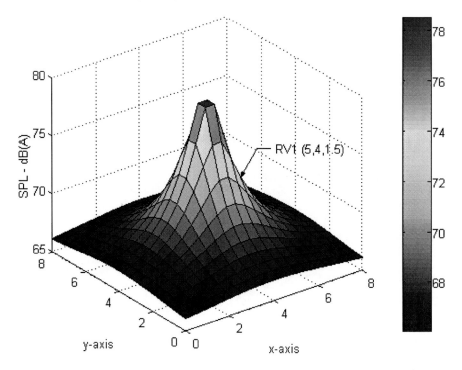

Figure 12. *SPL* distribution curve within a machine room using an optimally shaped three-layer sound absorber and an acoustical hood (*Case* III).

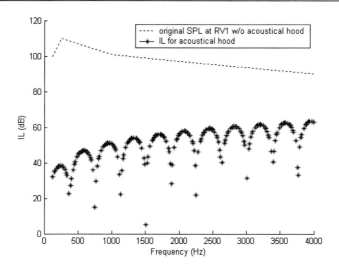

Figure 13. *IL* curve for an optimally shaped one-layer acoustical hood (*Case* I).

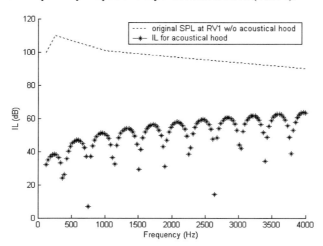

Figure 14. *IL* curve for an optimally shaped one-layer acoustical hood (*Case* II).

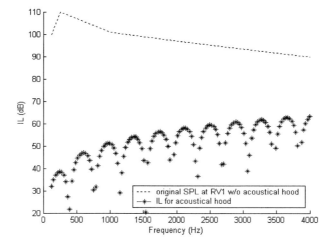

Figure 15. *IL* curve for an optimally shaped one-layer acoustical hood (*Case* III).

Table 4. Comparison of sound pressure levels at RV1 with respect to three kinds of cases [Case I: optimal one-layer sound absorber + optimal acoustical hood; Case II: optimal two-layer sound absorber + optimal acoustical hood; Case III: optimal three-layer sound absorber + optimal acoustical hood)

category of sound absorbers	Design parameters				SPL at RV1 dB(A)
Case I: One-layer sound absorber + acoustical hood	RT_1	RT_2	RT_3	RT_4	73.6
	26.78	0.01416	0.18	9916.2	
	RT_5	RT_6			
	0.004996	0.452			
Case II: Two-layer sound absorber + acoustical hood	RT_1	RT_2	RT_3	RT_4	73.3
	49.72	0.00309	49.99	0.00300	
	RT_5	RT_6	RT_7	RT_8	
	0.60	0.55	0.95	3000.3	
	RT_9	RT_{10}			
	0.004999	0.45			
Case III: Three-layer sound absorber + acoustical hood	RT_1	RT_2	RT_3	RT_4	73.0
	34.77	0.00467	18.08	0.00300	
	RT_5	RT_6	RT_7	RT_8	
	49.98	0.01993	0.50	0.84	
	RT_9	RT_{10}	RT_{11}	RT_{12}	
	0.021	0.35	0.79	3000.5	
	RT_{13}	RT_{14}			
	0.005	0.4424			

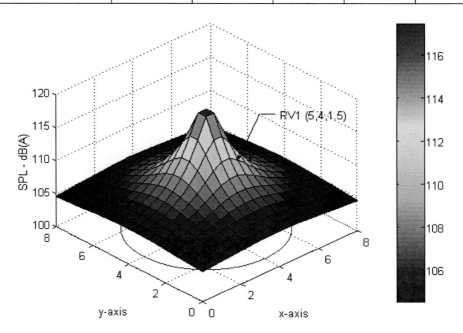

Figure 16. *SPL* distribution curve within a machine room using an optimally shaped one-layer sound absorber only.

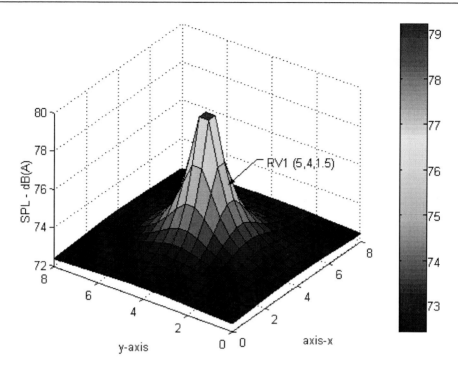

Figure 17. *SPL* distribution curve within a machine room using an optimally shaped one-layer acoustical hood only.

6.2. Discussion

To achieve sufficient optimization, the selection of the appropriate *GA* parameter set is essential. As indicated in Table 2 and Figs. 6~8, the best *GA* set with respect to a three-layer sound absorber at the targeted tone of 500 Hz reveals that the predicted maximal value of the sound absorption coefficient (α) is located at the desired frequency. Therefore, using the *GA* optimization to find a better design solution is reliable. Moreover, assessment of sound absorption coefficients with respect to three kinds of multi-layer sound absorbers shown in Figure 9 indicates that the sound absorber with more layers will broaden the curve of the sound absorption coefficient.

As can be seen in Figs. 10~12 and Table 4, the *SPL* at RV-1 with respect to three cases (*Case* I: one-layer sound absorber + acoustical hood; *Case* II: two-layer sound absorber + acoustical hood; and *Case* III: three-layer sound absorber + acoustical hood) can be improved from 122.5 dB(A) to 73.6 dB(A), 73.3 dB(A), and 73.0 dB(A). It would seem that a sound absorber with more layers will have a higher acoustical performance in lowering the reverberant sound within a space-constrain machine room. Additionally, as indicated in Figs. 13~15, the related *IL* of the optimally shaped acoustical hood is between 30~65 dB.

As indicated in Figure 16, the *SPL* at *RV*-1 reached at 111.6 dB(A) when only using a one-layer sound absorber. Moreover, the *SPL* at a region far from EQ-A will be reduced to below 105 dB(A). Similarly, as indicated in Figure 17, the *SPL* at RV-1 reached 76.0 dB(A) when only using an acoustical hood. Also, the *SPL* at a region far from EQ-A will be reduced to below 73 dB(A). Furthermore, as indicated in Figure 10, the *SPL* at RV-1 reached 73.6 dB(A) and the *SPL* at a region far from EQ-A will be reduced to below 66 dB(A) when both the one-layer sound absorber and the acoustical hood are used. Nevertheless, because of the mixed sound field, it would be more efficient for a sound field within a constrained space to eliminate both the reverberant and the direct sound energy simultaneously.

CONCLUSION

It has been shown that a one-layer acoustical hood and a multi-layer sound absorber in conjunction with a *GA* optimizer can be easily and efficiently optimized within a constrained space. As indicated in Table 2, six kinds of *GA* parameters (*pop, gen, chrm, pc, pm, elt*) play essential roles in the solution's accuracy during *GA* optimization. As indicated in Figs. 6~9, the sound absorption coefficient (α) is roughly maximized at the desired frequency. Therefore, the tuning ability established by adjusting the design parameters of the multi-sound absorbers and acoustical hood is reliable. In addition, as indicated in Figs. 10~12 and Table 4, the minimized *SPL* distribution curve within a space-constrained room has been assessed using an acoustical hood in conjunction with three kinds of sound absorbers (one-layer, two-layer, and three-layer). It would seem that sound absorber with more layers will perform better. Moreover, as indicated in Figs. 16 and 17, the *SPL* of *RV*-1 will be higher than *Case* I (one-layer sound absorber + acoustical hood) when using either a one-layer sound absorber or an acoustical hood. Nevertheless, because of the hybridized sound field of reverberant and direct noise within a constrained space, it would be more helpful for noise abatement to add optimally shaped acoustical hoods and multi-layer sound absorbers simultaneously.

Consequently, this approach used for optimally designing the shaped acoustical hood as well as multi-layer sound absorbers within a constrained space is easy and quite effective.

ACKNOWLEDGMENTS

The author acknowledges the financial support from the National Science Council (NSC 100-2622-E-235-001-CC3, ROC) and Chung Chou University of Science and Technology (CCUT –100– AC03).

REFERENCES

[1] Cheremisinoff, P. N, Cheremisinoff, P. P., *Industrial noise control handbook,* Ann Arbor Science, Michigan (1977).

[2] Alley, B. C., Dufresne, R. M., Kanji, N., Reesal, M. R., Costs of workers' compensation claims for hearing loss, *Journal of Occupational Medicine,* 31, pp. 134-138 (1989).

[3] Lan, T. S., Chiu, M. C., Optimal noise control on plant using simulated annealing, *Transactions of the Canadian Society for Mechanical Engineering,* 32(3-4), pp. 423-438 (2008).

[4] Chiu, M. C., Optimization of equipment allocation and sound-barriers shape in a multi-noise plant by using simulated annealing, *Noise & Vibration Worldwide,* 40(7), pp. 23-35 (2009).
[5] Roberts, N. J., Murray, W., Noise abatement device for a pneumatic tool, *United States Patent* 7845464 B2 (2010).
[6] Sanders, M., Composite sound barrier panel, United States Patent 7913812 B2 (2011).
[7] Oldham, D. J., Hilarby, S. N., The acoustical performance of small close fitting enclosures, part 1: theoretical methods, *Journal of Sound and Vibration,* 150, pp.261-281 (1991).
[8] Oldham, D. J., Hilarby, S. N., The acoustical performance of small close fitting enclosures, part 2: experimental investigation, *Journal of Sound and Vibration,* 150, pp.283-300 (1991).
[9] Chang, Y. C., Yeh, L. J., Chiu, M. C., Lai, G. J., Shape optimization on constrained single-layer sound absorber by using GA method and mathematical gradient methods, *Journal of Sound and Vibration*, 286(4-5), pp.941-961 (2005).
[10] Delany, M. E., Bazley, E. N., Acoustical properties of fibrous absorbent materials, *Applied Acoustics,* 13, pp. 105-116 (1969).
[11] Lee, F. C., Chen, W. H., Acoustic transmission analysis of multi-layer absorbers, *Journal of Sound and Vibration,* 248, pp.621-634 (2001).
[12] Munjal, M. L., *Acoustics of ducts and mufflers with application to exhaust and ventilation system design,* John Wiley & Sons, New York (1987).
[13] Beranek, L. L., Ver, I. L., *Noise and vibration control engineering,* John Wiley and Sons, New York (1992).
[14] Jinkyo, L., Swenson, G., Compact sound absorbers for low frequencies, *Noise Control Engineering Journal,* 38, pp.109-117 (1992).
[15] Holland, J. H., *Adaptation in natural and artificial System*, Ann Arbor, University of Michigan Press (1975).
[16] Jong, D., *An analysis of the behavior of a glass of genetic adaptive systems,* The University of Michigan Press (1975).

In: Handbook of Genetic Algorithms: New Research
Editors: A. Ramirez Muñoz and I. Garza Rodriguez

ISBN: 978-1-62081-158-0
© 2012 Nova Science Publishers, Inc.

Chapter 8

GENETIC ALGORITHM WITH OPTIMAL GENOTYPIC FEEDBACK: A NEW EVOLUTIONARY OPTIMIZATION APPROACH

S. R. Upreti[*], D. Joshi, F. Ein-Mozaffari and A. Lohi

Department of Chemical Engineering
Ryerson University, Toronto,
Ontario, Canada

ABSTRACT

Genetic algorithms or GAs are evolutionary optimization methods, which find applications in solving a variety of challenging optimization problems in chemical engineering. Different from traditional optimization methods, GAswork with populations of encoded optimization parametersetsto yield superior solutions, especially of problems that pose difficulties to conventional gradient search methods. However, GAssometime manifest slow or premature convergence, and reduced accuracies with the progression of genetic operations.

This paper addresses this problem by disseminating the genotypic information of the optimal parameter set in a GA population. After each iteration of the GA, randomly sized building blocks are extracted from the encoded parameter set that is found optimal. The blocks are then fed back or inserted into randomly selected members of the population for processing in the next iteration. Incorporated with the standard as well as an advanced GA, the new approach of optimal genotypic feedback (OGF) is successfully tested on 34 benchmark optimization functions.

The results demonstrate a significant improvement in the quality of results and computation times.When finally applied to the optimization problems of mixing characterization and minimum variance tuning of proportional controllers, OGF yields results, which are on a par with those obtained from a hybrid GA employing gradient search. Given its simplicity of implementation and quality of results, OGF is found to be a valuable enhancement to GA; better than the complex and computationally demanding add-on of gradient search.

---

[*] Corresponding author. Phone: (416) 979-5000 ext 6344. Fax: (416) 979-5083. Email: supreti@ryerson.ca

Keywords: optimization, genetic algorithm, gradient search, mixing characterization, minimum variance control

NOMENCLATURE

b - building block in binary encoding

c - chromosome, i.e. a binary encoded set of optimization parameters

c_{best} c - with maximum objective function value in the GA population

C_d - domain contraction factor (Upreti, 2004)

D_{min} - minimum domain size (Upreti, 2004)

J - objective function to be maximized by a GA

$\hat{J}$ - optimal objective function value

$\hat{J}_{ref}$ - true or best $\hat{J}$ known as the reference

N_{bits} - number of bits per optimization parameter

N_{gen} - number of GA generations per iteration of a GA

$\hat{N}_{iter}$ - maximum number of iterations in a GA

N_{obj} - number of objective function evaluations done by a GA to yield $\hat{J}$

N_{pop} - number of members (chromosomes) in a GA population

N_x - number of crossover sites per optimization parameter

p_i - bit-positions delimiting the substring extracted from c_{best}; $i = 1, 2$

p_m - probability of mutation

p_s - probability of selection

p_x - probability of crossover

u_i - binary numbers; $i = 1, 2, 3$

ABBREVIATIONS

AGA - Advanced Genetic Algorithm (Upreti, 2004)
GA - Genetic Algorithm
HGA - Hybrid GA with gradient search (Upreti and Ein-Mozaffari, 2006)
GS - Gradient Search
OGF - Optimal Genotypic Feedback
SGA - Standard GA

1. INTRODUCTION

Genetic algorithms (GAs) are evolutionary optimization methods that are well-suited for challenging problems plagued by non-linearities,discontinuities and multiple optima.To a given problem, a GA applies the principles of natural genetics and natural selection on a population of sets of optimization parameters (Holland, 1975). Typically encoded in binary representation as strings of bits (genes), the parameter sets (chromosomes) undergo the sequence of genetic operations, namely, selection, crossover and mutation (Goldberg, 1989).

During selection, stronger chromosomes having better fitnesses (objective function values) are preferentially selected, while the weaker ones are left out. The selected chromosomes then exchange genes in pairs during, what is called, crossover operation. It is equivalent to search by recombining parts of parent chromosomes in order to yield childchromosomes.In the next operation of mutation, the genes of chromosomes are altered with a low probability in equivalence to local search. The outcome is a new population with better-fit chromosomes. The genetic operations are iteratively applied on the new population for continued improvement in the quality of the optimal solution. Since a GA works on a population of optimization parameter sets, the algorithm is less likely to get trapped at a local optimum.The development in GAs includes messy GA having cut and splice crossover with variable length encoding (Goldberg et al., 1989), crossover and mutation with adaptive probabilities (Srinivas and Patnaik, 1994), simultaneous searches in sub domains (Andrea et al., 2001), switching between logarithmic and linear mappings in size-varying parameter domains (Upreti, 2004), new crossover and mutation operators (Deep and Thakur, 2007ab), and the use of optimization parameters in the decimal system itself (Davis, 1991; Janikow and Michalewicz, 1991; Michalewicz et al., 1991).

However, when high accuracy solutions are desired in a large search space, GAs face the obstacle of slow or premature convergence. They need longer chromosomes, and consequently a larger population to maintain suitable diversity of its members. With larger populations, the number of objective function evaluations increases, thereby increasing the computation time and slowing down the convergence to the optimal solution. If the population is not large enough then the lack of diversity, or the dominance of superior chromosomes causes premature convergence to suboptimal solutions. In this paper, we present a novel approach to feedback into a GA population, the genotypic information of its optimal chromosome. With a focus on speedy convergence and improvedsolution quality, the goal of this work is to come up with an enhanced form of GA.

2. THE NEW APPROACH: OPTIMAL GENOTYPIC FEEDBACK (OGF) IN GAS

The concept of OGF is the dissemination, into a newly generated GA population, of the genotypic information of the best available chromosome as several of its building blocks. The underlying expectation is that this feedback would further improve the population, and provide a better starting point for subsequent genetic operations. The effect could manifest as an improved quality of the final solution, and (or) a speedy convergence.

Figure 1 depicts the process of OGF. For the simplicity of presentation, the binary encoded parameter sets (chromosomes) are shown in six bits. Obtained after an iteration of a GA, the population of the parameter sets is shown on the left hand side. The best chromosome is at the bottom with its genes or bits in bold.

OGF begins with the extraction of three building blocks. Each building block carries a substring (of a given random length) of the best chromosome. Moreover, the substring is randomly positioned in the building block. In the next step, three different chromosomes of the population are randomly chosen as the recipients of the building blocks. When inserted into those chromosomes, the substrings replace the overlapping bits. Shown on the right hand side is the resulting population, which holds the best chromosome as well as its building blocks among other chromosomes. This population after OGF is ready to undergo the next iteration of the GA.

The dotted and solid lines of Figure 1 respectively denote the extraction and insertion of building blocks in OGF. Figure 2 presents their efficient implementation using bitwise operations. All numbers and chromosomes in the figure are unsigned integers. Extraction is effected by shifting the bits of the one's complement of zero (u_0, which has all 1s) separately to the left and right by random positions, p_1 and p_2. Together, they establish the length and placement of a substring to be later extracted from the best chromosome in the population. The resulting strings (u_1 and u_2) are "anded" to form u_3, whichwhen anded with the best chromosome, c_{best}, yields the building block, b. As shown in the figure, a substring '10' starting at the third bit-position from the right hand side is extracted from c_{best} and placed into b at the identical bit-position. Established by p_1 and p_2, the bit-position and substring length are both random. Inside b, the bits other than those in the extracted substring are all 0s.

The insertion of the building block b into a randomly chosen chromosome (c) is carried out as follows. As shown in the right hand column, by anding c with the one's complement of u_3 (i.e. $\sim u_3$), a number c_0 is obtained such that it has all the bits of c, but 0s at the sites that would receive the building block substring.

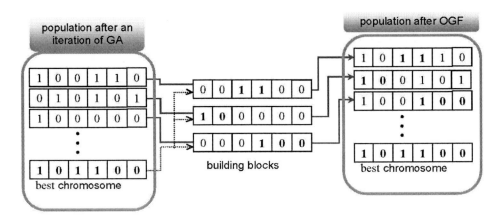

Figure 1. The process of Optimal Genotypic Feedback (OGF). For simplicity, the chromosomes are shown in six bits.

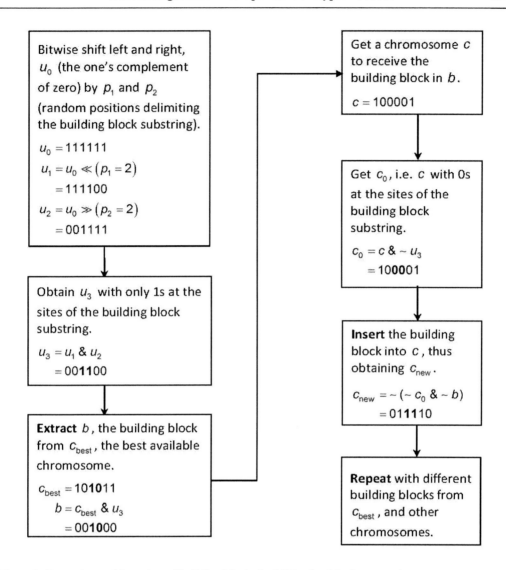

Figure 2. Extraction and insertion of building blocks in OGF using bitwise operations.

The one's complements of both c_0 and the building block, b, readily yield the desired chromosome c_{new} as shown in the figure. The procedures of extraction and insertion are repeated a specified number of times so that different building blocks from the best chromosome are introduced into several other chromosomes of the population. The OGF developed above was programmed in C++ to handle multi-parameter mapped binary encoding of chromosomes.

In this encoding, each chromosome is an ordered collection of substrings of binary encoded parameters of a set. The substrings are contiguously stored in an array of unsigned integersof smallest possible size. Thus, an additional unsigned integer is used only if all the bits of the existing unsigned integers in the arrayare used up.

Table 1. Inputs for the GA applications

Parameter	SGA and OGF-SGA (Case I)	AGA and OGF-AGA (Case II)	Mixing Characterization	Minimum Variance Control
C_d		0.75	0.75	0.75
D_{min}		10^{-4}	10^{-4}	10^{-4}
N_{bits}	10	10	10,10,7,7,7 respectively for a_1, a_2, d_1, d_2 and f	5
N_{gen}	10			
N_{pop}	$\sum_{i=0}^{i=N_x-1} N_{bit,i}$			
$\hat{N}_{iter}$	3000	3000	10000	1000
N_x	3	3	1	1
p_m	0.01	0.2	0.2	0.2
p_s	1			
p_x	0.9	0.6	0.98	0.6
ε			10^{-6}	10^{-6}

In the first step, the OGF was individually integrated with the following:

- **Case I** standard GA (SGA) [2]
- **Case II** an advanced GA (AGA) known for high quality and precision [6]

The motivation was to assess the impact of OGF on the two GAs, the second one being an improved variant using logarithmic and linear mappings on deviated parameters inside varying sizes of parameter domains. To that end, 34 challenging test functions from literature were selected. Their details are provided in the Appendix.

For the first case, each test function was minimized using OGF-SGA (i.e. OGF integrated with SGA), and SGA alone. Similarly for the second case, each test function was minimized using OGF-AGA (i.e. OGF integrated with AGA), and AGA alone. In the second step, the efficacy of OGF over gradient search in hybrid GA (or HGA) was assessed in the challenging optimization problems of mixing characterization, and minimum variance tuning of controllers. For that purpose, OGF was also combined with HGA. Table 1 lists the parameters of all GA applications above.

The maximum number of iterations was chosen to be sufficiently large in order to allow for convergence, whether true or suboptimal. The OGF was applied every iteration with the number of building blocks equal to the population size.

4. RESULTS AND DISCUSSION

We begin by presenting the impact of OGF on SGA and AGA in the first step.

4.1. Impact of OGF on SGA– Case I

For the first case, Table 2 presents the results obtained using OGF-SGA and SGA. The iteration number was incremented every N_{gen} generations. Thus, total of $N_{gen}N_{pop}$ objective function evaluations are carried out in one iteration of SGA, or OGF-SGA.

Table 2. Results of minimizing 34 test functions using SGA and OGF-SGA

Function No.	$\hat{J}_{ref}$	$\hat{J}\left[N_{obj} \times 10^{-2}\right]$ obtained from:	
		SGA	OGF-SGA
1	0	2.546364×10^{-3} [138]	2.546364×10^{-3} [40]
2	0	4.041497×10^{-6} [1814]	4.041497×10^{-6} [394]
3	−106.76	-1.070000×10^{2} [622]	-1.070000×10^{2} [20]
4	0	8.951425×10^{-3} [884]	8.951425×10^{-3} [746]
5	0	6.236181×10^{-6} [210]	6.236181×10^{-6} [232]
6	0.4	4.000014×10^{-1} [186]	4.000014×10^{-1} [30]
7	0	1.187166×10^{-1} [4022]	1.560737×10^{-1} [246]
8	−43.315	-4.499950×10^{1} [1628]	-4.499950×10^{1} [758]
9	0	1.001619×10^{-5} [482]	5.885723×10^{-6} [8]
10	−1	-9.999132×10^{-1} [58]	-9.999132×10^{-1} [12]
11	959.64	9.590007×10^{2} [1342]	9.590003×10^{2} [2354]
12	0	4.449941×10^{-4} [4]	4.449941×10^{-4} [4]
13	0	3.989878×10^{-6} [3556]	1.417646×10^{-6} [1540]
14	3	3.000615 [1894]	3.000615 [694]
15	0.1	1.000000×10^{-1} [588]	1.000000×10^{-1} [26]
16	0	4.459714×10^{-5} [5706]	7.64081×10^{-5} [2404]
17	0	1.343988×10^{-5} [8]	6.830052×10^{-6} [46]
18	−1.032	-9.999785×10^{-1} [3724]	-9.999820×10^{-1} [3072]
19	0	-2.017731×10^{-1} [1012]	-2.024008×10^{-1} [1616]
20	0	8.821027×10^{-7} [62]	4.939599×10^{-7} [802]
21	0	8.318477×10^{-8} [48]	5.895606×10^{-8} [5210]
22	−1.801	−1.800026 [2496]	−1.800026 [128]
23	0	1.351851×10^{-9} [4]	1.351851×10^{-9} [8]
24	−0.963	-4.871042×10^{-1} [248]	-4.871042×10^{-1} [20]
25	−45.77	4.500000×101 [6210]	4.500000×101 [2470]
26	0	$6.303304\times10–3$ [200]	$6.303304\times10–3$ [10]

Table 2. (Continued)

Function No.	$\hat{J}_{ref}$	$\hat{J}\left[N_{obj}\times 10^{-2}\right]$ obtained from: SGA	OGF-SGA
27	0	4.743326×10^{-5} [504]	4.743326×10^{-5} [500]
32	0	-6.234933×10^{-2} [5000]	-6.234933×10^{-2} [992]
33	0	1.977847×10^{-7} [14]	1.977847×10^{-7} [10]
34	-0.003	-3.791160×10^{-3} [602]	-3.791160×10^{-3} [164]
28	0	1.742554×10^{-4} [1956]	1.096691×10^{-4} [544]
29	-186.73	-1.864998×10^{2} [1062]	-1.865000×10^{2} [544]
30	-1	-9.999312×10^{-1} [1044]	-9.999904×10^{-1} [508]
31	0	1.194039×10^{-6} [16]	5.015361×10^{-7} [1652]

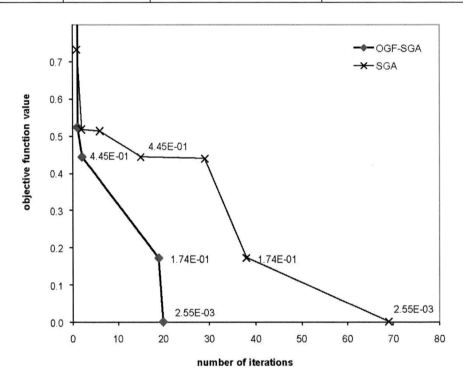

Figure 3. Objective function value versus iteration for OGF-SGA and SGA for Ackley function.

4.2. Impact of OGF on AGA – Case II

For the second case, Table 3 presents the results obtained using OGF-AGA and AGA. In both the GAs, one iteration has N_{gen} generations each for a logarithmic and the linear mapping (Upreti, 2004). Thus, $2N_{gen}N_{pop}$ objective function evaluations are carried out in one iteration of AGA, or OGF-AGA.

Table 3. Results of optimizing 34 test functions using AGA and OGF-AGA

Function No.	$\widehat{J}_{ref}$	$\widehat{J}\left[N_{obj} \times 10^{-2}\right]$ obtained from: AGA	$\widehat{J}\left[N_{obj} \times 10^{-2}\right]$ obtained from: OGF-AGA
1	0	1.952220×10^{-9} [10024]	1.940950×10^{-9} [4744]
2	0	1.354472×10^{-14} [5060]	1.798560×10^{-14} [4824]
3	−106.76	-1.070000×10^{2} [11408]	-1.070000×10^{2} [7148]
4	0	0.000000[508]	0.000000[484]
5	0	3.375080×10^{-14} [10364]	6.705750×10^{-14} [512]
6	0.4	4.000000×10^{-1} [11312]	4.000000×10^{-1} [1472]
7	0	5.734000×10^{-3} [6500]	3.670790×10^{-3} [6416]
8	−43.315	-4.500000×10^{1} [548]	-4.500000×10^{1} [536]
9	0	2.442491×10^{-14} [5384]	3.552714×10^{-15} [9344]
10	−1	-9.999983×10^{-1} [5312]	-9.999983×10^{-1} [5300]
11	959.64	9.590000×10^{2} [5828]	9.590000×10^{2} [3260]
12	0	2.760014×10^{-13} [10376]	1.671996×10^{-13} [3440]
13	0	3.907985×10^{-14} [9380]	3.552714×10^{-15} [8372]
14	3	3.000000[11312]	3.000000 [4472]
15	0.1	1.000000×10^{-1} [9080]	1.000000×10^{-1} [7556]
16	0	3.081855×10^{-10} [764]	0.000000[52]
17	0	7.127632×10^{-14} [10124]	8.482104×10^{-14} [9812]
18	−1.032	-1.000000×10^{-1} [6980]	-1.000000×10^{-1} [2228]
19	0	4.538592×10^{-13} [1208]	5.480061×10^{-13} [440]
20	0	1.110223×10^{-15} [11360]	6.661338×10^{-16} [9284]
21	0	0.000000[280]	1.604482×10^{-5} [244]
22	−1.801	−1.800026[10772]	−1.800026[5168]
23	0	0.000000[7184]	0.000000[736]
24	−0.963	-4.871042×10^{-1} [9356]	-4.871042×10^{-1} [1508]
25	−45.77	4.500000×10^{1} [56800]	4.500000×10^{1} [56380]
26	0	4.140776×10^{-10} [11360]	2.431797×10^{-10} [4048]
27	0	0.000000[400]	0.000000[364]
28	0	1.298961×10^{-13} [6392]	6.106227×10^{-14} [7352]
29	−186.73	−186.500000[488]	−186.450000[5396]
30	−1	−1.000000[1436]	−1.000000[9368]
31	0	0.000000[352]	0.000000 [316]
32	0	1.307637×10^{-6} [2536]	1.307637×10^{-6} [3256]
33	0	0.000000[352]	0.000000[316]
34	−0.003	-3.791237×10^{-3} [9560]	-3.791237×10^{-3} [4136]

As seen from Table 3, the quality and precision of results is far more superior than that in the first case (Table 2), although at the cost of increased number of N_{obj}. The main reason is that AGA helps reduce premature convergence because of its inherent characteristics of alternating linear and logarithmic mappings in size varying domains of deviation parameters (Upreti, 2004). It is desired to examine the improvement in AGA by integrating it with OGF.

It is observed from Table 3 that OGF-AGA reduced N_{obj} up to 95% in optimizing 29 out of the 34 test functions.

Looking for the improvement in $\hat{J}$, OGF-AGA accomplished that up to 100% for those 29 functions. In the remaining five functions where OGF-AGA needed more N_{obj} (No. 9, 28–30, and 32),the quality of $\hat{J}$ got improved except in Function 32 where $\hat{J}$ stayed the same as in AGA. In five functions with zero as reference $\hat{J}$ (No. 2, 5, 17, 19 and 21), the $\hat{J}$ values yielded by both OGF-AGA and AGA were insignificant.Thus, OGF-AGA always fared better than AGA either in terms of improvement in $\hat{J}$, or reduction in N_{obj}, or both.

Figure 4 shows the improvement of the objective function value to $\hat{J}$ in Ackley function when it is solved by AGA as well as OGF-AGA. Due to the efficacy of AGA, its $\hat{J}$ values with and without OGF are quite superior; in fact six orders of magnitude better than those obtained with the SGA applications. The effect of OGF on AGA is noticeable after 106 iterations leading to over 50% reduction in N_{obj} coupled with a slight betterment in $\hat{J}$.

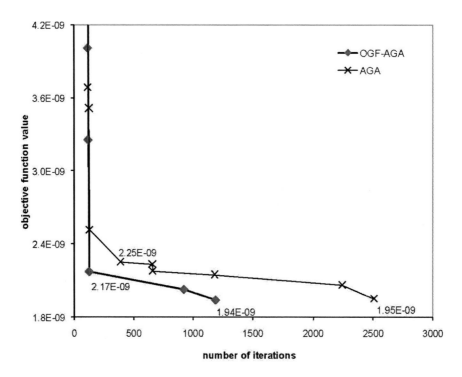

Figure 4.Objective function value versus iteration for OGF-AGA and AGA in Ackley function.

The above results demonstrate that the incorporation of OGF into SGA or AGA accelerates the convergence to the optimal solution in a meaningful manner. At regular intervals, OGF injects into the GA population, building blocks with substrings extracted from the strongest available chromosome, which has the maximum objective function value.

With the desirable traits in the building blocks, their injection alters the chromosomes so that they are either stronger (i.e. have higher fitnesses or objective function values), or more suitable to rapidly yield stronger chromosomes in subsequent genetic operations.

As a consequence, the iterative application of OGF helpsa GA population grow stronger much earlier than it would do otherwise. The effect is a healthier population, which gets progressively adjusted and steered effectively toward the optimal solution.

4.3. Reliability of OGF-AGA

Because of superior quality of results it delivered, OGF-AGAemerges as an enhanced form of GA in this work. To assess the reliability of OGF-AGA, we applied it 90 times with a different initialization seed to minimize each of the 34 benchmark functions. For each function, the standard deviation in the 90 optimal objective function values was calculated. Figure 5 plots the standard deviation for each function save Function 4, which had the deviation very close to zero.

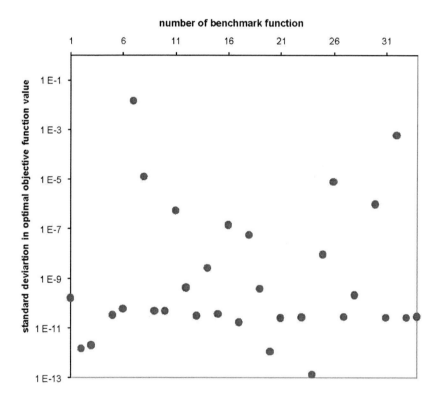

Figure 5 For a given benchmark function, the standard deviation in the optimal objective function values yielded by 90 OGF-AGA applications, each executed with a unique random initialization seed.

It is observed that except Function 7, the standard deviations are of the order 10^{-4} or less.

As a matter of fact, of all functions considered, Function 7 happens to be the most challenging as expected from the sharply discontinuous and constrained nature of the function to be minimized (see Appendix). However, for most of the functions (32 out of 34), the standard deviations are significantly low. This result demonstrates a strong level of confidence in the high-quality solutions provided by OGF-AGA developed in this work.

In the second step, we present the results of the application of OGF to the optimization problems of (i) mixing characterization, and (ii) minimum variance tuning of controllers. Similar to Cases I and II, the impact of OGF in both optimization problems was also examined by combining it with the hybrid GA. The hybrid GA or HGA is an amalgamation of AGA and gradient search (Upreti and Ein-Mozaffari, 2006).

4.4. Characterization of Mixing in Pulp Suspension

The characterization of mixing in pulp suspensions is a difficult optimization problem with the objective to find the identificationparameters (Ein-Mozaffari et al., 2003), which minimize the discrepancy between the model-predicted pulp concentration (y) and its experimental counterpart (y_{exp}). Thus, the objective function is given by

$$I = \sum_{i=0}^{i=N_s-1} (y_i - y_{exp,i})^2 \qquad (1)$$

in the discrete-time domain over N_s samples, y_i and $y_{exp,i}$, of y and y_{exp} (Kammer et al., 2005).

The mixing model is as follows:

$$y_i = \begin{cases} y_{exp,0} = u_0, & i \leq 0 \\ -\rho_6 y_{i-1} - \rho_7 y_{i-2} - \rho_8 y_{i-d_2} - \rho_9 y_{i-d_2-1} + \rho_1 u_{i-d_1} \\ \quad + \rho_2 u_{i-d_1-1} + \rho_3 u_{i-d_2} + \rho_4 u_{i-d_2-1} + \rho_5 u_{i-d_1-d_2}; & i = 1,2,\ldots,(N_s-1) \end{cases} \qquad (2)$$

where

$$u_i = u_0, \quad i \leq 0 \qquad (3)$$

is the input disturbance, and

$$\begin{aligned}
& d_i = 1 + \frac{T_i}{t_s}; \quad i = 1,2; \quad d_1 \leq d_2 \\
& \rho_1 = f(1-a_1), \quad \rho_2 = -a_2\rho_1, \quad \rho_3 = (1-f)(1-a_2) \\
& \rho_4 = -a_1\rho_3, \quad \rho_5 = 0, \quad \rho_6 = -a_1 - a_2 \\
& \rho_7 = a_1 a_2, \quad \rho_9 = -a_1\rho_8, \quad a_i = e^{-t_s/\tau_i}; \quad i = 1,2
\end{aligned} \qquad (4)$$

d_1, d_2, a_1, a_2 and f are the characterization parameters to be optimally determined in presence of the following constraints:

$$d_i = 1, 2, \ldots, d_{i,\max}; \quad i = 1, 2; \quad d_1 < d_2 \tag{5}$$

$$0 < a_i < 1; \quad i = 1, 2; \quad a_1 < a_2 \tag{6}$$

$$0 \leq f \leq 1 \tag{7}$$

In the above equations T_i s are τ_i s respectively are the integer time delays and constants, and τ_s is the sampling time. For each characterization parameter, Table 4 lists the minimum and maximum values enclosing the interval searched by the genetic operations.

Subject to the above conditions, the performance index to be maximized is

$$J = \frac{1}{1+I} \tag{8}$$

Table 5 shows the optimization results yielded by applications of HGA, OGF-HGA and OGF-AGA for Datasets 4, 5 and 6 in Upreti and Ein-Mozaffari (2006). It is observed that the fractional root mean-square-error (E) between the model-predicted and experimental outputs with OGF-HGA is either lower than or equal to that with HGA or OGF-AGA. The optimal value of a parameter agrees very well in all three applications. The number of objective function evaluations (N_{obj}) however is observed to generally increase in presence of OGF.

Table 4. Limits of search intervals for the optimization parameters in mixing characterization

Parameter	Value	Parameter	Value
$d_{1,\min}$	1	$d_{1,\max}$	5×10^2
$d_{2,\min}$	1	$d_{2,\max}$	10^3
$a_{i,\min}; \ i = 1, 2$	10^{-6}	$a_{i,\max}; \ i = 1, 2$	$(1-10^{-6})$
$f_{\min}$	0	$f_{\max}$	1

For Dataset 6, the net effect of OGF and gradient search is seen to provide the optimal results early with significantly fewer N_{obj}. It is interesting to find that the use of OGF alone (in OGF-AGA) delivers results that are almost of the same quality as of those obtained using the complicated gradient search.

Table 5. Results for mixing in pulp chest problem for three different datasets

	HGA	OGF-HGA	OGF-AGA
Dataset 4			
E	1.687870×10^{-2}	1.687870×10^{-2}	1.688060×10^{-2}
N_{obj}	1034410	1065343	1067256
d_1	9	9	9
d_2	28	28	27
a_1	8.227639×10^{-1}	8.227707×10^{-1}	8.121923×10^{-1}
a_2	9.928994×10^{-1}	9.928995×10^{-1}	9.929098×10^{-1}
f	1.848846×10^{-1}	1.848849×10^{-1}	1.796235×10^{-1}
Dataset 5			
E	2.653300×10^{-2}	2.652340×10^{-2}	2.653300×10^{-2}
N_{obj}	1528347	1734374	1294662
d_1	9	9	9
d_2	38	38	38
a_1	7.142029×10^{-1}	7.128597×10^{-1}	7.142088×10^{-1}
a_2	9.936603×10^{-1}	9.936492×10^{-1}	9.936603×10^{-1}
f	5.183089×10^{-1}	5.179999×10^{-1}	5.183071×10^{-1}
Dataset 6			
E	2.793870×10^{-2}	2.792280×10^{-2}	2.793870×10^{-2}
N_{obj}	1071160	635199	1452108
d_1	10	10	10
d_2	18	19	18
a_1	4.749037×10^{-1}	4.896894×10^{-1}	4.748997×10^{-1}
a_2	9.946615×10^{-1}	9.946559×10^{-1}	9.946614×10^{-1}
f	2.761970×10^{-1}	2.797227×10^{-1}	2.761969×10^{-1}

4.5. Minimum Variance Control Determination

The objective is find the proportional and integral controller gains (K_P and K_I), which minimize the variance in the process output (y)

$$V_y = \frac{1}{N-1} \sum_{k=0}^{N_s-1} y_k^2 \qquad (9)$$

over its N_s samples, y_k (Hanna et al., 2008).

The process output is given by

$$y_k = f_1 y_{k-1} + f_2 y_{k-2} + f_3 y_{k-L-1} + f_4 y_{k-L-2} + f_5 y_{k-L-3} + g_1 n_{k-L-1} + g_2 n_{k-L-2} + g_3 n_{k-L-3}, \quad k = 1, 2, 3, \ldots \quad (10)$$

where

$$\begin{aligned}
&f_1 = F + 1, & &f_2 = -F, & &f_3 = -G_1 K_P \\
&f_4 = G_1(K_P - K_I t_s) - G_2 K_P, & &f_5 = G_2(K_P - K_I t_s), & &g_1 = G_1 \\
&g_2 = G_2 - G_1, & &g_3 = -G_2, & &F = e^{-T/\tau} \\
&G_1 = K(1 - e^{-m/\tau}), & &G_2 = K(e^{-m/\tau} - e^{-T/\tau}) & &
\end{aligned} \quad (11)$$

In above equations, K is the steady-state gain, n is the input stochastic noise, L is that part of the time delay λ, which is an integral multiple of the sampling time T, and m is the other part of λ which is a fraction of T. The following constraints are present:

$$y_{min} < y_k < y_{max}; \quad k = 1, 2, 3, \ldots \quad (12)$$

$$K_{P,min} < K_P < K_{P,max} \quad (13)$$

$$K_{I,min} < K_I < K_{I,max} \quad (14)$$

Table 6 provides the process and existing controller parameters, and limits of search intervals for the optimization parameters.

Table 6. The process and existing controller parameters, and limits of search intervals for the optimization parameters in minimum variance control

Parameter	Control Loop: Montcalm	Roaster blower	Roaster feed
K	12.9	1	0.67
τ (s)	24.75	3.5	7
λ (s)	28	8.5	11
$K_{P,min}$	10^{-3}	10^{-3}	10^{-3}
$K_{P,max}$	10	10	10
$K_{I,min}$	10^{-3}	10^{-3}	10^{-3}
$K_{I,max}$	1	1	1
y_{min}	−4.5	−0.7	−0.4
y_{max}	4.5	0.7	0.4

Subject to the above conditions, the performance index to be maximized is

$$J = \frac{V_n}{V_y} \quad (15)$$

where V_n is the variance of the input stochastic noise.

Table 7 shows the optimization results yielded by applications of HGA, OGF-HGA and OGF-AGA to determine the minimum variance control parameters (controller gains) for Montcalm, roaster blower, and roaster feed control loops. The details of the three control loops are provided by Hanna et al. (2008).

Table 7. The results for the minimum variance control

	HGA	OGF–HGA	OGF–AGA
Montcalm loop			
I	2.649070×10^{-1}	2.649120×10^{-1}	2.649120×10^{-1}
N_{obj}	3435977737	3435976415	3435976391
K_P	3.739801×10^{-2}	3.739616×10^{-2}	3.739610×10^{-2}
K_I	1.000123×10^{-3}	1.000041×10^{-3}	1.000041×10^{-3}
Roster blower loop			
I	8.012760	8.012760	8.012760
N_{obj}	3435987948	3436008380	3436008344
K_P	6.857555×10^{-2}	6.857545×10^{-2}	6.857545×10^{-2}
K_I	1.000008×10^{-3}	1.000013×10^{-3}	1.000041×10^{-3}
Roster feed loop			
I	32.035950	32.035960	32.035960
N_{obj}	3436069545	3435983238	3435983210
K_P	3.341072×10^{-1}	3.341061×10^{-1}	3.341061×10^{-1}
K_I	1.000113×10^{-3}	1.000065×10^{-3}	1.000065×10^{-3}

The effect of OGF is observed to be very similar to that observed above in the problem of mixing characterization except that

1. the values of objective function (I) are identical up to six decimal places in each application on a control loop, and
2. the values of N_{obj} in the three applications are very close (the standard deviation in N_{obj} normalized by its average is of order 10^{-5} or less)

The above results further strengthen the high degree of parity between OGF and gradient search in terms of the quality of optimal results. This important upshot suggests that one could get more or less same performance enhancement using the relatively simple OGF than what could be delivered by more complicated and computationally intensive GS. It has to be emphasized that the complete implementation of OGF was carried out using bitwise operations as opposed to the arithmetic operations of gradient calculations and parameter corrections of GS.

CONCLUSION

A novel approach of optimal genotypic feedback (OGF) was developed with the aim to enhance the performance of genetic algorithms (GAs). From the best parameter set in the population, building blocks were regularly extracted and propagated into the population for its purposeful and progressive refinement. Integrated separately with standard GA, and an advanced GA, the OGF so developed was programmed and tested to minimize 34 benchmark functions.

With both GAs, the results demonstrated the effectiveness of OGF in significantly reducing in most cases, the number of objective function evaluations needed for convergence. When applied to the problems of mixing characterization and minimum variance control, OGF yielded results comparable to those obtained using gradient search with GA. With its inherent simplicity of implementation, the developed OGF is therefore promising in helping GAs attain optimal solutions with less computational overheads. The OGF integrated with GA emerges as a new evolutionary optimization algorithm with the potential to deliver solutions of high quality and precision.

ACKNOWLEDGMENTS

We acknowledge Natural Sciences and Engineering Research Council (NSERC), Canada for the financial support provided toward this work.

APPENDIX

Following are the 34 benchmark functions used to test the GAs developed in this work. The optimization problem is that of function minimization subject to the constraints on the parameters (x_i s) as specified.

1. Ackley function

$$f = -20\exp(-0.02\sqrt{\frac{1}{2}\sum_{i=1}^{2}x_i^2}) - \exp\left[\frac{1}{2}\sum_{i=1}^{2}\cos(2\pi x_i)\right] + 20 + e, \qquad (A.1)$$

$$-30 \le x_i \le 30, \quad i = 1, 2$$

2. Beale function

$$f = (1.5 - x_1 + x_1 x_2)^2 + (2.25 - x_1 + x_1 x_2^2)^2 + (2.625 - x_1 + x_1 x_2^3)^2,$$
$$-4.5 \leq x_i \leq 4.5, \quad i = 1,2 \tag{A.2}$$

3. Bird function

$$f = \sin(x_1)\exp(1 - \cos x_2)^2 + \cos(x_2)\exp(1 - \sin x_1) + (x_1 - x_2)^2,$$
$$-2\pi \leq x_i \leq 2\pi, \quad i = 1,2 \tag{A.3}$$

4. Bohachevsky function

$$f = x_1^2 + 2x_2^2 - 0.3\cos(3\pi x_1) - 0.4\cos(4\pi x_2) + 0.7,$$
$$-100 \leq x_i \leq 100, \quad i = 1,2 \tag{A.4}$$

5. Booth function

$$f = (x_1 + 2x_2 - 7)^2 + (2x_1 + x_2 - 5)^2; \quad -10 \leq x_i \leq 10; \quad i = 1,2 \tag{A.5}$$

6. Branin function

$$f = \left(x_2 - \frac{5x_1^2}{4\pi^2} + \frac{5x_1}{\pi - 6}\right)^2 + 10(1 - 8\pi)^{-1}\cos(x_1) + 10,$$
$$-5 \leq x_1 \leq 10 \text{ and } 0 \leq x_2 \leq 15 \tag{A.6}$$

7. Bukin function

$$f = 100\sqrt{|x_2 - 0.01x_1^2|} + 0.01|x_1 + 10|,$$
$$-15 \leq x_1 \leq -5 \text{ and } -3 \leq x_2 \leq 3 \tag{A.7}$$

8. Chichinadze function

$$f = x_1^2 + 2x_1 + 11 + 10\cos\left(\frac{\pi x_1}{2}\right) + 8\sin(5\pi x_1) - \frac{1}{10}\exp\left[-0.5(x_2 - 0.5)^2\right],$$
$$-30 \leq x_i \leq 30, \quad i = 1,2$$
$$\tag{A.8}$$

9. Dixon-Price function

$$f = (x_1 - 1)^2 + \sum_{i=2}^{m} i(2x_i^2 - x_{i-1})^2; \quad -10 \leq x_i \leq 10; \quad i = 1,2; \quad m = 2 \tag{A.9}$$

10. Easom function

$$-100 \leq x_i \leq 100$$
$$f = -\cos(x_1)\cos(x_2)\exp\left[-(x_1 - \pi)^2 - (x_2 - \pi)^2\right]; \quad -100 \leq x_i \leq 100; \quad i = 1,2 \tag{A.10}$$

11. Egg holder function

$$f = -(x_2 + 47)\sin\sqrt{|x_2 + 0.5x_1 + 47|} - x_1 \sin\sqrt{|x_1 - x_2 - 47|} \tag{A.11}$$
$$-512 \leq x_i \leq 512; \quad i = 1,2$$

12. Freudenstein-Roth function

$$f = \{-13 + x_1 + [(5 - x_2)x_2 - 2]x_2\}^2 + \{-29 + x_1 + (x_2 + 1)x_2 - 14x_2\}^2 \tag{A.12}$$
$$-10 \leq x_i \leq 10; \quad i = 1,2$$

13. Generalized penalized (GP) function

$$f = 0.1\left\{\sin^2(3\pi x_1) + \sum_{i=1}^{n-1} 2(x_i - 1)\left[1 + \sin^2(3\pi x_{i+1})\right] + (x_n - 1)^2\left[1 + \sin^2(2\pi x_n)\right]\right\} \tag{A.13}$$
$$+ \sum_{i=1}^{n} u(x_i, 10, 100); \quad -50 \leq x_i \leq 50; \quad i = 1,2,..,n; \quad n = 12$$

$$u(x, a, b) = \begin{cases} b(x-a)^4 & \text{if } x > a \\ -b(x-a)^4 & \text{if } x < a \\ 0 & \text{if } x = a \end{cases}$$

14. Goldstein and Price function

$$f = \left[1 + (x_1 + x_2 + 1)^2 (19 - 14x_1 + 3x_1^2 - 14x_2 + 6x_1 x_2 + 3x_2^2)\right] \times$$
$$\left[30 + (2x_1 - 3x_2)^2 (18 - 32x_1 + 12x_1^2 + 48x_2 - 36x_1 x_2 + 27x_2^2)\right] \tag{A.14}$$
$$-10 \leq x_i \leq 10; \quad i = 1,2$$

15. Giunta function

$$f = 0.6 + \sum_{i=1}^{2}\left[\sin\left(\frac{16}{15}x_i - 1\right) + \sin^2\left(\frac{16}{15}x_i - 1\right) + \frac{1}{50}\sin\left(4\frac{16}{15}x_i - 1\right)\right] \tag{A.15}$$
$$-1 \leq x_i \leq 1; \quad i = 1,2$$

16. Griewank function

$$f = \sum_{i=1}^{2}(\frac{x_i^2}{4000}) - \prod_{i=1}^{2}\cos(\frac{x_i}{\sqrt{i}}) + 1; \quad -600 \leq x_i \leq 600; \quad i = 1,2 \tag{A.16}$$

17. Himmelblau function

$$f = (x_1 + x_2^2 - 7)^2 + (x_1^2 + x_2 - 11)^2 + 0.1\left[(x_1 - 3)^2 + (x_2 - 2)^2\right] \tag{A.17}$$
$$-6 \leq x_i \leq 6; \quad i = 1,2$$

18. Hump function

$$f = 4x_1^2 - 2.1x_1^4 + \frac{x_1^6}{3} + x_1 x_2 - 4x_2^2 + 4x_2^4; \quad -5 \leq x_i \leq 5; \quad i = 1,2 \tag{A.18}$$

19. Leon function

$$f = 100(x_2 - x_1^2)^2 + (1 - x_1)^2; \quad -1.2 \leq x_i \leq 1.2; \quad i = 1,2 \tag{A.19}$$

20. Levy function

$$f = \sin^2(\pi y_1) + (y_1 - 1)^2\left[1 + 10\sin^2(\pi y_2)\right] + (y_2 - 1)^2 \tag{A.20}$$
$$y_i = 1 + 0.25(x_i - 1); \quad -10 \leq x_i \leq 10; \quad i = 1,2; \quad n = 2$$

21. Matyas function

$$f = 0.26(x_1^2 + x_2^2) - 0.48 x_1 x_2; \quad -10 \leq x_i \leq 10; \quad i = 1,2 \tag{A.21}$$

22. Michalewicz function

$$f = -\sum_{i=1}^{2}\sin(x_i)\left[\sin\left(\frac{ix_i^2}{\pi}\right)\right]^{2p}; \quad 0 \leq x_i \leq \pi; \quad i = 1,2; \quad p = 10 \tag{A.22}$$

23. Power sum function

$$f = \sum_{k=1}^{4}\left(b_k - \sum_{i=1}^{4}x_i^k\right)^2; \quad 0 \leq x_i \leq 4; \quad i = 1,2,3,4 \tag{A.23}$$
$$b_1 = 8, \quad b_2 = 18, \quad b_3 = 44, \quad b_4 = 114$$

24. Pen holder function

$$f = -\exp\left(-\left|\cos(x_1)\cos(x_2)\exp\left|1-\left[\frac{(x_1^2+x_2^2)^{0.5}}{\pi}\right]\right|\right|\right)^{-1}$$

$$-11 \leq x_i \leq 11; \quad i = 1,2$$

(A.24)

25. Paviani function

$$f = \sum_{i=1}^{10}\left[\ln^2(x_i-2)+\ln^2(10-x_i)\right]-\left[\prod_{i=1}^{10}x_i\right]^{0.2}; \quad 2 \leq x_i \leq 10; \quad i=1,2,..,10$$

(A.25)

26. Quintic function

$$f = \frac{x_1^4}{4} - \frac{x_1^2}{2} + \frac{x_1}{10} + \frac{x_2}{2}; \quad -10 \leq x_i \leq 10; \quad i = 1,2$$

(A.26)

27. Rastrigin function

$$f = 20 + \sum_{i=1}^{2}\left[x_i^2 - 10\cos(2\pi x_i)\right]; \quad -5.12 \leq x_i \leq 5.12; \quad i = 1,2$$

(A.27)

28. Rosenbrock function

$$f = 100(x_2 - x_1^2)^2 + (x_2 - 1)^2; \quad -5 \leq x_i \leq 10; \quad i = 1,2$$

(A.28)

29. Shubert function

$$f = \prod_{j=1}^{2}\sum_{i=1}^{5} i\cos[(i+1)x_j + i]; \quad -10 \leq x_j \leq 10; \quad j = 1,2$$

(A.29)

30. Test tube holder function

$$f = -4\left|\sin(x_1)\cos(x_2)\exp\left|\cos\left(\frac{x_1^2+x_2^2}{200}\right)\right|\right|$$

$$-9.5 \leq x_1 \leq 9.4, \quad -10.9 \leq x_2 \leq 10.9$$

(A.30)

31. Three-hump camel back function

$$f = 2x_1^2 - 1.05x_1^4 + \frac{x_1^6}{6} + x_1 x_2 + x_2^2; \quad -5 \leq x_i \leq 5; \quad i = 1,2$$

(A.31)

32. Weierstrass function

$$f = \sum_{i=1}^{2}\sum_{k=0}^{k}\left\{a^k \cos\left[2\pi b^k(x_i+0.5)\right]\right\} - 2\sum_{k=0}^{k}\left\{a^k \cos\left[\pi b^k\right]\right\} \quad \text{(A.32)}$$

$$-0.5 \leq x_i \leq 0.5; \quad i=1,2; \quad a=0.5, \quad b=3, \quad k=20$$

33. Zakharov function

$$f = \sum_{i=1}^{2} x_i^2 + \left(\sum_{i=1}^{2}\frac{ix_i}{2}\right)^2 + \left(\sum_{i=1}^{2}\frac{ix_i}{2}\right)^4; \quad -5 \leq x_i \leq 10; \quad i=1,2 \quad \text{(A.33)}$$

34. Zettle function

$$f = (x_1^2 + x_2^2 - 2x_1)^2 + 0.25x_1; \quad -5 \leq x_i \leq 5; \quad i=1,2 \quad \text{(A.34)}$$

REFERENCES

Andrea J., Siarryb, P. and Dognona, T, 2001, An improvement of the standard genetic algorithm fighting premature convergence in continuous optimization, *Advances in Engineering Software*, 32:49–60.

Davis, L., 1991, *Handbook of Genetic Algorithm*,(Van Nostrand Reinhold, New York, USA).

Deep, K., and Thakur, M., 2007b, A newmutationoperator for realcodedgeneticalgorithms, *Appl. Math. Comp.* 193:211–230.

Deep, K.,and Thakur, M., 2007a, A new crossover operator for real coded genetic algorithm. *Appl. Math. Comp.* 188:895–911.

Ein-Mozaffari, F., Kammer, L.C., Dumont, G.A., and Bennington, C.P.J., 2003, Dynamic modeling of agitated pulp stock chests. *TAPPI J.* 2: 13–17.

Goldberg, D.E., 1989, *Genetic Algorithms in Search, Optimization andMachine Learning*,(Addison-Wesley, New York, USA), p.124.

Goldberg, D.E., Korb, B. and Deb, K., 1989,Messy genetic algorithms: motivation, analysis and first results,*Complex System*, 3:493–530.

Hanna, J., Upreti, S.R., Lohi, A. and Ein-Mozaffari, F., 2008, Constrained minimum variance control using hybrid geneticalgorithm − An industrial experience, *J. Proc. Cont.* 18: 36–44.

Holland, J.H., 1975, *Adaptation in Natural and Artificial Systems*,(University of Michigan Press, Ann Arbor, USA).

Janikow, C.Z. and Michalewicz, Z., 1991, An experimental comparison of binary and floating point representation in genetic algorithms. *Proceedings of the Fourth International Conference on Genetic Algorithms*, San Francisco, Morgan Kaufman, 31–36.

Kammer, L.C., Ein-Mozaffari, F., Dumont, G.A. and Bennington, C.P.J.,2005, Identification of channelling and recirculation parameters ofagitated pulp stock chests, *J. Proc. Cont.* 15: 31–38.

Michalewicz, Z., Janikow C.Z., and Krawczyk, J.B., 1992, A modified genetic algorithm for optimal control problem. *Comp. Math. Appl.* 23:83–94.

Srinivas, M. and Patnaik, L.M. 1994, Adaptive probabilities of crossover and mutations in GAs, *IEEE Trans SystemsMan and Cybernetics*, 24:656–667.

Upreti, S.R. and Ein-Mozaffari, F.,2006, Identification of dynamic characterizationparameters of agitated pulp chests usinghybrid genetic algorithm, *Chem. Eng. Res. Des.* 84: 1–10.

Upreti, S.R., 2004, A new robust technique for optimal control of chemical engineering processes, *Comp. Chem. Eng.* 28:S1325–S1336.

In: Handbook of Genetic Algorithms: New Research
Editors: A. Ramirez Muñoz and I. Garza Rodriguez

ISBN: 978-1-62081-158-0
© 2012 Nova Science Publishers, Inc.

Chapter 9

GEOMETRICAL OPTIMIZATION IN GAS TURBINES BY APPLYING COMPUTATIONAL FLUID DYNAMICS AND GENETIC ALGORITHMS

A. Gallegos-Muñoz[*], V. Ayala-Ramírez and J. A. Alfaro-Ayala

Faculty of Engineering, University of Guanajuato,
Comunidad de Palo Blanco, Carretera Salamanca-Valle de Santiago Salamanca, Gto., México

ABSTRACT

A methodology that uses Genetic Algorithms (GA) with a computerized vision tool and Computational Fluid Dynamics (CFD) to optimize the geometric parameters of gas turbine components, is presented. The proposed methodology applies the results of the CFD simulation where the temperature and velocity contours are use to create the population of individuals. Each population generated is composed by the geometrical parameters that represent a feasible geometry. The set of parameters corresponding to the individual genotype were decoded and a script for the CFD software (Fluent ®) that describes the geometrical shape was created. The optimization process considers an initial set of individuals and a CFD simulation is created to obtain the temperature and velocity contours according to the desirable properties of the thermal behavior and subsequent populations were generated by applying selection, crossover and mutation genetic operators to the best individuals. The composition of the new population was created with 2 elite individuals, 6 individuals obtained from the application of the genetic operators and 2 new random individuals. A morphometric analysis computes several geometric properties of the temperature and velocity profiles to provide information to decide the best individuals. For the case of the transition piece of the gas turbine, the temperature and velocity profiles required at the outlet must be uniform because a non-uniformity in the temperature profile affects the useful life of the blades and nozzles of the first stage. However, a diminution in the average value of the turbine inlet temperature (TIT) produces a reduction of the thermal efficiency and power of the gas

---

[*] Tel.: (464) 647 9940. E-mail: gallegos@.ugto.mx.

turbine. Then, by applying a genetic algorithm and CFD simulation to optimize the geometry of the transition piece it is possible to analyze the geometrical parameters that are not intuitive for a human designer.

NOMENCLATURE

C_μ	Specific heat
C	Circumferential direction.
e	Energy
F	External forces
g_i	Gravity
G_k	Turbulent kinetic energy generation
G_b	Turbulent kinetic energy generation due to floatability
k	Thermal Conductivity, Turbulent kinetic energy
P	Pressure
q_k	Heat flux
R	Gas constant, Radial direction.
T	Temperature
u_i, v	Velocity
Y_M	Fluctuation of expansion in turbulence
y^+	Dimensional parameter.

GREEK SYMBOLS

ε	Dissipation rate
μ	Viscosity
Φ	Viscous heating dissipation
μ_t	Turbulent viscosity
ρ	Density
σ_k	Turbulent Prandtl number by k
σ_ε	Turbulent Prandtl number by ε
τ_{ij}	Stress tensor

1. INTRODUCTION

The optimization methods to diagnose power systems have been applied to estimate the performance status of a gas turbine [1]. In this case, the Genetic Algorithms (GA) were applied to estimate the design point of component parameters like turbine entry temperature and efficiency, determining the optimal operating conditions and design parameters of the engine under the maximum power condition [2], to optimize the aerodynamic design of turbine blades [3, 4] or as an strategy to optimize the turbine stages and to minimize the adverse effects of the three-dimensional flow features on the turbine performance [5]. On the other hand, coupling Computational Fluid dynamics (CFD) with GA permit the prediction of aerodynamic characteristics and shape optimization of airfoil using a multi-objective Pareto-based Genetic Algorithm [6, 7], determining the optimal geometric parameters of the microchannels in micro heat exchangers [8, 9], to find the geometry most favorable to simultaneously heat exchange while obtaining a minimum pressure drop in heat exchangers [10], to design of optimal hydraulic shapes for axial water turbines [11], optimization of the chord length and installation angle of the blades in wind turbines to obtain good aerodynamics performance at low wind speed [12], to find the optimum settings for nitric oxide emission minimization in the bubbling fluidized bed boiler studying the combustion process [13], to determine the convective heat transfer coefficient of an engine air-cooling system in different air velocity conditions finding the optimal geometric configuration of the finned cylinder [14] and, actually, for shape optimization of transonic airfoils using an efficient evolutionary algorithm to increase the robustness and convergence rate of the genetic algorithm [15].

In the present study, the CFD simulation and the GA are applied to obtain the optimal geometry of the transition piece reaching a uniform temperature and velocity distribution at the center of the outlet section in the transition piece, reducing the effect by hot streak over the vanes and blades of the first stage in the gas turbine.

2. CFD MODEL

The thermal and fluid dynamics analysis requires the mathematical representation of the physical domain, transforming it to a computational domain to obtain a solution with CFD. The governing equations applied to the computational model include the equation of mass, momentum, energy and the turbulence k-ε model in steady state.

Continuity equation

$$\frac{\partial(\rho u_i)}{\partial x_i} = 0 \tag{1}$$

Momentum equation

$$\frac{\partial(\rho u_i u_j)}{\partial x_i} = -\frac{\partial P}{\partial x_i} + \frac{\partial \tau_{ij}}{\partial x_j} + \rho g_i + F_i \tag{2}$$

where P is the total pressure, τ_{ij} is the stress tensor, ρg_i and F_i are the gravitational body force and external body forces, respectively.

Energy equation

$$\rho\left(u_i \frac{\partial e}{\partial x_i}\right) = -\frac{\partial q_k}{\partial x_k} - P\frac{\partial u_k}{\partial x_k} + \varphi \tag{3}$$

where φ is the viscous heating dissipation and the heat flux is given by Fourier's law.

$$q_k = -k\frac{\partial T}{\partial x_k} \tag{4}$$

Turbulence model (k-ε)

The standard turbulence model is a semi-empirical model proposed by Launder and Spalding [16], based on the model of the equations of transport for the turbulent kinetic energy (k) and the dissipation rate (ε). This model is appropriate to study the turbulence in practical engineering flows due to its reasonable accuracy for a wide range of turbulent flows, the equations are defined as

$$\frac{\partial}{\partial x_i}(\rho k u_i) = \frac{\partial}{\partial x_j}\left[\left(\mu + \frac{\mu_t}{\sigma_k}\right)\frac{\partial k}{\partial x_j}\right] + G_k + G_b - \rho\varepsilon - Y_M \tag{5}$$

$$\frac{\partial}{\partial x_i}(\rho \varepsilon u_i) = \frac{\partial}{\partial x_j}\left[\left(\mu + \frac{\mu_t}{\sigma_\varepsilon}\right)\frac{\partial \varepsilon}{\partial x_j}\right] + C_{1\varepsilon}\frac{\varepsilon}{k}(G_k) - C_{2\varepsilon}\rho\frac{\varepsilon^2}{k} \tag{6}$$

In these equations, G_k represents the generation of turbulent kinetic energy due to the mean velocity gradients, G_b is the generation of turbulent kinetic energy due to buoyancy, Y_M represents the contribution of the fluctuating dilatation in compressible turbulence to the overall dissipation rate, $C_{1\varepsilon}$ and $C_{2\varepsilon}$ are constants. The turbulent (or eddy) viscosity μ_t is computed by combining k and ε as follows:

$$\mu_t = \rho C_\mu \frac{k^2}{\varepsilon} \tag{7}$$

Equation of state

The density varies according to ideal gas law depending on the local pressure and temperature. The compressibility effect is accommodated during the simulation.

$$P = \rho R T \tag{8}$$

R is the gas constant.

2.1. Computational Domain

The geometry model and mesh were created using GAMBIT® and the governing equations are transformed to algebraic equations inside computational domain. Figure 1 shows the structured grid used in the modeling *(a)* inlet and *(b)* outlet. The use of quadrilateral cells reduces the amount of finite volume elements needed for the grid independence (if these were compared with triangular cells). The quadrilateral cells provide stability during the numerical solution and reduce the solution time. Close to the wall, the enhance wall function grid has been used to ensure that the results were independent of the mesh based on the y^+, getting a field of y^+ from 6.6 to 90, the grid is about 5.1 million cells.

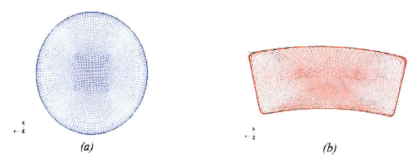

Figure 1. Structured mesh for the transition piece (a) inlet (b) outlet.

2.2. Boundary Conditions

Figure 2 shows the transition piece used for the modeling and Table 1 shows a summary of the boundary conditions in the transition piece. Velocity and temperature contours are set on the surface of the circular section, pressure on the exit mouth and natural convection on the outside wall of the transition piece were considered as boundary conditions. The assumption of the solid wall of the transition piece is modeled with a hypothesis of negligible thermal resistance by conduction, there are no film holes of cooling on the surface and the thermal properties of the material were considered.

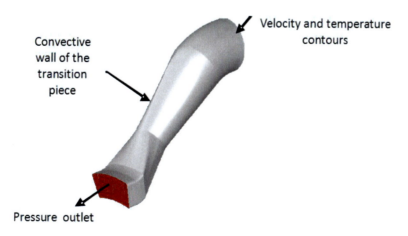

Figure 2. Boundary conditions of the transition piece.

The simulation was performed using air as the working fluid. This assumption was used due to the quantity of predominant air added after the reacting flow in the combustor liner. This secondary or dilution air after the reaction in the combustor liner is about 72% of all air added in the combustor [17].

Table 1. Boundary conditions

Boundary condition	Magnitude
Velocity contour	--
Temperature contour	--
Turbulent Intensity inlet	5 [%]
Hydraulic Diameter inlet	0.3302 [m]
Pressure outlet.	965 675 [Pa]
Turbulent Intensity outlet	5 [%]
Hydraulic Diameter outlet	0.471705 [m]
Convection coefficient	10 [W/m^{2}K]
Temperature	273 [K]

The boundary condition for the inlet of the transition piece was obtained from Alfaro-Ayala et al. [18], in which reacting flow is considered with realistic operation conditions in the combustor, where non-premixed combustion was adopted in the process whereas a chemical equilibrium model was used for the reacting system, getting temperature and velocity contours that were supplied to the computational model, see Figure 3.

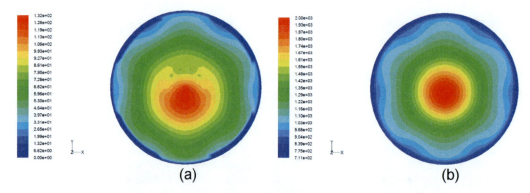

Figure 3. Contours of (a) velocity magnitude [m/s] and (b) static temperature [K] obtained from combustor [18].

2.3. Script for the CFD Software

An automatic process to generate the geometry and set the boundary conditions was made to optimize the gas turbine transition piece in order to reduce time in the procedure. Some of the commands used in the script for the CFD software (Fluent ®) that describes the geometrical shape, are shown in Table 2.

Table 2. Commands written in the script to generate geometry and boundary conditions

Command	Description
vertex create coordinates x y z	It is used to create a vertex, it just needs the coordinates x, y, z.
edge create straight "vertex. 1" "vertex. 2"	An edge (line) is created for two vertexes. In this case vertex.1 and vertex.2
edge create center 2points "vertex.3" "vertex.4" "vertex.5" minarc arc	The command creates an arc, it just needs one vertex for the center (vertex.3) and two vertex (vertex.4 and vertex.5) that belong to the arc.
edge create threepoints "vertex.27" "vertex.28" "vertex.20" arc	Another form to create an arc can be done for three points that belong to the arc.
edge split "edge.22" vertex "vertex.10" connected	Split edges is very common when a geometry is being made, in this case the edge.22 is divided in two edges for the vertex.10
face create wireframe "edge.24" "edge.26" "edge.27" "edge.28"	Creating a face can be done for some of edges, in this case, the face is formed for edge.24, edge.26, edge.27 and edge.28.
volume create wireframe "edge.40" "edge.41" "edge.42" "edge.43" "edge.44" "edge.45" "edge.46" "edge.47" "edge.48"	A volume is created for some faces or some edges. In this case a volume is created from edge.40 to edge.48.
physics create "salida" b-type "PRESSURE_OUTLET" face "face.10"	The boundary conditions can be set for a command also, this example shows that the face.10 has a pressure outlet and is called outlet.
physics create "pared" btype "WALL" face "face.4" "face.8"	Another boundary condition used is applied to the faces 4 and 6 as a wall and is called wall.

Also, some commands were set in the script to mesh the geometry. Such as in the procedure to create the geometry and the boundary conditions, all edges, faces and volumes must be identified with the *assignable number* (example vertex.*1,* face.*20*, volume.*4*, etc.) to give the number of nodes to the edges, then mesh the faces with their element and type and then mesh the volumes. The assignable number is given automatically in the pre-processor (GAMBIT ®) while vertex, edge, faces and volume are created. These commands can be seen in Table 3.

Table 3. Commands written in the script to mesh the geometry

Command	Description
edge mesh "edge.1" successive ratio1 1 intervals 15	This is the command to make the mesh. It is applied to the edge.1, and it has 15 nodes spaced with the same interval. To chance the space between nodes we can use a different successive ratio (different of one)
face mesh "face.1" map size 1	To mesh a face with quadrilateral cells, this command uses the map element in the face.1 with a size 1. Note: There are some different forms to mesh a face.
Volume mesh "volume.1" cooper source "face.1" "face.5" size 1	To mesh a volume, we use the cooper source to expand the mesh from face.1 to face.5 with a size of one, in order to keep the same way of mesh.

The procedure to generate the geometry, set the boundary conditions and mesh of the gas turbine transition piece by the commands, is described in Figure 4. The use of these commands should be careful due to the assignable number should be found by the user. Many times, an assignable number in a line of the command is wrong and then it fails, the sequence is broken and the goal of generate the geometry and mesh of the individuals can be impossible to achieve. The sequence of the commands is simple. At first, all the commands for the vertex followed for the commands of edges, faces and volumes. This is the first step, the full geometry. Then the second step is the commands of the boundary conditions and the last the commands of mesh. This sequence is just to have a good order as if it were the graphical interface.

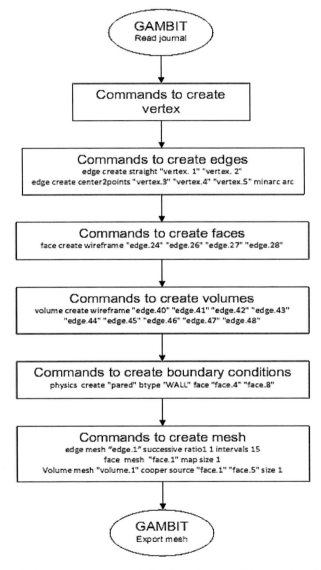

Figure 4. Diagram applied to create geometry, define boundary conditions and mesh for the gas turbine transition piece.

3. GENETIC ALGORITHM MODEL

To obtain the optimal geometry of the transition piece, a simple genetic algorithm (SGA) in an interactive way, was implemented. The individual phenotype GA was composed of fourteen geometrical parameters shown in Figure 5. For each parameter, *16* bits for its binary representation were used. So, the individual genotype has a length of *128* bits. Each parameter in the phenotype was encoded into a normalized interval [0,1]. In this representation, zero value is used for the minimum value of the feasible interval for that parameter and one value is assigned when the parameter takes the upper bound of the same interval, resulting in a normalized representation of the individual that simplifies its handling by the genetic operators. Using this approach, the optimization problem search space becomes $[0,1]^p$, with p being the number of parameters encoded in each individual.

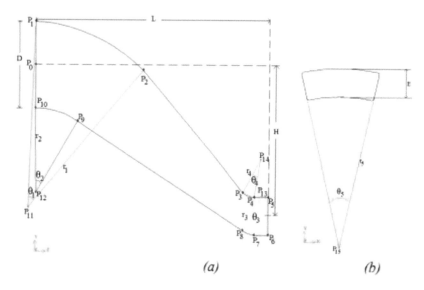

Figure 5. Geometry;(a) side view and (b) outlet of the transition piece.

In order to evaluate the fitness score of each potential solution encoded by a given computational individual, the set of parameters corresponding to the individual genotype were decoded and a script for FLUENT® describing the geometrical shape induced by the set of parameters was generated. Using this tool, a CFD simulation was obtained. The CFD results obtained following this procedure were analyzed in two steps; i) choice of a set of individuals that exhibit the best performance according to characteristics defined in the literature [18-21], and ii) a computer program in CFD analyzes the behavior of the temperature and velocity contours corresponding to the individual under study. The image obtained in this analysis is evaluated with respect to the desirable properties of the profile at the outlet. In particular, the elliptic form of the hottest region was analyzed at the outlet transition piece.

Using a small population (10 individuals) as initial population a random set of parameters was created in order to cover the most of the search space. Subsequent populations were generated by applying the selection, crossover and mutation genetic operators to the best individuals. The composition of a new population was taken as follows: 2 elite individuals, 6 individuals resulting from the application of the genetic operators and 2 new random individuals.

3.1. Interactive Process of Optimization

Figure 6 shows the sequence of activities that were performed in the optimization process. The GA is responsible of generating the population of computational individuals. As said before, each individual corresponds to a feasible geometric configuration of the transition piece. These individuals are translated into a FLUENT script in order to simulate its thermal behavior. The contours of the CFD simulation are converted into images. A morphometric analysis module computes several geometric properties of the temperature profile to provide information to decide which the best individuals are. This analysis computes geometric indexes like the elliptic form of the hottest region in the image, its displacement relative to the ideal position and the TIT variation in the graphs. According to the results and comparing the temperature profiles, a decision is taken to generate new individuals that allow obtaining a uniform temperature at the outlet of the transition piece. These individuals are then used to generate the new population of the genetic algorithm by applying on them the genetic operators (crossover and mutation).

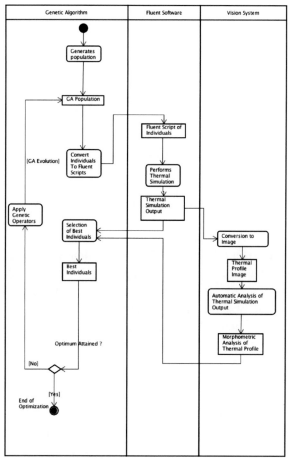

Figure 6. Sequence of the optimization process.

4. RESULTS

In order to improve the temperature distributions at the outlet of the transition piece that affect the vanes and blades due to the non-uniform heat load, the Genetic Algorithms were applied in the geometric parameters to obtain a reduction in the peak temperature values and a uniform profile at the outlet of the transition piece. These new geometric parameters are showed in Table 4, and Figure 7 shows a comparison of the geometries.

Table 4. Geometric parameters (initial and final)

Parameter	Initial (Original)			Final (Genetic algorithm)		
	x [mm]	y [mm]	z [mm]	x [mm]	y [mm]	z [mm]
P_0	0	0	0	0	0	0
P_1	0	20.6981	0	0	20.6981	0
P_2	0	- 2.6030	56.5934	0	10.2424	58.8650
P_3	0	- 60.7445	109.0252	0	60.2443	109.4082
P_4	0	- 62.9943	115.1897	0	62.9943	115.1897
P_5	0	- 62.9943	122.3891	0	-62.9943	122.3891
P_6	0	- 80.9927	122.3891	0	-80.9927	122.3891
P_7	0	- 80.9927	115.1897	0	-80.9927	115.1897
P_8	0	- 78.7429	109.0252	0	-84.4566	103.3830
P_9	0	- 26.9975	22.3180	0	-27.1576	22.0180
P_{10}	0	- 20.6981	0	0	-20.6981	0.0000
P_{11}	0	- 68.1175	- 3.4744	0	-60.9778	- 4.0137
P_{12}	0	- 61.0854	0.6484	0	-62.1054	1.2008
P_{13}	0	- 62.8974	118.3011	0	-60.0278	113.3067
P_{14}	0	- 44.8990	118.3011	0	-48.3899	122.9203

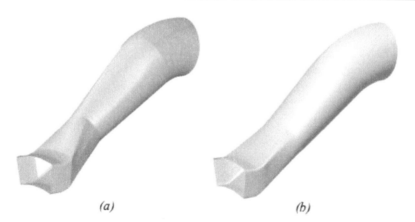

Figure 7. (a) Initial geometry and (b) Optimal geometry obtained by genetic algorithms.

Figure 8 shows a comparison of the velocity and temperature distribution in the radial direction of the original and genetic algorithm profiles. In the original geometric, the average velocity was of 53.54 and the peak velocity is about 1.16 located at 55% in the radial direction and the average value of turbine inlet temperature (TIT) was 1209.28 K, with peak temperature value of 1.08 located between 40 to 100% in the radial direction. The velocity distribution using the genetic algorithm has an average velocity of 49.27 m/s, where the peak velocity is 1.11 located at 50% in the radial direction. The peak temperature is located from 40% to 100% in the radial direction and the peak value of temperature is between 1.0 and 1.02 and the average value of turbine inlet temperature (TIT) is 1197.8 K. Also, a peak reduction of 3.62% in the velocity profile and 5.55% in the temperature profile can be seen.

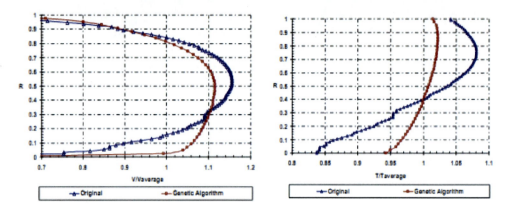

Figure 8. Radial velocity and temperature profile.

Figure 9 shows a comparison of the velocity and temperature circumferential profile of the original and genetic algorithm profiles. The profiles are symmetric, where the average velocity is 55.58 m/s with peak velocity of 1.1 located at 53% in the circumferential direction to original geometry. The peak temperature is 1.05 located at 55% with average value of 1240.13 K. Using genetic algorithm, the velocity at the outlet of the transition piece is symmetric, with an average velocity of 51.4 m/s located at 50% from the circumferential direction. The temperature profile is symmetric and the peak temperature is between 1.0 and 1.015 and it is located from 20% to 80% of the circumferential direction. The average value is 1194.6 K. As it can be seen the circumferential profile does not present a change on its shape, only the peak value is decreased. Also, a peak reduction of 3.63% in the velocity profile and 3.33% in the temperature profile can be seen.

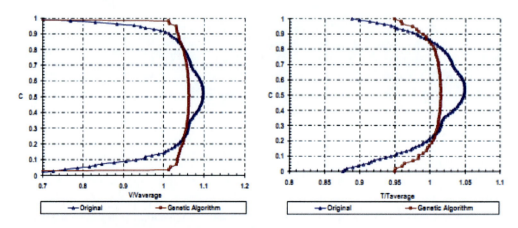

Figure 9. Circumferential velocity and temperature profile.

Figure 10 shows the velocity and temperature contours at the exit mouth with the original geometry. It can be seen that there is asymmetry in the radial direction and symmetry in the circumferential direction. The asymmetric radial profile is attributed to the curved shape geometry between the inlet and outlet of the transition piece and to the difference in density of the hot streak and the lack of cooling system in the wall of the transition piece. This phenomenon is one of the principal causes that affect the turbine blades due to the difference heat load Bai-Tao AN et. al. [20].

The new temperature and velocity contours are showed in Figure 11. It can be seen that the temperature and velocity distribution changes due to the new shape of the transition piece, provoking that the new temperature distributions decrease about 60 K (difference between maximum temperatures), it is attributed to a good mixture of the flow paths and a better uniformity of the velocity at the outlet of the transition piece.

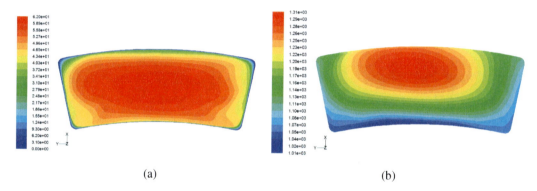

Figure 10. Contours of (a) velocity magnitude (m/s) and (b) static temperature (K) obtained at the outlet.

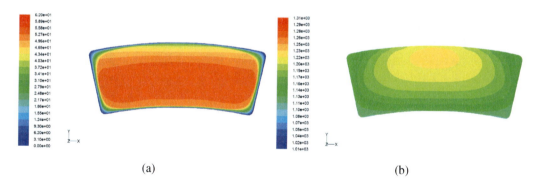

Figure 11. Contours of (a) velocity magnitude (m/s) and (b) static temperature (K) obtained at the outlet of the transition piece, applying a genetic algorithm.

CONCLUSION

The determination of the temperature and velocity profiles at the outlet of the transition piece will give a better approach to the numerical solution of heat transfer, fluid dynamics and structural analysis in components like vanes and blades, because these profiles are used as boundary condition at the inlet of the first stage of the gas turbine.

Applying an optimization method like a genetic algorithm is possible to improve the geometry of the transition piece, through interactive processes between CFD simulation and optimization method to define the geometrical parameters. This optimization permits the exploration of large search spaces and the inspection of geometrical parameters that are not intuitive for a human designer. The evolution process filters the best individuals according to the genetic operators: selection, crossover and mutation. This interactive process represented an efficient method to optimize the transition piece.

With the new geometry of the transition piece, the average value of the turbine inlet temperature (TIT) reduces to about 2.32% (28.29 K) and the average velocity reduces about 7.73% (4.217 m/s). The peak value of the temperature is decreased about 5.5% (11.4 K) in the radial direction whereas the velocity is 3.62% (4.27 m/s). The peak value of the temperature in the circumferential direction is decreased about 3.33% (45.51 K) whereas velocity is 3.63% (4.171 m/s). For all cases, the peak values of the profiles are reduced but the radial temperature profile has the maximum change on its shape, it is more uniform at the outlet and the peak temperature is reduced more than other profiles. However, it is important to include parameters like thermal efficiency and power (to maintain the same operation conditions of the gas turbine) in the optimization process to obtain better results.

REFERENCES

[1] Li Y.G. (2008). A Genetic Algorithm Approach to Estimate Performance Status of Gas Turbine. *Proceedings of the ASME Turbo 2008: Power for Land, Sea and Air.* Berlin, Germany, 431-440.

[2] E. Haghighi, B. Borzou, A. Ghahremani, M. Behshad (2009). Optimization of a Regenerative Gas Turbine Engine With Isothermal Heat Addition With Genetic Algorithm. *Proceeding of the ASME Turbo Expo 2009: Power for Land, Sea and Air.* Orlando, FL USA, 265-274.

[3] O. Oksuz, I. Akmandor (2008). Axial Turbine Blade Aerodynamics Optimization Using a Novel Multi-Level Genetic Algorithm. *Proceedings of ASME Turbo Expo 2008: Power for Land, Sea and Air.* Berlin, Germany, 2361-2374.

[4] O. Oksuz I. Akmandor (2008). Multi-Objective Aerodynamics Optimization of Axial Turbine Blades Using a Novel Multi-Level Genetic Algorithm. *Proceedings of ASME Turbo Expo 2008: Power for Land, Sea and Air.* Berlin, Germany, 2375-2388.

[5] M. Arabnia, W. Ghaly (2009). A Strategy for Multi-Point Shape Optimization of Turbine Stages in Three-Dimensional Flow. *Proceeding of the ASME Turbo Expo 2009: Power for Land, Sea and Air.* Orlando, FL. USA, 489-502.

[6] K. Park, B.S. Kim, J. Lee and K.S. Kim (2009). Aerodynamics and Optimization of Airfoil Under Ground Effect. *Proceeding of World Academy of Science, Engineering and Technology*, 332-338.

[7] K. Park, J.W. Han, H.J. Lim, B.S. Kim and J Lee (2008). Optimal Design of Airfoil with High Aspect ratio in Unmanned aerial Vehicles. *Proceeding of World Academy of Science, Engineering and Technology, Vol. 2*, 66-72.

[8] K. Foli, T. Okabe, M. Olhofer, Y. Jin and B. Sendhoff (2006). Optimization of Micro Heat Exchanger: CFD, Analytical Approach and Multi-Objetive Evolutionary Algorithms. *Int. J. of Heat and Mass Transfer, Vol 49*, 1090-1099.

[9] S. Baodong, W. Lifeng, L. Jianyung and C. Heming (2011). Multi-Objetive Optimization Design of a Micro-Channel Sink Using Adaptive Genetic Algorithm. *Int. J. Numerical Methods for Heat and Fluid Flow, Vol 21 (3)*, 353-364.

[10] R. Hilbert, G. Janiga, R. Baron and D. Thévenin (2006). Multi-Objetive Shape Optimization of a Heat Exchanger Using Parallel Genetic Algorithms. *Int. J. of Heat and Mass Transfer, Vol 49*, 2567-2577.

[11] T. Rovsnikova and L. Slovenia (2004). Optimization Method for the Design of Axial Hydraulic Turbines. IMechE, *Proc. Instn. Mech. Eng., Vol 218, part A: J. Power and Energy*, 43-50.

[12] J. Zhang, Z. Zhou and Y. Lei (2011). Optimization for High Performance Wind Turbine Blades at Low Wind Speed-Based on Parallel Genetic Algorithm. *Acta Energiae Solaris Sinica, Vol 32 (8)*, 1275-1280.

[13] A. Saario, A. Oksanen and M. Ylitalo (2006). Combination of Genetic Algorithm and Computational Fluid Dynamics in Combustion Process Minimization. *Combustion Theory and Modeling, Vol 10 (6)*, 1037-1047.

[14] B. Najafi, H. Najafi and M.D. Idalik (2011). Computational Fluid Dynamics Investigation and Multi-Objective Optimization of an Engine Air-Cooling System Using Genetic Algorithm. *Journal of Mechanical Engineering Science; Proceeding of the Institution of Mechanical Engineers, Part C, Vol 225 (6)*, 1389-1398.

[15] A. Jahangirian and A. Shahrokhi (2011). Aerodynamics Shape Optimization Using Efficient Evolutionary Algorithms and Unstructured CFD Solver. *Computers and Fluids, Vol 46 (1)*, 270-276.

[16] B. E. Launder, D. B. Spalding. *Lectures in Mathematical Models of Turbulence*. London, England: Academic Press; 1972.

[17] P. B. Meherwan. *Gas Turbine Engineering Handbook*. Second edition. Gulf Professional Publishing; 2002.

[18] J. Arturo Alfaro-Ayala, A. Gallegos-Muñoz A., J. M. Riesco-Avila, A. Campos-Amezcua, M.P. Flores-López, A. Mani-González (2011). Analysis of the Flow in the Combustor-Transition Piece Considering the Variation in the Fuel Composition. *Journal of Thermal Science and Engineering Applications, Vol. 3 (2)*, 1-11.

[19] J. A. Alfaro-Ayala, A. Gallegos-Muñoz, A. Zaleta-Aguilar, A. Campos-Amezcua and M. Zdzislaw (2009). Thermal and Fluid Dynamic Analysis of the Gas Turbine Transition Piece. *Proceeding of the ASME Turbo Expo 2009: Power for Land, Sea and Air*. Orlando, FL. USA, 1387-1396.

[20] A. Bai-Tao, L. Jian-Jun, J. Hong-De (2008). Numerical Investigation on Unsteady Effects of Hot Streak on Flow and Heat Transfer in a Turbine Stage. *Proceedings of ASME Turbo Expo 2008: Power for Land, Sea and Air*. Berlin, Germany, 1735-1746.

[21] Yan Xiong, J. Lucheng, Z. Zhedian, Yue Wang and Yunhan Xiao (2008). Three-Dimensional CFD Analysis of a Gas Turbine Combustor for Medium/Low Heating Value Syngas Fuel. *Proceedings of ASME Turbo Expo 2008: Power for Land, Sea and Air*. Berlin, Germany, 1769-1778.

In: Handbook of Genetic Algorithms: New Research
Editors: A. Ramirez Muñoz and I. Garza Rodriguez

ISBN: 978-1-62081-158-0
© 2012 Nova Science Publishers, Inc.

Chapter 10

STUDY OF THE INFLUENCE OF FOREST CANOPIES ON THE ACCURACY OF GPS MEASUREMENTS BY USING GENETIC ALGORITHMS

P. J. García Nieto[1], C. Ordóñez Galán[2], J. Martínez Torres[3], M. Araújo[2] and E. Giráldez[2]

[1]Department of Mathematics, University of Oviedo, Oviedo, Spain
[2]Department of the Natural Resources and Environment, University of Vigo, Vigo, Spain
[3]Centro Universitario de la Defensa, Academia General Militar, Universidad de Zaragoza, Zaragoza, Spain

ABSTRACT

The present paper analyzes the influence of the forest canopy on the precision of the measurements performed by global positioning systems (GPS) receivers. The accuracy of a large set of observations is analyzed in the present research. These observations were taken with a GPS receiver at intervals of one second during a total time of an hour in twelve different points placed in forest areas characterized by a set of forest stand variables (tree density, volume of wood, Hart-Becking index, etc.). The influence on the accuracy of the measurements of other variables related to the GPS signal, such as the Position Dilution of Precision (PDOP), the signal-to-noise ratio and the number of satellites, was also studied. The analysis of the influence of the different variables on the accuracy of the measurements was performed by using genetic algorithms. The results obtained show that the variables with the highest influence on the accuracy of the GPS measurements are those related to the forest canopy, that is, the forest stand variables. The influence of these variables is almost equally important without significant statistical differences. As was expected, those observations recorded in areas covered by an important forest canopy have larger errors than those obtained in areas with less canopy cover. Finally, conclusions of this study are exposed.

Keywords: Forest environments; GPS measurements; Machine learning; Genetic algorithms; Multivariate adaptive regression splines (MARS).

1. INTRODUCTION

The *global* positioning *system* (GPS) is a space-based global navigation satellite system that provides reliable location and time information in all weathers, at all times and anywhere on or near the Earth when and where there is an unobstructed line of sight to four or more GPS satellites. It is maintained by the United States Government and is freely accessible by anyone with a GPS receiver. GPS was created and realized by the U.S. Department of Defense (DOD) and was originally run with 24 satellites. It was established in 1973 to overcome the limitations of previous navigation systems [1-2]. GPS consists of three parts: the space segment, the control segment, and the user segment. GPS satellites broadcast signals from space, which each GPS receiver uses to calculate its three-dimensional location (latitude, longitude, and altitude) plus the current time [3].

GPS was initially developed to be primarily employed in open spaces. However, in practice many users have to operate GPS receivers in less favourable conditions such as forest cover. The occurrence of an overhead canopy may degrade the positional precision by one order of magnitude [4]. This research paper studies the impact of forest canopy on quality and accuracy of GPS measurements by means of the genetic algorithms (GA) technique, in combination with the multivariate adaptive regression splines (MARS) method [5-7] with success. A genetic algorithm (GA) is a search heuristic that mimics the process of natural evolution. This heuristic is routinely used to generate useful solutions to optimization and search problems. Genetic algorithms belong to the larger class of evolutionary algorithms (EA), which generate solutions to optimization problems using techniques inspired by natural evolution, such as inheritance, mutation, selection, and crossover. In a genetic algorithm, a population of strings (called chromosomes or the genotype of the genome), which encode candidate solutions (called individuals, creatures, or phenotypes) to an optimization problem, evolves toward better solutions. Traditionally, solutions are represented in binary as strings of 0s and 1s, but other encodings are also possible. The evolution usually starts from a population of randomly generated individuals and happens in generations. In each generation, the fitness of every individual in the population is evaluated, multiple individuals are stochastically selected from the current population (based on their fitness), and modified (recombined and possibly randomly mutated) to form a new population. The new population is then used in the next iteration of the algorithm. Commonly, the algorithm terminates when either a maximum number of generations has been produced, or a satisfactory fitness level has been reached for the population. If the algorithm has terminated due to a maximum number of generations, a satisfactory solution may or may not have been reached. Genetic algorithms find application in bioinformatics, phylogenetics, computational science, engineering, economics, chemistry, manufacturing, mathematics, physics and other fields. MARS technique is a non-parametric regression technique and can be seen as an extension of linear models that automatically models non-linearities and interactions. MARS technique starts with a model which consists of just the intercept term (which is the mean of the response values). MARS technique then repeatedly adds basis function in pairs to the model. At each step it finds the pair of basis functions that gives the maximum reduction in sum-of-squares residual error (it is a greedy algorithm). The two basis functions in the pair are identical except that a different side of a mirrored hinge function is used for each function.

On one hand, it must be highlighted that the aim of the present research consists of finding, by means of genetic algorithms, those variables that have a higher influence on horizontal and vertical accuracies. On the other hand, the goodness of the variables selected will be evaluated by means of Pearson's correlation coefficient R^2 of a multivariate adaptive regression splines (MARS) model [6]. This paper is in line with previous research focused on the use of GPS under forest canopy [7-11].

2. THE AIM OF THE PRESENT RESEARCH

As input variables, the dasometric characteristics of the fields (i.e. arithmetic mean diameter, average height, crown height, stand density, etc.) together with the GPS signal variables (position dilution of precision, X accuracy, Y accuracy, vertical accuracy, horizontal accuracy, etc.) are taken into account. As output variables, horizontal and vertical accuracies were calculated for each sample through the following expressions:

$$H_{acc} = \sqrt{(E_i - E_{true})^2 + (N_i - N_{true})^2} \qquad (1)$$

$$V_{acc} = |Z_i - Z_{true}| \qquad (2)$$

where H_{acc} and V_{acc} indicate horizontal and vertical accuracies, respectively. E_i, N_i and Z_i are the measured positions at the i second, and E_{true}, N_{true} and Z_{true} are the true positions along the easting, northing and ellipsoidal height directions, respectively. In this sense, the aim of the present work is to find, by means of genetic algorithms, those variables that have a higher influence on H_{acc} and V_{acc}. The performance of the model is evaluated by Pearson's correlation coefficient [6]:

$$R^2 = \frac{(SSY - SSE)}{SSE} \qquad (3)$$

where $SSY = \sum_{i=1}^{n}(y_i - \bar{y})^2$ and $SSE = \sum_{i=1}^{n}(y_i - \hat{y}_i)^2$ for a sample of data y_i being $\bar{y}$ the mean of data y_i and $\hat{y}_i$ the fitted values from the regression analysis.

3. MATHEMATICAL MODEL

3.1. Genetic Algorithms

The genetic algorithms (GAs) are based upon Darwin's Theory of Evolution. The genetic algorithms are modelled on a relatively simple interpretation of the evolutionary process. However, it has proven to be a reliable and powerful optimization technique in a wide variety of applications. Holland [8] in 1975 was the first to propose the use of genetic algorithms for problem-solving. The GA uses the current population of strings to create a new population whereby the strings in the new generation are on average better than those in the current population; the selection depends on their fitness value. The selection process determines which string in the current will be used to create the next generation. The crossover process determines the actual form of the string in the next generation. Weak individuals are discarded and only the strongest survive. In this way, how do they work?

- Initialization: Initially many individual solutions are randomly generated to form an initial population. The population size depends on the nature of the problem, but typically contains hundreds or even thousands of possible solutions. Traditionally, the population is generated randomly, covering the entire range of possible solutions (the search space). Occasionally, the solutions may be "seeded" in areas where optimal solutions are likely to be found.
- Evaluation: An evaluation function is applied in order to know the goodness of each of the solutions of the population.
- Stop criterion: The GA will stop when the optimum solution is found or after a certain number of iterations/generations. If the stop criterion is not accomplished then a new iterative loop is carried out.
- Selection: During each successive generation, a proportion of the existing population is selected to breed a new generation. Individual solutions are selected through a fitness-based process, where fitter solutions (as measured by a fitness function) are typically more likely to be selected. Certain selection methods rate the fitness of each solution and preferentially select the best solutions. Other methods rate only a random sample of the population, as this process may be very time-consuming. The fitness function, f, maps a chromosome representation into a scalar value so that represents the data type of the elements of an dimensional chromosome:

$$f : \Gamma^{n_x} \to \Re \qquad (4)$$

- Crossover: In genetic algorithms, crossover is a genetic operator used to vary the programming of a chromosome or chromosomes from one generation to the next. It is analogous to reproduction and biological crossover, upon which genetic algorithms are based. Crossover operators can be divided into three main categories based on the arity (i.e. the number of parents used) of the operator. This gives rise to three main classes of crossover operators: (1) asexual, where an offspring is generated from one parent; (2) sexual, where two parents are used to produce one or two offspring - the operator employed in the present research - and (3) multi-recombination, where more than two parents are used to produce one or more offspring.
- Mutation: A genetic operator, used to maintain genetic diversity from one generation of a population of algorithm chromosomes to the next. It is analogous to biological mutation. Mutation is used in support of crossover to ensure that the full range of allele is accessible for each gene. Mutation is applied at a certain probability, p_m, to

each gene of the offspring, $\tilde{x}_i(t)$, to produce the mutated offspring $x_i(t)$. The mutation probability, also referred to as the mutation rate, is usually a small value, $p_m \in [0,1]$, to ensure that good solutions are not distorted too much. Given that each gene is mutated at probability p_m, the probability that an individual will be mutated, taking into account that the individual contains n_x genes, is given by:

$$Prob(\tilde{x}_i(t) \text{ is mutated}) = 1 - (1 - p_m)^{n_x} \qquad (5)$$

- Replacement: the least-fit population is replaced with new individuals.

3.2. Multivariate Adaptive Regression Spline

Multivariate adaptive regression splines (MARS) is a form of regression analysis introduced by Jerome Friedman in 1991 [6]. It is a non-parametric regression technique and can be seen as an extension of linear models that automatically models non-linearities and interactions. Its main purpose is to predict the values of a continuous dependent variable, from a set of independent explanatory variables. The MARS model is not only able to predict the value of this dependent variable, but also reports on the importance of each input variable to the continuous dependent variable. MARS builds models of the form:

$$\hat{f}(x) = \sum_{i=1}^{M} c_i B_i(x) \qquad (6)$$

The model is a weighted sum of basis functions $B_i(x)$. Each c_i is a constant coefficient. Each basis function $B_i(x)$ takes one of the following three forms: (1) a constant 1. There is just one such term, the *intercept*; (2) a *hinge* function. A hinge function has the form $\max(0, x - const)$ or $\max(0, const - x)$. MARS automatically selects variables and values of those variables for knots of the hinge functions; and (3) a product of two or more hinge functions. These basis functions can model interaction between two or more variables. The importance of the variable is a measure of the effect that the input variables have on the observed response.

- Nsubsets: this criterion counts the number of model subsets in which each variable is included. Variables that take part in more subsets are considered the most important. By subsets, we mean the subsets of terms generated by the pruning pass. There is one subset for each model size, and the subset is the best set of terms for that model size. Only those subsets that are smaller than or equal in size to the final model are used for estimating variable importance.

- Generalized cross validation (GCV): the GCV is the mean squared residual error divided by a penalty dependent on the model complexity. The GCV criterion for n objects is defined by Eq. (7), where $C(M)$ is a complexity penalty that increases with the number of basis functions in the model, and which is defined as $C(M) = (M+1) + dM$, M is the number of basis functions and the parameter d is a penalty for each basis function included in the model [6-7]:

$$GCV(M) = \frac{\frac{1}{n}\sum_{i=1}^{n}\left(y_i - \hat{f}_M(\bar{x}_i)\right)^2}{\left(1 - C(M)/n\right)^2} \quad (7)$$

- Residual sum of squares (RSS): this is the residual sum of squares measured in the training data. It is a measure of the discrepancy between the data y_i and the estimation model $f(x_i)$. A small RSS indicates a tight fit of the model to the data.

$$RSS = \sum_{i=1}^{n}\left(y_i - f(x_i)\right)^2 \quad (8)$$

4. RESULTS

The data used in the present research were collected using two double-frequency GPS receivers (Hiper-Plus, Topcon Positioning Systems, Inc., Livermore, CA USA) observing GPS pseudorange and carrier phase. The GPS experimental data was collected over 4 days for periods of 5-6 hours between 20th and 23rd August 2007. The antenna heights ranged from 1.45 to 1.60 m and the logging rate was 1 second. The collection of observations lasted for at least 1.5 hour and the process was repeated 3 times per day. The GPS data was revised to ensure continuity and was cut to obtain 12 data sets of 1 hour (3 data sets per day). The coordinates of the control point were 42°41'08.79872''N, 6°38'03.210587'' (latitude-longitude WGS84) and the ellipsoidal height was 933.829 m. These coordinates were calculated by differential correction using the data of the base station named *ponf*, which is the nearest reference station of the Regional GNSS Network (http://gnss.itacyl.es/). This control point was projected by setting up the following UTM coordinates: (m) 693814.623, 4728635.531 (Easting, Northing - Datum ETRS89; zone 29N). This position was used to calculate the "true positions" of experimental points.

The input variables employed in the present research may be outlined as follows:

- In order to characterize each tree mass, the trees in each parcel in a radius of 10 metres around the observation point were measured and the following dasometric parameters were calculated: arithmetic mean diameter (d_m), mean height (h_m), slenderness coefficient (SLC), crown height (h_c), dominant height (H_0), treetop

height (Tt_h), tree density (N), basal area (G), quadratic mean diameter (d_g), Hart-Becking index (HBI), wood volume (V) and biomass (W).
- In addition to parameters reflecting plant cover, a number of variables were recorded in every second measurement, so that the accuracy of observation was conditioned independently of the plant cover (GPS signal variables). These variables are listed as follows:

> $PDOP_p$: position dilution of precision ($PDOP$) for each point under the forest canopy.
> $PDOP_r$: position dilution of precision ($PDOP$) for the reference point.
> σ_{Xr_acc}: error X for the reference point.
> σ_{Yr_acc}: error Y for the reference point.
> σ_{XYr_acc}: error XY for the reference point.
> σ_{Zr_acc}: error Z for the reference point.
> E_p: mean elevation angle for the satellites transmitting the signal for points under the the forest canopy.
> E_r: mean elevation angle for the satellites transmitting the signal received by the reference point.
> $DLLSNCA_p$: indicator of the signal-noise ratio in CA code (in dB*Hz) for a point under the forest canopy.
> $DLLSNCA_r$: indicator of the signal-noise ratio in CA code (in dB*Hz) for the reference point.
> nCA_p: number of satellites receiving CA code for a point under the forest canopy.
> nCA_r: number of satellites receiving CA code for the reference point.
> $DLLSNL1_p$: indicator of the signal-noise ratio in P code for $L1$ (in dB*Hz) for a point under the forest canopy.
> $DLLSNL1_r$: indicator of the signal-noise ratio in P code for $L1$ (in dB*Hz) for the reference point.
> $nL1_p$: number of satellites receiving code in the $L1$ carrier for a point under the forest canopy.
> $nL1_r$: number of satellites receiving code in the $L1$ carrier for the reference point.
> $DLLSNL2_p$: indicator of the signal-noise ratio in P code for $L2$ (in dB*Hz) for a point under the forest canopy.
> $DLLSNL2_r$: indicator of the signal-noise ratio in P code for $L2$ (in dB*Hz) for the reference point.

- $nL2_p$: number of satellites receiving code in the $L2$ carrier for a point under the forest canopy.
- $nL2_r$: number of satellites receiving code in the $L2$ carrier for the reference point.

In order to find those input variables with a greater influence over H_{acc} and V_{acc}, a variable selection by means of GA was performed. The goodness of the variables selected in all the members of the population of each generation was evaluated by mean of Pearson's correlation coefficient R^2 of a MARS model. Figure 1 represents a flowchart that summarizes this process. Table 1 shows the parameters used by the GA technique. It must be remarked that clones are not allowed in our procedure. This assumption is made to ensure that the population has sufficient genetic diversity, which is necessary to enable the algorithm to complete the specified number of generations. However, even this selection does not guarantee that there are no repetitions: only the offspring of couples are tested for clones. As may be observed in Table 1, the optimum generation for H_{acc} was achieved after 237 generations and for V_{acc} after 274. In our case, the stop criterion causing the finalization of the iterations of the models was that the mean of the R^2 of the best 10 models of a generation does not improve more than a 0.05% the R^2 of any of the 10 previous generations.

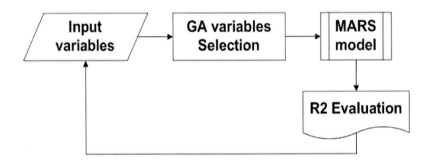

Figure 1. Flowchart of the process.

Table 1. Input variables for the genetic algorithm model

Parameters	Values
Size of the population	2,000 individuals
Probability of mutation	10%
Probability of crossover	50%
Clones allowed	No
Percentage of each population kept	20%
Cardinality of the smallest subset that is wanted	1
Cardinality of the largest subset that is wanted	20
Generations in which the optimum is achieved	237 (for H_{acc}) and 274 (for V_{acc})

Figure 2 (a) and Table 2 show the summarized information of the best model of the last iteration for the output variable H_{acc}. The R^2 of the MARS model performed using as input variables the corresponding variables of this individual of the population was 0.8024. In the same way, Figure 2 (b) and Table 3 represent the information of the best model of the last iteration for the output variable V_{acc}. The R^2 of this MARS model carried out using as input variables the corresponding variables of this individual of the population was 0.6739.

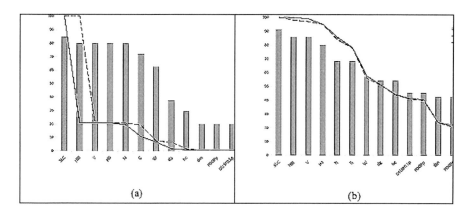

Figure 2. (a) Classification of importance of those variables that have influence over the output variable H_{acc} using nsubsets, GCV and RSS parameters, and (b) classification of importance of those variables that have influence over the output variable V_{acc} using nsubsets, GCV and RSS parameters.

Table 2. Input variables for the genetic algorithm model Classification of importance of those variables that have influence over the output variable H_{acc} by means of the parameters Nsubsets, *GCV* and *RSS*

Variables	Nsubsets	GCV	RSS
SLC	34	100.0000	100.0000
HBI	32	100.0000	21.0223
V	32	20.9356	21.0223
H_0	32	20.9356	21.0223
N	32	20.9356	19.5735
G	29	19.4836	10.8318
W	25	7.3603	6.6852
d_g	15	6.6084	1.5564
h_c	12	1.5064	0.7243
d_m	8	0.6828	0.6898
$PDOP_p$	8	0.6801	0.6566
$DLLSNL1_p$	8	0.6614	0.6233
$nL2_r$	8	0.6573	0.6146
$DLLSNCA_p$	7	0.3777	0.5965

Table 3. Classification of importance of those variables that have influence over the output variable V_{acc} by means of the parameters Nsubsets, *GCV* and *RSS*

Variables	Nsubsets	GCV	RSS
SLC	32	100	100.0000
HBI	30	98	98.1291
V	30	97	98.5717
H_0	28	95	96.7914
N	24	84	85.8379
G	24	78	79.1626
W	20	56	56.1396
d_g	19	51	51.08429
h_c	19	44	45.0488
$DLLSNL1_p$	16	41	41.1493
$PDOP_p$	16	40	40.9767
d_m	15	24	25.7437
$PDOP_r$	15	21	22.0236
$DLLSNCA_p$	13	17	17.5290
$nL2_r$	7	12	12.1380

Conclusion

The genetic algorithms (GAs), in combination with the multivariate adaptive regression splines (MARS) model, have been shown to be a suitable tool in the modelling and analysis of singular problems, such as the analysis of the influence of a forest canopy on the accuracy of GPS measurements. The results obtained show that the variables with the biggest influence on the accuracy of the GPS measurements are those related to the forest overstorey, that is, the forest stand variables. The influence of these variables is almost equally important without significant statistical differences. To fix ideas, the best-fitted MARS models have values of the Pearson's correlation coefficient R^2 of 0.8024 and 0.6739 R^2 for the variables H_{acc} and V_{acc}, respectively.

The GA plus MARS [6] quite accurately reproduces the mechanism of the prediction of the GPS measurements and the associated errors due to the forest canopy. Therefore, the key step in an engineering analysis is choosing appropriate mathematical models. These models will clearly be selected depending on what phenomena are to be predicted, and the selection of mathematical models that are *reliable* and *effective* in predicting the quantities sought is of the utmost importance.

In this article, the influence of dasometric and GPS parameters on the accuracy of GPS measurements under a forest canopy has been studied in depth for both H_{acc} and V_{acc}

variables. The results of this research are in line with previous studies [7-12] and have proved that for both parameters, the most important variables are, in this order, SLC, HBI, V, d_m, H_0, N and G. Other variables that were found to be important for both H_{acc} and V_{acc} are: d_g, W, $PDOP_p$, $DLLSNL1_p$, h_c, d_m, $nL2_r$, $PDOP_r$ and $DLLSNCA_p$. Please note that one of the most important findings of this research is that the variables that have an influence on H_{acc} and V_{acc} are practically the same and that many of them appear in the MARS models in the same order for both output variables. Finally, the methodology developed in this research paper can be applied to other, similar problems with equal success.

ACKNOWLEDGMENTS

The authors wish to acknowledge the computational support provided by the Department of Mathematics at University of Oviedo and Department of Natural Resources and Environment at University of Vigo. Furthermore, authors would like to express their gratitude to the Department of Education and Science of the Principality of Asturias for its partial financial support (Grant reference FC-11-PC10-19). Finally, we would like to thank Anthony Ashworth for his revision of English grammar and spelling of the manuscript.

REFERENCES

[1] Bao-Yen Tsui, J. *Fundamentals of Global Positioning System Receivers: A Software Approach*. New York: Wiley-Interscience; 2004.

[2] Kaplan, E.D. and Hegarty, C. *Understanding GPS: Principles and Applications*. New York: Artech House Publishers; 2005.

[3] El-Rabbany, A. *Introduction to GPS: The Global Positioning System*. New York: Artech House Publishers; 2006.

[4] Sigrist, P., Coppin, P. and Hermy, M. (1999). Impact of forest canopy on quality and accuracy of GPS measurements. *International Journal of Remote Sensing*, 20(18), 3595-3610.

[5] Engelbrecht, A.P. *Computational Intelligence: An Introduction*. New York: Wiley; 2007.

[6] Friedman, J.H. (1991). Multivariate adaptive regression splines. *Annals of Statistics*, 19(1), 1-141.

[7] Hastie, T., Tibshirani, R. and Friedman, J. *The Elements of Statistical Learning: Data Mining, Inference, and Prediction*. New York: Springer; 2009.

[8] Holland, J.H. *Adaptation in Natural and Artificial Systems*. Ann. Arbor: University of Michigan Press; 1975.

[9] Cordero, R., Mardones, O. and Marticorena, M. (2006). Evaluation of forestry machinery performance in harvesting operations using GPS technology." In: Ackerman PA, Längin DW and Antonides MC (Editors): Precision forestry in plantations, semi-

natural and natural forests. *Proceedings of the International Precision Forestry Symposium*, Stellenbosch University, South Africa, 163-173.

[10] Gegout, J.C. and Piedallu, C. (2005). Effects of Forest Environment and Survey Protocol on GPS Accuracy. *Photogrammetric Engineering and Remote Sensing*, 71(9), 1071-1078.

[11] Hasegawa, H. and Yoshimura, T. (2007). Estimation of GPS positional accuracy under different forest conditions using signal interruption probability. *Journal of Forest Research*, 12(1), 1-7.

[12] Martin, A.A., Owende, P.M. and Ward, S.M. (2001). The effects of peripheral canopy on DGPS performance on forest roads. *International Journal of Forest Engineering*, 12 (1), 71-79.

Reviewed by Prof. Dr. Pedro Arias, Departament of Natural Resources and Environment, University of Vigo, 36310 Vigo, Spain. (Email: parias@uvigo.es).

In: Handbook of Genetic Algorithms: New Research
Editors: A. Ramirez Muñoz and I. Garza Rodriguez

ISBN: 978-1-62081-158-0
© 2012 Nova Science Publishers, Inc.

Chapter 11

ROUNDNESS EVALUATION BY GENETIC ALGORITHMS

Michele Lanzetta and Andrea Rossi*

Department of Mechanical, Nuclear and Production Engineering
University of Pisa, Via Diotisalvi 1, 56122 Pisa, Italy

ABSTRACT

Roundness is one of the most common features in machining, and various criteria may be used for roundness errors evaluation. The minimum zone tolerance (MZT) method produces more accurate solutions than data fitting methods like least squares interpolation. The problem modeling and the application of Genetic Algorithms (GA) for the roundness evaluation is reviewed here. Guidelines for the GA parameters selection are also provided based on computation experiments.

Keywords: Minimum zone tolerance (MZT), roundness error, genetic algorithm, CMM

1. INTRODUCTION

In metrology, the inspection of manufactured parts involves the measurement of dimensions for conformance to product specifications (Figure 1). In series or lot production, feedbacks from statistical analyses performed on multiple products allow process control for quality improvement.

Product specifications are associated with tolerances, which represent the acceptable limits for measured parts. Tolerances come from manufacturing requirements, e.g. assemble parts that fit, or from functional requirements for use or operation of the final product, e.g. rotation of a wheel, power of an engine.

---

* Tel.: +39 050 2218122; fax: +39 050 2218140.

Metrology involves the acquisition or sampling of individual points by manual instruments, like analog or digital calipers, micrometers and dial gages, or by automated tools like coordinate measuring machines (CMM) or vision systems. Automated tools are equipped with software for post processing of data and are able to measure complex surfaces.

Measurements can be linear, such as size, distance, and depth and in two or three dimensions, such as surfaces and volumes. In addition to dimension tolerance there are form tolerances for two or three dimensional geometric primitives, like straightness (for an edge, an axis), flatness (for a plane), circularity or roundness (for a circle, an arc) or cylindricity (for a peg, a bar, a hole).

The simplest way to assess form tolerances is finding the belonging geometric primitives by interpolation of individual acquired data points. Not always linear regression represents the most accurate estimation of the form error. Overestimates represent a waste for the rejection of acceptable parts, while inversely underestimated form errors may produce defective parts.

The estimation of form errors by non linear methods is an optimization problem where metaheuristics, such as genetic algorithms, ant colony systems or neural networks can provide more accurate results with respect to linear methods, subject to proper modeling of the mathematical problem.

The application of metaheuristics for the estimation of form error is an active research field and final solutions are still far to come, particularly regarding the processing time due to the problem compexity compared to interpolation methods, which provide results in fractions of the second.

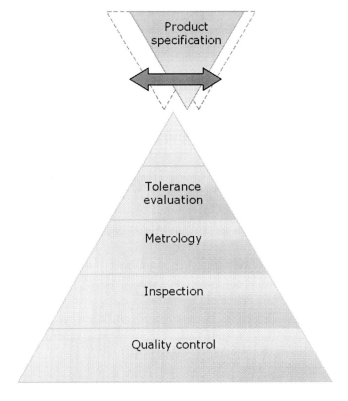

Figure 1. Adjustment between manufacturing tolerances and product specifications.

In the remainder the application of genetic algorithms for roundness evaluation will be discussed. The approach presented can be extended to other types of form errors.

2. ROUNDNESS ERROR

Roundness is the property of being shaped like or approximately like a circle or cylinder. In manufacturing environments, variations on circular features may occur due to: imperfect rotation, erratic cutting action, inadequate lubrication, tool wear, defective machine parts, chatter, misalignment of chuck jaws, etc. The out of roundness of circular and cylindrical parts can prevent insertion, produce vibrations in rotating parts, irregular rotation, noise etc.

Form tolerance is evaluated with reference to an ideal geometric feature, i.e. a circle in the case of roundness.

The most used criteria to establish the reference circle are: the Least- Squares method (LSQ), the Maximum Inscribed Circle (MIC), the Minimum Circumscribed Circle (MCC) and the Minimum Zone Tolerance (MZT). The use of a particular interpolation or data fitting method depends on the required part application, e.g. MIC and MCC can be used when mating a peg into a hole is involved to assess interference. The LSQ is one of the methods used by the coordinate measuring machines for rapidity and because it is efficient in computation with a large number of measured points. The roundness error determined by the LSQ is larger than those determined by other methods, such as the MZT.

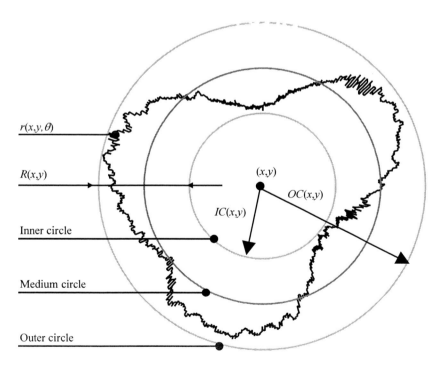

Figure 2. Roundness profile of a real profile with 1800 equally-spaced CMM sample points and reference circles: inner, medium, and outer circles for a given associated derived center (x,y).

3. LITERATURE

The MZT can be considered the best estimation of the roundness error because its definition meets the standard definition of the roundness error, as reported in ISO 1101 [1]. The MZT determines two concentric circles that contain the measured profile and such that the difference in radii is the least possible value as it is shown in Figure 2, where c_1 and c_2 are two possible centers of two concentric circles that include the measured points and where Δr_1 and Δr_2 are their difference in radii. So, once found, the MZ error can be considered the roundness error itself and the related MZ center. However, the MZT is a non linear problem and several methods to solve this problem have been proposed in the literature: computational geometry techniques and the solution of a non linear optimization problem. The first approach is, in general, very computationally expensive, especially, when the number of data points is large. One of these methods is based on the Voronoi diagram [2]. The second approach is based on an optimization function but the inconvenience is that this function has several local minima.

Some examples are: the Chebyshev approximation [3], the simplex search / linear approximation [4] [5], the steepest descent algorithm [6], the particle swarm optimization (PSO) [7] [8], the simulated annealing (SA) [9], and genetic algorithms (GAs) [10] [11] [12] [13].

Xiong [14] develops a general mathematical theory, a model and an algorithm for different kinds of profiles including roundness where the linear programming method and exchange algorithm are used. As limaçon approximation is used to represent the circle, the optimality of the solution is however not guaranteed. Performance of methods have been reviewed in [15].

A strategy based on geometric representation for minimum zone evaluation of circles and cylinders is proposed by Lai and Chen [16]. The strategy employs a non-linear transformation to convert a circle into a line and then uses a straightness evaluation schema to obtain minimum zone deviations for the feature concerned. This is an approximation strategy to minimum zone circles.

M. Wang et al. [17] and Jywe et al. [18] present a generalized non-linear optimization procedure based on the developed necessary and sufficient conditions to evaluate roundness error. To meet the standards the MZ reference circles should pass through at least four points of the sample points. This can occur in two cases: a) when three points lie on a circle and one point lies on the other circle (the 1-3 and the 3-1 criteria); b) when two points lie on each of the concentric circles (the 2-2 criterion). In order to verify these conditions the computation time increases exponentially with the dataset size. Gadelmawla [19] use a heuristic approach to drastically reduce the number of sample points used by the min-max 1-3, 3-1 and 2-2 criteria.

Samuel and Shunmugam [20] establish a minimum zone limaçon based on computational geometry to evaluate roundness error; with geometric methods, global optima are found by exhaustively checking every local minimum candidate. Moroni and Petro [15] propose a technique to speed up the exhaustive generation of solutions (*brute force algorithm*) which starts with a single point and increases one sample point at each step in order to generate all the possible subsets of points, until the tolerance zone of a subset cover the whole dataset (*essential subset*).

A mesh based method with starting center on the LSC, where the convergence depends on the number of mesh cross points, representing a compromise between accuracy and speed, is proposed by Xianqing et al. [21].

The strategy to equally-spaced points sampled on the roundness profile is generally adopted in the literature. Conversely, in previous works the authors developed a cross-validation method for small samples to assess the kind of manufacturing signature on the roundness profile in order to detect critical points such as peaks and valleys [22] [23]. They use a strategy where a next sampling increasing the points near these critical areas of the roundness profile.

In [24], some investigations proved that the increase of the number of sample points is effective only up to a limit number. Recommended dataset sizes are given for different data fitting methods (LSQ, MIC, MCC, MZT) and for three different out-of-roundness types (oval, 3-lobing and 4-lobing). Similar works are [25] and [26] in which substantially the same results are given.

A sampling strategy depends on the optimal number of sample points and the optimum search-space size for best estimation accuracy, particularly with datasets which involve thousands of sample points available by CMM scanning techniques. The sampling strategy tailored for a fast genetic algorithm to solve the MZT problem can be defined as *blind* according to the classification in [27] if it is not manufacturing specific. By sampling strategy not only the number and location of sample points on the roundness profile but also their use by the data fitting algorithm is concerned.

4. MZT MODEL

In the MZT method, the unknown are the (x,y) coordinates of the associated derived center of the minimum zone reference circles of the roundness profile (MZCI [28]). MZCI is formed by two concentric circles enclosing the roundness profile, the inner minimum zone reference circle and the outer minimum zone reference circle, having the least radial separation. The difference between the inner minimum zone reference circle and the outer minimum zone reference circle is the minimum zone error (*MZE*). MZE is the target parameter of our optimization algorithm as a function of (x,y).

Given an extracted circumferential line $r(x,y,\theta)$, with $\theta \in (0, 2\pi]$, of a section perpendicular to the axis of a cylindrical feature, the *roundness error* $R(x,y)$ is defined by:

$$R(x, y) = OC(x, y) - IC(x, y), (x,y) \in E_{r(x,y,\theta)} \qquad (1)$$

where $OC(x,y)$ and $IC(x,y)$ are the radii of the reference circles of center (x,y), and $E_{r(x,y,\theta)}$ is the area enclosed by $r(x,y,\theta)$:

$$OC(x, y) = \max_{\theta \in (0,2\pi]} r(x, y, \theta) \qquad (2)$$

$$IC(x, y) = \min_{\theta \in (0,2\pi]} r(x, y, \theta) \qquad (3)$$

As a CMM scans the roundness profile by sampling a finite number, n, of equally-spaced points θ_i of the extracted circumferential line ($\theta_i = i \times \frac{2\pi}{n}$, $i=1,...,n$), the $OC(x,y)$ and $IC(x,y)$ are evaluated by:

$$OC(x,y) = \max_{\theta_i = i \times \frac{2\pi}{n},\, i=1,...,n} r(x,y,\theta_i) \qquad (4)$$

$$IC(x,y) = \min_{\theta_i = i \times \frac{2\pi}{n},\, i=1,...,n} r(x,y,\theta_i) \qquad (5)$$

Figure 2 shows the mentioned features for a given (x,y).

MZE is evaluated by applying the MZT data-fitting method to solve the following optimization problem:

$$MZE = \min_{(x,y) \in E_{r(x,y,\theta_i)}} R(x,y) = \begin{cases} \min\left[\max_{\theta_i = i \times \frac{2\pi}{n},\, i=1,...,n} r(x,y,\theta_i) - \min_{\theta_i = i \times \frac{2\pi}{n},\, i=1,...,n} r(x,y,\theta_i)\right] \\ \text{subject to } (x,y) \in E_{r(x,y,\theta_i)} \end{cases} \qquad (6)$$

where $E_{r(x,y,\theta_i)}$ is a restricted area in the convex envelopment of the n equally-spaced sample points, i.e. the *search space*.

5. GENETIC ALGORITHMS FOR ROUNDNESS EVALUATION

GAs were proposed for the first time by Holland [29] and constitute a class of search methods especially suited for solving complex optimization problems [30].

Genetic algorithms are widely used in research for non-linear problems. They are easily implemented and powerful being a general-purpose optimization tool. Many possible solutions are processed at the same time and evolve with both elitist and random rules, so to quickly converge to a local optimum which is very close or coincident to the optimal solution.

Genetic algorithms constitute a class of implicit parallel search methods especially suited for solving complex optimization or non-linear problems. They are easily implemented and powerful being a general-purpose optimization tool. Many possible solutions are processed concurrently and evolve with inheritable rules, e.g. the elitist or the roulette wheel selection, so to quickly converge to a solution which is very close or coincident to the optimal solution.

Genetic algorithms maintain a population of center candidates (the *individuals*), which are the possible solutions of the MZT problem. The center candidates are represented by their *chromosomes*, which are made of pairs of x_i and y_i coordinates. Genetic algorithms operate on the x_i and y_i coordinates, which represent the inheritable properties of the individuals by means of genetic operators. At each generation the genetic operators are applied to the selected center candidates from current population in order to create a new generation. The selection of individuals depends on a fitness function, which reflects how well a solution fulfills the requirements of the MZT problem, e.g. the objective function.

Sharma et al. [2] use a genetic algorithm for MZT of multiple form tolerance classes such as straightness, flatness, roundness, and cylindricity. There is no need to optimize the algorithm performance, choosing the parameters involved in the computation, because of the small dataset size (up to 100 sample points).

Wen et al. [11] implement a genetic algorithm in real-code, with only crossover and reproduction operators applied to the population. Thus in this case mutation operators are not used. The algorithm proposed is robust and effective, but it has only been applied to small samples.

In a genetic algorithm for roundness evaluation the center candidates are the individual of the population (*chromosomes*). The search space is an area enclosed by the roundness profile where the center candidates of the initial population are selected for the data fitting algorithm. The area is rectangular because the crossover operator changes the x_i and y_i coordinates of the parents to generate the offspings [10]. After crossover, the x_i and y_i coordinates of parents and offsprings are located to the rectangle vertexes.

In order to find the MZ error the search space must include the global optima solution i.e. the MZ center. Therefore the centre of the rectangular area is an estimation of the MZ center evaluated as the mean value of the x_i and y_i coordinates of the sampled points [10], [11], [12], [13]. In [11] the search-space is a square of fixed 0.2 mm side, in [13] it is 5% of the circle diameter and center. In [10], the side is determined by the distance of the farthest point and the nearest point from the mean center. In [12] it is the rectangle circumscribed to the sample points.

Optimal sampling and genetic algorithm parameters are listed in Table 1.

Table 1. Algorithm parameters according to [13]

Optimization Geometric and algorithm GA genetic parameters	Symbol	Value	Comment
sample size	n	500	number of equally spaced sample points
search space	E	0.5	initial population randomly selected within
population size	P_s	70	set of chromosomes used in evolving epoch
selection			elitist selection
crossover	P_c	0.7	one point crossover of the *pc×pop* parents' genes (i.e. coordinates) at each generation
mutation	P_m	0.07	*pm×pop* individuals are modified by changing one gene (i.e. coordinate) with a random value
stop criterion	N	100	the algorithm computes N generations after the last best roundness error evaluated rounded off to the fourth decimal digit (0.1 µm)

6. GA Operation

Selection: during this operation, a solution has a probability of being selected according to its fitness. Some of the common selection mechanisms are: the roulette wheel procedure, the Tournament selection, and the elitist selection. With this latter, the individuals are ordered on the basis of their fitness function; the best individuals produce offspring. The next generation will be composed of the best chromosomes chosen between the set of offspring and the previous population.

Crossover: this operator allows to create new individuals as offspring of two parents by inheriting genes from parents with high fitness. The possibility for this operator to be applied depends on the crossover probability. There are different crossover types: the one point and multiple point crossover, and other sophisticated ones. In the proposed GA was used the arithmetic crossover mechanism, which generates offspring as a component-wise linear combination of the parents.

Mutation: a new individual is created by making modifications to one selected individual. In genetic algorithms, mutation is a source of variability, and is applied in addition to selection and crossover. This method prevents the search to be trapped only in local solutions. The relative parameter is the mutation probability, that is the probability that one individual is mutated.

Stop criterion: the algorithm has an iterative behavior and needs a stop condition to end the computation. Possible criteria include: overcoming a predefined threshold for the fitness function or iteration number or their combinations. In the proposed GA, the stop criterion is controlled by the number N of the iterations if no improvement in the solution occurs.

7. Testing Algorithms

To analyze the behavior of an algorithm with the MZT method, dataset with known MZ error are available. These datasets are generated with NPL Chebyshev best fit circle certified software [31].

The use of certified software has the following benefits

- it produces randomly distributed error makes the results more general, because the results achieved with the genetic algorithm are not manufacturing signature specific;
- the circle center is known, so it allows estimating the circular profile center is computed as an average value of the measuring points coordinates [11] the average MZ center found by the algorithm

Datasets produced by certified software have a user-selected center and radius.

For performance assessment, the algorithm is usually executed on a dataset several tens of times for each test, and the average MZ error and the average computation time are computed and compared with the nominal MZ error.

An example of execution in shown in Figure 3, which displays the average and standard error.

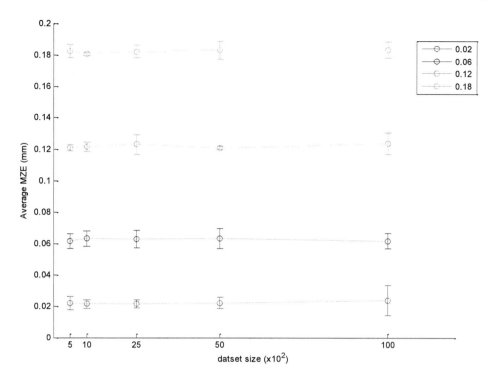

Figure 3. Average MZE with error bars for large datasets and different roundness errors.

In a typical experimental approach, a first raw estimation of the circular profile center is necessary as a starting point for a more accurate roundness evaluation.

A first estimation of the circular profile center is computed as an average value of the measuring points coordinates in expression (1).

SUMMARY

The application of a genetic algorithm for the roundness evaluation of circular profiles using the MZT method has been described.

The GA approach described and the parameters provided may solve most roundness measurement needs, both for small and large datasets (up to several thousands datapoints). The listed literature may serve as a guideline for optimal GA parameters estimation on new problems.

ACKNOWLEDGMENTS

Part of this work has been derived from Alessandro Meo and Luca Profumo's project work for the class of automation of production processes of the master degree in automation engineering at the school of engineering at Pisa University.

REFERENCES

[1] International Organization for Standardization, Geneva, Switzerland, ISO 1101, *Geometrical Product Specifications (GPS)–tolerances of form, orientation, location and run out*, 2nd ed.; December 2004.

[2] Sharma, R., Rajagopal, K. and Anand, S., A genetic algorithm based approach for robust evaluation of form tolerances, *Journal of Manufacturing Systems*, 2000, 19-n1, 46–57.

[3] Dhanish, P.B. and Shunmugam, M.S., An algorithm for form error evaluation – using the theory of discrete and linear chebyshev approximation, *Computer Methods in Applied Mechanics and Engineering*, Vol. 92, 1991, 309–324.

[4] Weber, T., Motavalli, S., Fallahi, B., Cheraghi, S.H., A unified approach to form error evaluation, *Journal of the International Societies for Precision Engineering and Nanotechnology* 26, 2002, 269–278.

[5] Murthy, T.S.R. and Abdin, S.Z., Minimum zone evaluation of surfaces, *International Journal of Machine Tool Design Research*, 20, 1980, 123–136.

[6] Zhu, L.M., Ding, H. and Xiong, Y.L., A steepest descent algorithm for circularity evaluation, *Computer Aided Design*, 2003, 35, 255–265.

[7] Yashpal Kovvur, Hemant Ramaswami, Raj Bardhan Anand and Sam Anand, Minimum-zone form tolerance evaluation using particle swarm optimisation, *Int. J. Intelligent Systems Technologies and Applications*, Vol. 4, Nos. 1/2, 2008.

[8] Mao, J., Cao, Y., Yang, J., Implementation uncertainty evaluation of cylindricity errors based on geometrical product specification (GPS), *Measurement*, 42(5), June 2009, 742-747.

[9] Shakarji, C.M. and Clement, A., Reference Algorithms for Chebyshev and One-Sided Data Fitting for Coordinate Metrology, *CIRP Annals - Manufacturing Technology*, 53(1), 2004, 439-442.

[10] Rohit Sharma, Karthik Rajagopal, and Sam Anand, A Genetic Algorithm Based Approach for Robust Evaluation of Form Tolerances, *Journal of Manufacturing Systems* Vol. 19/No. I 2000.

[11] Wen, X., Xia, Q., Zhao, Y., An effective genetic algorithm for circularity error unified evaluation, *International Journal of Machine Tools and Manufacture* 46, 2006, 1770–1777.

[12] Yan, L., Yan, B., Cai, L., Hu, G., Wang, M., Research on roundness error evaluation of shaft parts based on genetic algorithms with transfer-operator, 9th International Conference on Electronic Measurement and Instruments, 2009. ICEMI '09, 2-362 - 2-366. doi:10.1109/ICEMI. 2009.5274566.

[13] Rossi, A.; Antonetti, M.; Barloscio M.; Lanzetta, M.: Fast genetic algorithm for roundness evaluation by the minimum zone tolerance (MZT) method, *Measurement*, vol. 44, n. 7, August 2011, ISSN 0263-2241, DOI: 10.1016/j.measurement.2011.03.031, 1243-1252 (10).

[14] Xiong, Y.L., Computer aided measurement of profile error of complex surfaces and curves: theory and algorithm, *International Journal Machine Tools and Manufacture*, 1990, 30, 339–357.

[15] Giovanni Moroni, Stefano Petro, Geometric tolerance evaluation: *A discussion on minimum zone fitting algorithms, Precision Engineering,* Volume 32, Issue 3, July 2008, Pages 232-237, ISSN 0141-6359, DOI: 10.1016/j.precisioneng.2007.08.007.
[16] Lai, J. and Chen, I., Minimum zone evaluation of circles and cylinders, *International Journal of Machine Tools and Manufacturing,* 1995, 36(4), 435–51.
[17] Wang, M., Cheraghi, S.H. and Masud, A.S.M., Circularity error evaluation: *theory and algorithm, Precision Engineering,* 1999, 23(3), 164–76.
[18] Jywe, W.-Y, Liu G.-H., Chen C.-K., The min-max problem for evaluating the form error of a circle, *Measurement,* 1999, 26, 273-282.
[19] Gadelmawla, E.S., Simple and efficient algorithm for roundness evaluation from the coordinate measurement data, *Measurement,* 2010, 43, 223-235.
[20] Samuel, G.L. and Shunmugam, M.S., Evaluation of circularity from coordinate and form data using computational geometric techniques, *Precision Engineering,* 2000, 24, 251–263.
[21] Xianqing, L., Chunyang, Z., Yujun, X., Jishun, L., *Roundness Error Evaluation Algorithm Based on Polar Coordinate Transform, Measurement,* In Press, Accepted Manuscript, Available online 23 October 2010.
[22] Rossi, A., A form of deviation-based method for coordinate measuring machine sampling optimization in an assessment of roundness, Proc Instn Mech Engrs, Part B: *Journal of Engineering Manufacture,* 2001, 215, 1505–1518.
[23] Rossi, A., A minimal inspection sampling technique for roundness evaluation, In *first CIRP International Seminar on PRogress in Innovative Manufacturing Engineering* (Prime), Sestri Levante, Italy, June 2001.
[24] Chajda, J., Grzelka, M., Gapinski, B., Pawłowski, M., Szelewski, M., Rucki, M., *Coordinate measurement of complicated parameters like roundness, cylindricity, gear teeth or free-form surface,* 8 International conference advanced manufacturing operations.
[25] Gapinski, B. and Rucki, M., Uncertainty in CMM Measurement of Roundness, *AMUEM* 2007 - International Workshop on Advanced Methods for Uncertainty Estimation in Measurement Sardagna, Trento, Italy, 16-18 July 2007.
[26] Gapinski, B., Grezelka, M., Rucki, M., Some aspects of the roundness measurement with cmm, *XVIII IMEKO World Congress Metrology for a Sustainable Development September,* 17 – 22, 2006, Rio de Janeiro, Brazil.
[27] Colosimo, B.M., Moroni, G. and Petro, S., A tolerance interval based criterion for optimizing discrete point sampling strategies, *Precision Engineering,* Volume 34, Issue 4, October 2010, 745-754, ISSN 0141-6359, DOI: 10.1016/j.precisioneng.2010.04.004.
[28] International Organization for Standardization, Geneva, Switzerland, ISO/TS 12181–1: *Geometrical Product Specifications (GPS) – Roundness — Part 1: Vocabulary and parameters of roundness,* 2003.
[29] Holland, J., Adaptation in Natural and Artificial System, *Ann. Arbor.,* MI: The University of Michigan Press, 1975.
[30] DeJong, K.A., Analysis of the behavior of a class of genetic adaptive systems, *PhD thesis* (University of Michigan, USA), 1975.
[31] National Physical Laboratory (UK), *Data Generator for Chebyshev Best-Fit Circle,* http://www.npl.co.uk/mathematics-scientific-computing/ software – support – for - metrology/, last accessed January 2011.

In: Handbook of Genetic Algorithms: New Research
Editors: A. Ramirez Muñoz and I. Garza Rodriguez
ISBN: 978-1-62081-158-0
© 2012 Nova Science Publishers, Inc.

Chapter 12

INFLUENCE OF BUILDING CODES IN THE BEHAVIOR OF A MODIFIED ELITIST GENETIC ALGORITHM APPLIED TO THE OPTIMIZATION OF STEEL STRUCTURES

J. J. del Coz Díaz[1], P. J. García Nieto[2], M. B. Prendes Gero[1] and A. Bello García[1]

[1]Department of Construction and Manufacturing Engineering,
University of Oviedo, 33204 Gijón, Spain
[2]Department of Mathematics, University of Oviedo, Oviedo, Spain

ABSTRACT

In this work an elitist genetic algorithm (GA) developed by the authors and implemented in an advanced analysis program of three-dimensional steel structures (named ESCAL3D) is evaluated in order to compare the optimization results over a typical portal-frame structure. The minimum weight that satisfies the ultimate limit states of different applicable building codes (MV-103 Spanish code, Eurocode-3 and AISC-LRFD) was checked in order to obtain the influence of several parameters, such as the population's size, the number of generations, the function's evaluation, etc. Finally, the cost and weight improvements obtained using this GA for the different building codes as well as the computational effort are discussed, giving place to the conclusions exposed in this study.

Keywords: Elitist genetic algorithm; Steel structures; Minimum weight; Ultimate limit states; Building codes.

1. INTRODUCTION

The continuous development of genetic algorithms (GA) has led to obtain the so-called elitist GA [1]. The purpose of this last one is to avoid that the best individual of a population fails when it obtains offspring in the following generation. With this aim, they copy the best individual from the present population in the new one, normally achieving a speed increase in the obtaining of the optimal individual. The elitist GA implemented in this study does not save a single individual, but a percentage of the best individuals according to the elite probability, thus achieving a greater speed of convergence. In this work this algorithm is applied to steel structures, with the aim of obtaining individuals of minimum weight in the structural elements, but which fulfil the safety factors established by three selected building codes (MV-103 Spanish code [2], Eurocode-3 [3] and AISC-LRFD [4]). In order to solve this problem, a modified objective function has been defined which considers the constraints of these coefficients. In addition, it is necessary to take into account the following remarks [5]:

- The codification of the design variables has been modified, achieving the same probability of initial selection in all of them.
- A selection operator denominated 'aptitude' has been implemented, considering the dispersion of the individuals in the population.
- A crossover operator denominated 'phenotype crossover' has been developed so that it only exchanges the section type assigned to the structural elements, without modifying it previously.

Besides, the design of a two-dimensional portal-frame and a three-storey steel building have been implemented in order to check the performance of the developed elitist algorithm and the suggested modifications, with the three building codes indicated above. The first structure, two-dimensional portal-frame (Figure 1), with thirteen nodes and twelve beams collected in three groups, is subjected to a total distributed load of 4250 N/m (1250 N/m of dead load and 3000 N/m of snow load) on the roofing beams. The overall dimensions of the structure are 20 m span and 7.7 m high.

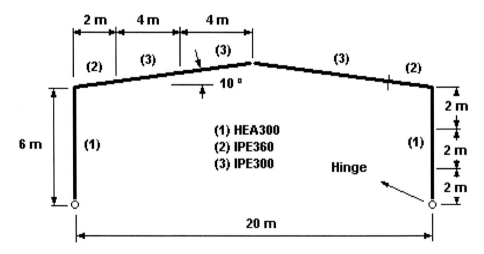

Figure 1. Detail of a portal-frame structure: overall dimensions and initial sections.

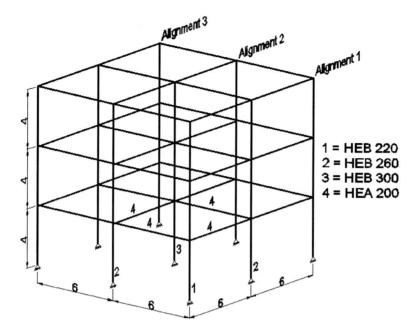

Figure 2. Detail of a three-storey structure: overall dimensions and initial sections.

The second structure, three-storey building, is totally symmetric with a separation of 4 metres between storeys and 6 metres between the columns (see Figure 2 below). Three simple load hypotheses have been considered: the weight of the structure, a live load due to the wind with a maximum value of 18,835 N in the middle column and middle floor and a live load by use with a maximum value of 17,658 N/m in the second alignment. The steel properties used are: Young's modulus $= 2.1 \times 10^{11}$ Pa, specific weight = 7,850 kg/m³ and yield stress $= 275 \times 10^{6}$ Pa.

2. ELITIST GENETIC ALGORITHM

The elitist genetic algorithm developed has been implemented in the advanced structural analysis program ESCAL3D [6], with these modifications [7]:

1. The design variables are encoded by means of chains of bits, according to a computational procedure in which all the sections have the same probability of initial selection.

$$\lambda = 2^n \qquad (1)$$

where λ is the number of sections of the commercial catalogue represented and n is the integer of bits necessary to represent λ sections of this catalogue.

2. A reproductive operator denominated 'aptitude' is implemented with the end of considering the population's dispersion: A new function denominated 'aptitude

function' is defined on the basis of the modified objective function and the value of this one is obtained for all the individuals. Then, the aptitudes less than the average value as well as their individuals are removed and the probability of rejection is calculated over the surviving individuals. This value is the inverse one of the aptitude. On this way, the new population is created from the best individuals of the previous population, avoiding isolated elements and increasing the speed of the GA in the search of the optimal individual.
3. A new crossover denominated *phenotype* crossover is implemented. In this operator, the crossover point is located between two phenotypes or design variables from two individuals termed parents. Two new strings, the children, are created swapping all the characters from the selected position to the overall length of the parents' strings. With it, the crossover interchanges real sections without modifying them (see Figure 3 below).

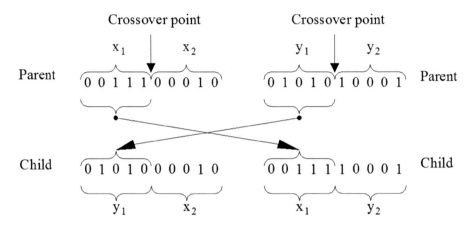

Figure 3. Phenotype crossover in binary representation.

3. FORMULATION OF THE PROBLEM

Using the algorithm pointed out previously, the function used as a measure of effectiveness of the design is denominated the *objective function, merit function* or *cost function*. This function may be formulated from a simple objective $f_1(x)$ or multiple objectives as shown in Eq. (2).

$$F(x) = \{f_1(x), f_2(x), ..., f_p(x)\} \qquad (2)$$

The aim of structural optimization, and in particular GAs, is to obtain the sections of the structural elements that minimize an objective, subject to certain limits or constraints. These constraints may be classified into two types: explicit and implicit. Explicit constraints are analyzed without a simulation system. On the contrary, implicit constraints require analysis and verification of the designs, such as for instance the allocation of areas to the sections. Several methods of adjusting the constraints exist [8]. In this work, it is used the penalizing of the solutions that violate one or more constraints. Nevertheless, there are difficulties if penalty functions are applied to dependent problems.

In general, the problems covered by GA are of the restricted optimization type, and for this reason the optimization problem must be transformed into non restrictive problems. In this case, the penalty is based on the transformation method represented below:

$$\text{Minimize} \quad \overline{F}(x,r) = F(x) + \overline{P}(r, G(x), H(x)) \tag{3}$$

where $\overline{F}(x,r)$ is the modified objective function, $F(x)$ is the objective function and $\overline{P}(r, G(x), H(x))$ is the penalty term defined as a function of the penalty coefficient r and the constraint functions $G(x)$ and $H(x)$.

The goal of the implemented elitist GA is to obtain steel structures of minimum weight that fulfil the safety factors established by the applicable building code. This may be mathematically expressed according to Eq 9.

$$\overline{F}(x,r) = \rho \cdot \sum_{s=1}^{n_{bar}} x_s \cdot L_s + \sum_{s=1}^{n_{bar}} \left[r_1 \cdot G_s(x) + r_2 \cdot H_s(x) + \ldots + r_{n_c} \cdot T_s(x) \right] \tag{4}$$

where the *modified objective function* $\overline{F}(x,r)$ is the result of add two terms, the weight of the structure that depends on the density of the material ρ, the area of the section x_s and the length L_s of the number of structural elements n_{bar} that compose the structure and the penalty factor defined or the limit values that the number of safety factors n_c established by the applicable building code $G_s(x)$, $H_s(x)$, ..., $T_s(x)$ may reach in each structural element.

Of the terms that define the modified objective function, the easiest to obtain is the weight, since it is directly defined on the basis of the geometric data of the structure, the characteristics of the material assigned to the bars and the properties of the sections assigned to these. In contrast, in order to obtain the second term it is necessary to define the penalty coefficient and to carry out an analysis of the structure. In this way, the stresses and moments that define the safety factors and therefore the constraints of the problem will be obtained. The analysis of the structure and the verification of the safety factors are carried out by means of the program named ESCAL3D, which is able to obtain the safety factors established by Spanish, American and European Building Codes.

In them, the safety factors are defined as the quotient between the calculated value and the maximum allowed value of: axial stress, shear stress, bending stress, Von Mises stress, compressive buckling stress, compressive and bending buckling stresses, torsion buckling stress and torsion and bending buckling stresses. The *safe bar* concept was considered for the definition of the penalty coefficient. A safe bar is the one whose safety factors are equal to or lower than one ($G_s(x) \leq 1$ $H_s(x) \leq 1$ ... $T_s(x) \leq 1$) [7].

In addition, if the coefficient is far below unity the bar is considered *oversized*. The penalty coefficient is defined as the value which multiplied by the safety factor calculated in a bar, increases this coefficient if it is different from one and maintains it constant if it is equal to one. The sum of the penalized coefficients of all structural elements will be the penalty term of the modified objective function. Different algorithms for the penalty coefficient were developed by the authors [5, 9], and the best function for the GA is indicated next.

$$r(c) = \begin{cases} 0 & if \quad c = 0 \\ e^{2-c} & if \quad 0 < c < 1 \\ 1 & if \quad c = 1 \\ e^c & if \quad c > 1 \end{cases} \quad c = G_s(x) \qquad (5)$$

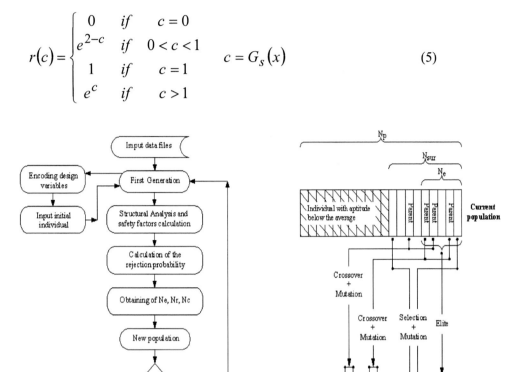

Figure 4. Elitist GA flowchart (left) and generation scheme of a new population (right).

The flowchart of the implemented elitist GA is represented in Figure 4. It can be seen how each new population is formed by three types of individuals obtained from the surviving population:

- *Elite* individuals, selected between the best individuals of the current population without mutation and whose number N_e is the result of multiplying the elite probability by the population size N_p.
- Crossover *individuals*, selected between the surviving individuals as a function of probability of rejection with mutation and whose number N_c is the result of multiplying the crossover probability P_c by the population size N_p.
- *Random* individuals, whose number is equal to the difference between the total number of individuals in the population and the sum of elite and crossover individuals.

4. COMPARATIVE BETWEEN STANDARD RULES: RESULTS AND CONCLUSIONS

These sections comprised the series IPN, IPE, HEA, HEB y HEM [9], which belong to the Spanish market [11] but nevertheless are also used in the European market. Although the American code uses different denominations, the above-mentioned series are equivalent to the American forms W, M, S y HP, which can be divided into five groups according to their resistance capacity.

Of these five groups, numbers 4 and 5, according to the AISC-LRFD code, are suitable for columns or compression components, as are the series HEA, HEB y HEM. This similarity makes it possible to use the Spanish sections with the American building code, since the error which is committed is very small.

4.1. Portal-Frame Structure

In order to check the suggested improvements, and to investigate the effect of tuning GA parameters (N_p, P_{mut}, P_c y P_e) on the performance of the developed GA, the domain of each parameter has been set as follows:

- The population size N_p ranges from 20 to 100.
- The elite probability P_e is 30%.
- The crossover probability P_c is 70%.
- The mutation probability P_{mut} is 0.5%.

The study carried out on the portal-frame structure (Figure 1) and with sections can be summarised as follows:

1. The modified objective function shows a very little variation in its value, when the population size ranges from 20 to 100. Besides, the values for the AISC code are the biggest ones whereas the values for the Eurocode-3 are the smallest, as consequence of the different safety factors evaluated in each rule (see Figure 5 below).
2. The weight of the structural elements shows important variations when the size of population changes. In general, the tendency is to diminish when the population increases, but some oscillations exist. The best building code, from the minimum structural weight point of view for a population size of hundred individuals, is the MV-103 code, next the Eurocode-3 and, finally, the AISC-LRFD code. This is one of the most important results of this work as it is shown in Figure 6.
3. The number of new generations is very similar between the codes, around fifty, and their differences are smaller as the size of population increases (see Figure 7 below).
4. Finally, the number of evaluations in the modified objective function is shown in Figure 8, and we can appreciate that it increases with the size of population, with small differences between rules.

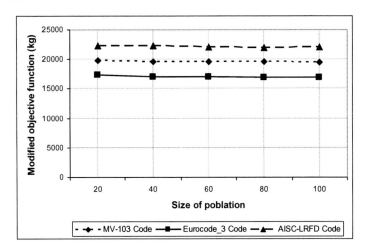

Figure 5. Building code's influence over the modified objective function.

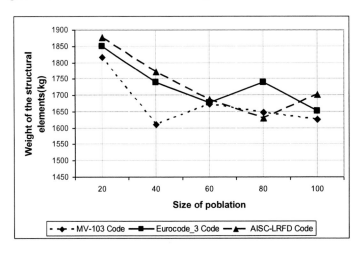

Figure 6. Building code's influence over the weight of the structural elements.

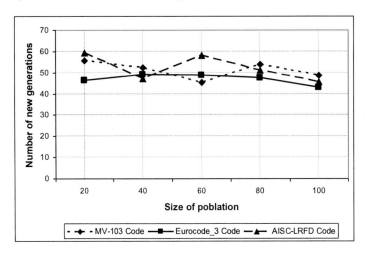

Figure 7. Building code's influence over the number of generation of new generations.

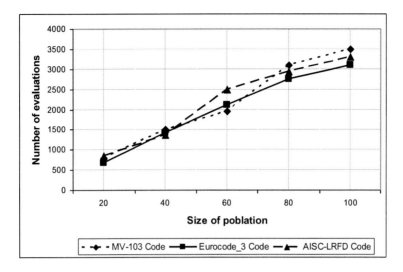

Figure 8. Building code's influence over the number of evaluations of the modified objective function.

4.2. Three Storey Building

In the three dimensional structure (see Figure 2) the results are similar to the previous ones, how it is showed in Figures 9 and 10. In the first one, the modified objective function shows a soft fall when the population size ranges from 20 to 100 and a better behaviour of the Eurocode_3. In the second the number of evaluations increases with the size of population, without important differences between the codes. Al last, is important to compare the best optimum individuals obtained with the three building codes. In the case of the Eurocode_3 and the AISC-LRFD code the individuals are obtained with the combination of parameters N_p = 120; P_m = 2%; P_c = 70%; P_e = 20%. In the case of the AISC-LRFD code, the individual was obtained with the combination of parameters N_p = 80; P_m = 2%; P_c = 70%; P_e = 20%.

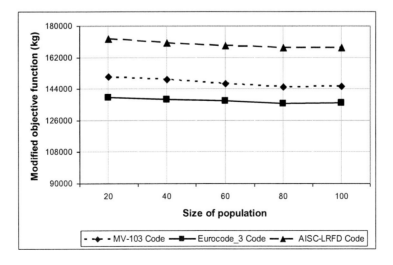

Figure 9. Building code's influence over the modified objective function in the three storey structure.

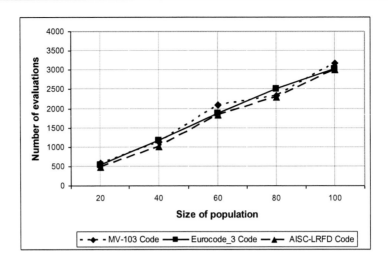

Figure 10. Building code's influence over the number of evaluations of the modified objective function in the three storey building.

Table 1. Sections of the optimum individuals

Groups of structural elements	Initial individual	Building code		
		AISC-LRFD	Eurocode_3	MV-103
Group I	HEB220	IPN320	IPE300	IPE300
Group II	HEB260	IPE300	HEA240	HEA240
Group III	HEB300	HEM160	HEM140	HEM140
Group IV	HEA200	IPE330	IPE330	IPE330
Total weight (kg)	17865.75	16158.3	15844.42	15844.42

The results reflected in Table 1 again show that the AISC-LRFD code is the one which results in the heaviest structure, whilst the Eurocode_3 and MV-103 code produce similar optimum individuals. The obtained weight difference is of scarcely 314 kg, that is to say 1.98% of surplus weight. This is due to the fact that groups of structural elements such as IV have the same cross-section in the three building codes, and that groups such as III have a very similar cross-section in the three building codes. This small percentage difference between weights, despite there being more extra weight in the AISC-LRFD structure, makes it possible to affirm that the elitist generic algorithm is suitable for the optimization of metallic structures using the three building codes analysed: MV-103, Eurocode_3 and AISC-LRFD.

CONCLUSIONS

The implemented elitist GA is a robust method of optimization of little mathematical complexity that is adequate for designers. It does not need prior information about the objective function or the constraint functions and can work with complex structures under different load and constraint conditions. In addition, it permits the use of commercial sections catalogues as design variables and is able to apply the engineer experience in creating groups of identical section, selecting these variables and their relation with the structural members. An increase in the size of the population leads to a soft decrease in the weight of the structural elements and of the coefficients. However, populations of more individuals lead to an excessive increase in the number of evaluations. For this reason a working range of populations of between 80 and 100 individuals is established.

An analysis of the different building codes using sections used in Spain, shows that the AISC-LRFD code results in the heaviest structures whilst the European code results in the lightest. The computational cost or number of evaluations is very similar in the three codes when working with populations of less than 100 individuals, approximately 6 hours of work. The results obtained with the three building codes are very similar. It is therefore possible to affirm that the elitist genetic algorithm is suitable for the optimization of metallic structures with regard to the three building codes analysed: AISC-LRFD, MV-103 and Eurode_3.

ACKNOWLEDGMENTS

The authors wish to acknowledge the computational support provided by the Department of Mathematics and Construction and Manufacturing Engineering at University of Oviedo. Furthermore, we would like to thank Anthony Ashworth for his revision of English grammar and spelling of the manuscript.

REFERENCES

[1] Davis L., *Handbook of Genetic Algorithms*. New York, V. Nostrand-Reinhold; 1991.
[2] *Basic Standard: NBE EA-95. Steel structures in building,* In: Ed. Madrid V.; 1996.
[3] EC3 Eurocode *3: Design of steel structures. Part 1-1. General. General rules and rules for buildings*, In: AENOR: Spanish association for normalization and certification, Madrid; 1996.
[4] AISC, *Specification for Structural Steel Buildings*. American Institute of Steel Construction, Chicago; 2005.
[5] Prendes Gero, M.B., Bello García, A. and Coz Díaz, J.J. (2005). A modified elitist genetic algorithm applied to the design optimization of complex steel structures. *Journal of Constructional Steel Research*, 61, 265–80.
[6] Coz Díaz, J.J., Ordieres Meré, J.B., Suárez Domínguez, F.J., Felgueroso Fdez., D., Álvarez Fdez., M. and Bello García, A. (1998). Aprendizaje Interactivo mediante el programa de análisis estructural avanzado ESCAL3D. *Journal of Constructional Steel Research*, 46, 273–274.
[7] Prendes Gero, M.B. (2002). *Design and making optimization of metallic buildings applying Genetic Algorithms*, Ph.D. Dissertation Thesis, Department of Construction and Manufacturing Engineering, University of Oviedo.
[8] Álvarez, L.F. (2000). *Design optimization based on genetic programming. Approximation model building for design optimization using the response surface*

methodology and genetic programming, Ph.D. Dissertation Thesis, Department of Civil and Environmental Engineering, University of Bradford, UK.

[9] Prendes Gero, M.B., Bello García, A. and Coz Díaz, J.J. (2006). Design optimization of 3D steel structures:Genetic algorithms vs. classical techniques. *Journal of Constructional Steel Research*, 62, 1303–1309.

[10] Mahfouz, S.Y., Toropov, U.U. and Westbrook, R.K. (1998). Improvementes in the performance of a genetic algorithm: application to steelwork optimum design. In: *Proceedings of 7th AIAA/USAF/NASA/ISSMO*, Symposium on Multidisciplinary Analysis and Optimization, 2037-2045.

[11] ENSIDESA (Madrid), *ENSIDESA Compendium: Calculus rules. Measuring of structural elements*, Madrid, ENSIDESA; 1997.

In: Handbook of Genetic Algorithms: New Research
Editors: A. Ramirez Muñoz and I. Garza Rodriguez © 2012 Nova Science Publishers, Inc.
ISBN: 978-1-62081-158-0

Chapter 13

GENETIC ALGORITHMS FOR SINGLE MACHINE SCHEDULING PROBLEMS: A TRADE-OFF BETWEEN INTENSIFICATION AND DIVERSIFICATION

Veronique Sels[1,*] *and Mario Vanhoucke*[1,2,3,†]

[1]Faculty of Economics and Business Administration
Ghent University
Tweekerkenstraat 2, 9000 Gent
Belgium
[2]Operations and Technology Management Centre
Vlerick Leuven Gent Management School
Reep 1, 9000 Gent
Belgium
[3]Department of Management Science and Innovation
University College London
Gower Street, London WC1E 6BT
United Kingdom

Abstract

In order to increase the efficiency of genetic algorithms, these algorithms are often hybridized with other heuristic procedures, such as local search algorithms or other meta-heuristics. The main purpose of this hybridization is to intensify the search process of the genetic algorithm in order to accelerate the search for high quality solutions. However, diversity is also a crucial component of the genetic algorithm that will guarantee a uniform sample of the search space. Therefore, hybrid algorithms require a careful trade-off between the diversification and intensification strategy. In this chapter, we discuss several techniques that take this important balance into account. To be more precise, we will show that the definition of a clever, often restricted, neighborhood increases the effectiveness of the embedded local search algorithm or meta-heuristic. In addition, we will discuss how the extension from a single population to

---

[*]E-mail address: veronique.sels@ugent.be
[†]E-mail address: mario.vanhoucke@ugent.be

multiple populations and the use of a distance measure to define these populations can be an important stimulator to add diversity to the search process. These techniques are illustrated by means of a commonly known machine scheduling problem.

Keywords: Genetic algorithms, single machine scheduling, intensification, diversification.

1 Introduction

In this chapter, we will discuss some new research endeavors related to the use of genetic algorithms. In order to illustrate these new approaches, we study a widely investigated branch of the operational research domain, i.e. the scheduling of production systems. Scheduling, in general, can be seen as the allocation of limited resources to tasks in order to optimize a certain objective function. Machine scheduling, in particular, is the processing of a number of jobs on a number of machines. Machine scheduling problems can be classified using the "$\alpha|\beta|\gamma$"-scheme introduced by [20]. The α-field represents the machine environment, the β-field describes the job characteristics and problem restrictions and the γ-field defines the optimality criterion. These different aspects give rise to a multitude of different machine scheduling problems. One of the most basic machine scheduling environments that is of significant theoretical and practical interest is the single machine scheduling (SMS) problem. Production processes often have one or more critical machines that have a huge impact on the rest of the production line. Focusing on those single machines can be sufficient to control the entire production process. Moreover, the SMS problem often occurs as a component in solving other scheduling environments such as flow shops or job shops. For these reasons, the single machine scheduling problem has received considerable attention in the literature.

The single machine scheduling problem addressed in this chapter aims at scheduling a number of jobs n on a single machine. This machine is assumed to be continuously available and can process at most one job at a time. Job preemption is not allowed and each job j (index $j = 1, 2, \ldots, n$) has a processing time p_j, a release time r_j and a due date d_j. In addition, precedence constraints between the jobs may be present. If $j \rightarrow k$, then job j has to be completed before job k can start. These precedence constraints specify partial orders for the jobs [38]. Moreover, the set of jobs can be partitioned into $|F|$ subsets or families (index $f, f = 1, 2, \ldots |F|$) based on their specifications [25]. If this partitioning is made, a family setup time s_{fg} is incurred whenever there is a switch from processing a job from one family f to a job from another family g [45]. We assume that this setup can be executed in absence of a job and that it depends on the sequence in which jobs are scheduled. The completion time of a job, C_j, is the time the job finishes processing. The lateness of a job j is the difference between its completion time and its due date, i.e. $L_j = C_j - d_j$. If this difference is negative, the job is said to be early, if it equals zero, the job is on time, otherwise the job is said to be late. The objective function to be optimized in this study is the minimization of the maximum lateness value ($L_{max} = \max_j L_j$), which measures the worst violation of the due dates. Using the classification scheme of Graham et al., the problems under study can be represented by:

- $1|r_j|L_{max}$ = single machine scheduling problem (1) with release times (r_j) and the objective of minimizing the maximum lateness (L_{max});

- $1|r_j, prec|L_{max} = 1|r_j|L_{max}$ with precedence relations between the jobs (*prec*);

- $1|r_j, s_{fg}|L_{max} = 1|r_j|L_{max}$ with family setup times (s_{fg}).

A simplified example of a single machine maximum lateness scheduling problem with release times and due dates is given in part (a) of Figure 1, the extension to family setup times and to precedence constraints are given in parts (b) and (c), respectively. The corresponding data is given in Table 1. These problems were discussed in the papers of [53–55].

Table 1. Example of a single machine scheduling problem with 6 jobs

	p_j	r_j	d_j
job 1	8	0	20
job 2	10	12	21
job 3	5	2	10
job 4	15	20	45
job 5	4	15	30
job 6	6	35	50

2 Genetic Algorithms

2.1 General introduction

The genetic algorithm (GA) is a well-known search technique for solving optimization problems based on the principles of genetics and natural selection. The method was initially developed by John Holland in the 1970s [23]. Holland was inspired by Charles Darwin's theory of evolution Òthe survival of the fittestÓ. A standard GA starts with a population of (randomly) generated solutions, represented by chromosomes. The purpose of the algorithm is to improve these solutions iteratively. A single iteration is referred to as a generation. The population of one generation consists of solutions surviving from the previous generation plus the newly generated solutions, while keeping the population size constant. Each solution of the population is characterized by its fitness value. This fitness value is related to the associated value of the objective function. In general, the fittest solutions from the population are selected to serve as parents in the reproduction phase. The reproduction phase uses a crossover operator to combine the characteristics of the parents in the hope to create fitter solutions. However, in order to maintain diversity in the population, a proportion of the new solutions is mutated, by randomly altering elements of the chromosome. This ensures an extensive search of the solution space. Furthermore, the new solutions can then be locally improved by means of a local search algorithm. Dependent on their fitness value, the obtained solutions are included in the population of the next generation, and the algorithm is iteratively repeated until some stopping criterion is met.

Genetic algorithms have been applied to a wide variety of scheduling problems, including the single machine scheduling problem [66]. The following paragraph and the table

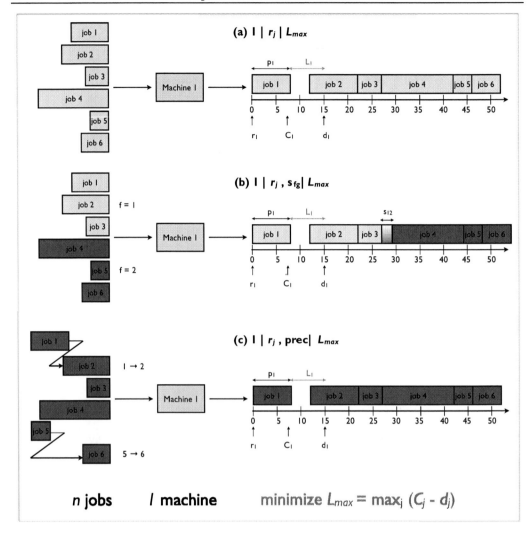

Figure 1. The single machine maximum lateness problem (a), with family setup times (b) and precedence constraints (c).

below (Table 2) will give an overview of GA applications in SMS. A distinction is made based on the objective function, which can be a regular or non-regular measure of performance.

Regular performance measures are non-decreasing (non-increasing) functions of the job completion times in the case of minimization (maximization) problems [16]. Some commonly used regular performance measures are given in the following overview. [48] developed a GA for the problem of sequencing jobs on a single machine with distinct due dates and sequence-dependent setup times to minimize *total tardiness*. The same problem was discussed in the work of [60], [2] and [15]. [10] investigated a single machine total tardiness problem with family setup times and developed a hybrid GA with improving heuristics. The single-batch-processing machine problem with non-identical job sizes to minimize the *makespan* was addressed by [27] as well as by [14]. Similarly, [30] de-

Table 2. Overview of the literature on genetic algorithms for SMS problems

Performance measure	References
Regular *e.g. total (weighted) tardiness, makespan, (weighted) number of late jobs, maximum lateness, ...*	[48], [60], [2], [15], [10], [27] [14], [30], [12], [13], [37], [21], [64], [29], [56], [57], [36], [8] and [24]
Non-regular *e.g. earliness-tardiness penalties, variance of completion time, ...*	[32], [6], [41], [62], [65], [63], [66], [59], [22], [35], [33], [67], [7], [44] and [11].
Regular + non-regular	[31], [34] and [26].

veloped a hybrid genetic algorithm for the problem with incompatible job families. The study of [12] extended the problem with arbitrary job release times. [13] compared different local search heuristics, including the GA, for the single-batch-processing machine problem with family sequence-independent setup times with the objective of minimizing *the number of late jobs*. The paper of [37] proposed a modified GA for the single machine scheduling problem with release times to minimize the *mean completion time* of the jobs. [21] examined the application of a GA for a single-batch-processing machine problem with sequence-independent setup times with the objective to minimize the *total flow time*. The paper of [64] considered the problem of minimizing the *maximum lateness* with dynamic job arrivals on a batch-processing machine.

When priority weights are assigned to the jobs, the objective functions to be minimized are somewhat different. [29] aimed at minimizing the *total weighted late work* using different scheduling rules and search algorithms, among which a GA. The minimization of the *weighted number of late jobs* on a single machine with release times was examined by [56]. [57] modified the GA of the former authors to deal with stochastic release dates and to compute robust schedules. [36] compared different meta-heuristics, such as simulated annealing, tabu search and a genetic algorithm, to solve the single machine *total weighted tardiness* problem with sequence-dependent setup times. The work of [8] and [24] discussed the single machine scheduling problem with the objective of minimizing of the *total weighted completion time* in the presence of release times.

Several researchers have also developed GAs for scheduling SMS problems with non-regular performance measures. For these performance measures, it can be advantageous to allow inserted idle time. A popular non-regular measure of performance is the minimization of the *earliness and tardiness penalties*. In the paper of [32], the objective of the GA was to find a schedule for the single machine scheduling problem with distinct due dates that minimizes the total earliness and tardiness penalties. A similar problem was solved by the genetic algorithm of [6], who considered the introduction of artificial chromosomes to increase performance. [41] developed a hybrid heuristic that combines the GA with some local search heuristics. [62] extended the problem with distinct ready times. To include a penalty for setup times, [65] adjusted the general earliness-tardiness penalty function. [63] assumed their earliness costs to be linear and their tardiness costs to be quadratic. [66] and [59] addressed the earliness-tardiness problem with family sequence-independent setups and an unrestricted common due date. Likewise, the studies of [22] and [35] also involved a common due date, but the authors did not consider setup times. The problem of minimizing the total weighted earliness and tardiness penalties was solved by the parallel GA of [33], the genetic dynamic programming algorithm of [67] and the genetic algorithm with dominance properties of [7]. A more complicated objective function was used in the real-life study of [44]. Their hybrid GA tried to minimize the *sum of setup, inventory and backlog costs* for a problem with sequence-dependent setup times and a fixed planning horizon. Another non-regular performance measure is the minimization of the *completion time variance* as studied in the paper of [11].

A lot of papers have attempted to optimize more than one optimality criterion at the same time. These multi-criteria scheduling problems consider regular and/or non-regular performance measures simultaneously. [31] for example, developed a GA that minimizes the flow time in combination with the number of tardy jobs and the maximum earliness. [34] considered the problem with sequence-dependent setups to minimize the number of tardy jobs together with the makespan. Similarly, the survey of [26] proposed a GA for the bi-criteria scheduling problem of minimizing the number of tardy jobs and the maximum earliness.

In this chapter, we present genetic algorithms for the single machine maximum lateness problems discussed in the previous section. These GAs are developed by means of comparing the different genetic operators described in the papers above. As such, the experience gained in other SMS problems is used to build effective GAs for the problems under study.

2.2 Genetic operators

An outline of a standard genetic algorithm is given in Figure 2. The following paragraphs describe the different components of the genetic algorithm in more detail, applied to the three single machine scheduling problems discussed in this paper.

Solution representation and schedule generation An important decision in building a genetic algorithm is the representation of a solution. This decision has an influence on the development of the genetic operators such as crossover, mutation and local search. The most natural way to represent a schedule in (single) machine scheduling is to use permutation encoding. In permutation encoding, every chromosome is a permutation of the n jobs in which each job appears exactly once. As such, it is ensured that each chromosome corresponds to a unique schedule and each schedule corresponds to a unique chromosome [41]. Moreover, permutation encoding is especially useful when the individual fitness of the genes depends on the positions of the genes in the chromosome. This feature holds for the objective function under study. For the example SMS problem of Table 1 scheduled in part (a) of Figure 1, the corresponding permutation encoding is

$$1\ 2\ 3\ 4\ 5\ 6.$$

The permutation of jobs needs to be translated into a feasible schedule. [56] describe

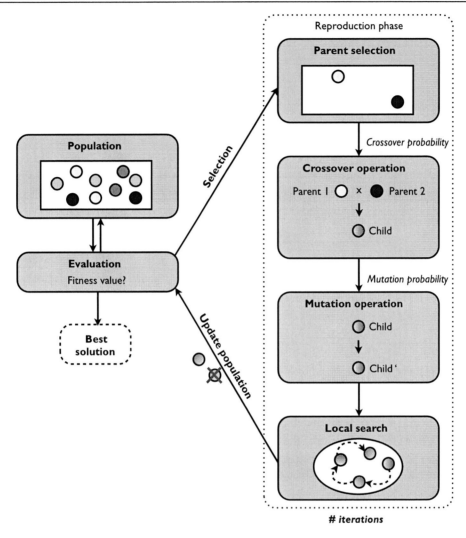

Figure 2. General outline of a standard genetic algorithm.

three construction engines to build feasible schedules. The *list engine* simply keeps the jobs in the order of the chromosome and assigns the earliest possible starting time to them. As such, semi-active schedules are built, in which no job can be completed earlier without changing the order of processing. The list engine is easy to implement and uses a minimal amount of computation time. The *first in first out* or *FIFO engine* builds active schedules in which no job can be completed earlier without delaying other jobs in the schedule. The engine maintains a list of idle periods and sequences every job as early as possible, at the earliest possible time slot. This may result in a different job order in the schedule. The updating of the list of idle periods can be computationally expensive, which is the major drawback of this construction engine. In the *non-delay engine*, the machine is never kept idle while there is a job waiting to be processed. A job is sequenced each time it is possible with no delay. If more than one job is available for processing, the order of scheduling is determined by the order that the jobs appear in the chromosome. In order to illustrate the differences between the construction engines, the permutation above for the example of Table 1 is translated in three different schedules using these engines in Figure 3.

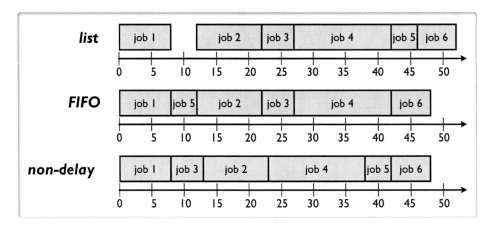

Figure 3. Illustration of the construction engines.

Population The initial solutions of the population can be generated randomly or with some construction heuristic. An example of a good construction heuristic for the single machine maximum lateness problem is the well-known earliest due date (EDD) rule, which is optimal for the problem without release times ($1||L_{max}$). The EDD rule sequences jobs in non-decreasing order of their due dates. However, when release times are present, the EDD-rule needs to be adjusted to take these release times into account. One such adjustment is made in the Schrage algorithm [40], where available jobs are scheduled according to their due date. For the example described above, this would result in the sequence 1 3 2 5 4 6. A good initial solution is very important since it not only seeds the population with known good solutions resulting in an improved quality of the future chromosomes, it also speeds up the search process [47]. An important parameter is the size of the population. If the population is too small, there is a risk of failing to cover the solution space adequately. However, if the population is too large, this will create a high computational burden without making acceptable progress. Our experience has taught us that the size of the population depends on the number of jobs. The larger the problem instance, the larger the population should be. Moreover, the intelligence of the heuristic also has an influence on the population size. According to [63], smaller populations can be used if local search and/or population initialization is used, without compromising the solution quality.

Evaluation: Using the objective function L_{max}, all solutions are evaluated based on their maximum lateness value and are assigned a proportional fitness value.

Selection When the initial population is constructed, the algorithm has to select parent solutions that will generate new solutions for the next generation. Parents are selected according to their fitness value. The better the fitness value, the higher the chance that the parent will be selected. The selection methods can be generally classified in two classes [43]. *Proportionate-based selection* selects parents according to their fitness value relative to the fitness value of the other solutions in the population. An example of such selection method is roulette wheel selection (RWS). *Ordinal-based selection* on the other hand, selects parents according to their rank within the population. Examples are tournament selection (TS) and ranking selection (RS). Although ordinal-based selection methods are normally preferred over proportional-based methods due to scaling problems [43], it seems to be that roulette wheel selection is more commonly used in single machine scheduling. More information on the selection methods described can be found in [47].

Crossover In a next step, the selected parents are recombined to create a new solution or offspring. This is done by the crossover operator. Crossover is executed with a certain probability. This probability can be kept constant during the search process, but is often lowered towards the end (degressive rate). The rates strongly depend on the population size: small populations require high rates whereas large populations require small rates. Crossover operators for permutation encoding can be roughly classified in three classes: a class that preserves the relative order of the jobs, a class that respects the absolute position of the jobs and a class that tends to preserve the adjacency information of the jobs. As we want to minimize the maximum lateness, the relative order and/or the absolute position of each job is more relevant to the total fitness of the schedule than the adjacency information. For that reason, the crossover operators implemented in our algorithm belong to one of the first two classes. These include (linear) order crossover (OX), (uniform) order-based crossover (OBX), cycle crossover (CX), position-based crossover (PBX) and partially mapped crossover (PMX). The working of these crossover operators is explained below and illustrated in Figure 4. These crossover operators are also discussed in the paper of [28], who give an overview of crossover operators in the single machine total weighted tardiness scheduling problem.

- Order crossover (OX): Two crossover points are selected at random and the job symbols between these points are copied to the same positions of the offspring. Those job symbols are deleted from the second parent and the remaining job symbols are inherited from the second parent, beginning with the first position following the second crossover point.

- Order-based crossover (OBX): A number of positions are chosen at random in the first parent and the corresponding job symbols of the unselected positions are placed in the position they appear in the second parent. The chosen symbols of the first parent are copied into the empty positions of the offspring by preserving their absolute order in the first parent.

- Cycle crossover (CX): A cycle is defined by the corresponding positions of symbols between parents. The symbols in the cycle of the first parent are copied to the same positions of the offspring, and these symbols are deleted from the other parent. The remaining symbols of the other parent are copied at their corresponding positions in the offspring.

- Position-based crossover (PBX): A number of positions from one parent are selected randomly and copied to the offspring. The symbols that correspond with these selected positions are deleted from the second parent, and the unselected symbols are inserted in the offspring at the unselected positions in the order they appear in the second parent.

- Partially mapped crossover (PMX): Two crossover points are selected at random and the job symbols of the first parent are copied into the offspring. The remaining symbols are copied from the second parent, but identical symbols are replaced according to the map defined by the crossover points.

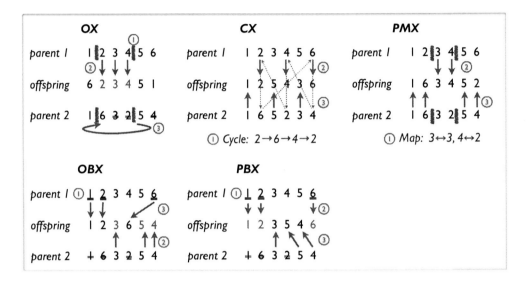

Figure 4. Illustration of the crossover operators.

Mutation After a number of generations, the chromosomes become more homogeneous and the population starts to converge. The mutation operation helps the algorithm to escape from the local optimum and serves as a tool to introduce diversity into a population. This ensures an extensive search of the search space. Mutation also occurs with a certain probability. In contrast to crossover, the mutation rate is often progressive. Low rates are used in the beginning of the search and higher rates are applied at the end. Additionally, this rate also depends on the population size in a similar fashion as the crossover rate. Commonly used mutation operators for permutation encoding are the swap mutation (SM), i.e. the swapping of two or more jobs, the insertion mutation (InsM), i.e. the reallocation of a job in the sequence, and the inversion mutation (InvM), i.e. the inversion of a sequence of jobs. These mutation operators are displayed in Figure 5.

Local searches Before introducing the offspring solutions into the population, a local improvement technique can be used to further improve these solutions. A local search algorithm iteratively searches the solution space by moving from one solution to another. It replaces the current solution by a better neighboring solution until no more improvements can be found or until some stopping criterion is met. A meta-heuristic may be used instead, in order to escape from local optima. As will be discussed in the remainder of this paper,

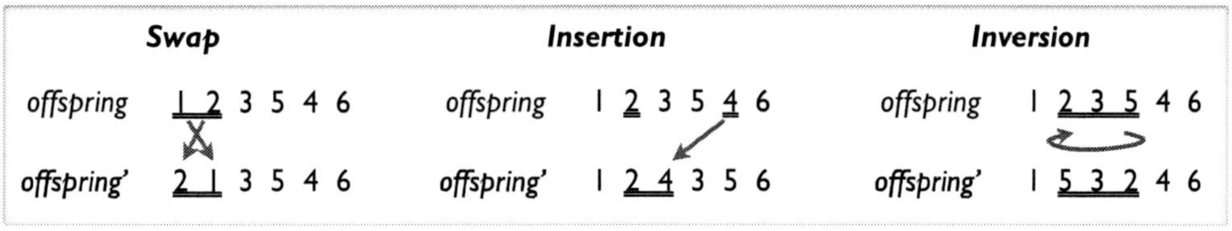

Swap	Insertion	Inversion
offspring 1 2 3 5 4 6	offspring 1 2 3 5 4 6	offspring 1 2 3 5 4 6
offspring' 2 1 3 5 4 6	offspring' 1 2 4 3 5 6	offspring' 1 5 3 2 4 6

Figure 5. Illustration of the mutation operators.

the inclusion of a local search algorithm or another meta-heuristic search procedure will result in a careful trade-off between diversification and hybridization.

Update of the population   Once the new solutions are created, they have to be inserted in the population. In an incremental replacement strategy, the new solutions will replace the least fit solutions of the current population. As such, it is ensured that the best solutions survive to the next generation. The algorithm can be stopped after a specified number of iterations, after a certain time limit or when it is guaranteed that the optimal solution is found.

3   Trade-off Between Intensification and Diversification

A meta-heuristic, such as the genetic algorithm described above, is built upon two main strategies: diversification and intensification. Diversification (or exploration) is used to ensure an extensive search of the solution space. In a standard genetic algorithm this is obtained by the different genetic operators and parameters, such as mutation and, to a lesser extent, crossover. Intensification (or exploitation) is used to reduce the search space around promising solutions. This can be achieved by hybridizing the genetic algorithm with a local search algorithm or another (meta-)heuristic. In order to construct an effective genetic algorithm, a careful trade-off between both strategies should be made. In the next paragraphs, we will discuss how this trade-off was scrutinized for the single machine scheduling problems under study.

3.1   Intensification

As already mentioned above, intensification can be obtained by the hybridization of the GA. Hybridization can be seen as the merging of two or more optimization techniques into a single algorithm. These optimization techniques may be simple local improvement algorithms or more advanced meta-heuristics.

When using a local search algorithm, an important feature to be decided is how the neighborhood is defined and, more importantly, how this neighborhood is explored. For the permutation encoding used for the three single machine scheduling problems under study, the most commonly used neighborhoods are the insertion and the swap NH [42]. An *insertion NH* generates a neighbor by deleting a job from the sequence and inserting it at another position in that sequence. Examples of local searches that are based on the insertion NH are:

- Insertion (INS): insert a job before or after every other job in the permutation;

- Backward Insertion (BWI): insert a job before every earlier job in the permutation.

- Largest cost insertion (LCI): insert the job with the largest L_{max}-value before or after every other job in the permutation;

In a *swap NH*, a neighboring solution is generated by interchanging two (or more) arbitrary or adjacent jobs of the sequence. Four local searches that use this NH are:

- Randomized pairwise swap (RPS): randomly swap two jobs in the permutation;

- Adjacent pairwise swap (APS): swap pairs of adjacent jobs in the permutation;

- General pairwise swap (GPS): swap a job with any other possible job in the permutation;

- Three swap (ThreeS): choose the best permutation for each adjacent trio of jobs.

When the problem is characterized by family setup times, the formation or destruction of family batches is sometimes desirable. A batch is a sequence of jobs that belong to the same family. For that reason, a third NH can be introduced, in which batches are combined or split (*batch NH*) [3,49,50,61]. The combination of batches will save setup times and as it is observed by [3], if this setup time occurs early in the sequence this saving can reduce the maximum lateness. Splitting batches can reduce lateness as well, if this results in a job with an early due date scheduled earlier in the sequence. The four batch NH local searches examined in this study are:

- Forward batch combining (FBC): combine each batch with the first next batch of the same family;

- Backward batch combining (BBC): combine each batch with the first previous batch of the same family;

- Batch splitting last job (BSLJ): split off last job of each batch and insert it before the first job of the next batch of the same family;

- Batch splitting first job (BSFJ): split off first job of each batch and insert it after the last job of the previous batch of the same family.

An illustration of all local searches is given in Figure 6. In this illustration, a possible neighboring solution is given for each local search.

In order to prevent local optimality, the inclusion of a meta-heuristic that sometimes allows small deteriorations in the objective function value may be desirable. An example of such a meta-heuristic is the tabu search algorithm. The tabu search algorithm (TS), introduced by Glover [18], is a well-known meta-heuristic for finding near optimal solutions in combinatorial optimization problems. The TS can be seen as a local improvement technique that tries to prevent local optimality by allowing non-improving moves. The algorithm starts with a feasible initial solution and tries to iteratively improve the solution by examining a set of candidate moves in the insertion, swap or batch neighborhood of that

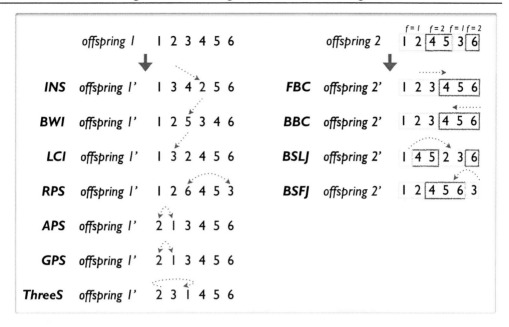

Figure 6. Illustration of the local search neighborhoods.

solution. The neighboring solutions are evaluated based on their objective function value and the current solution moves to its best neighboring solution, even if - in contrast to a local search algorithm - it results in a worse objective value. The best move is put on the tabu list, so that it cannot be made undone for some number of iterations in order to avoid a cyclic search. In addition, the move is recorded in a frequency list, which acts as a long-term memory to detect the most frequently executed moves. The number of iterations that a move is set tabu is determined by the size of the tabu list. The size of the list must be large enough to prevent cycling and to emphasize exploration and small enough to ensure the intensification of the search. In addition, an aspiration criterion is defined to deal with the situation where the moves are tabu. A commonly used aspiration criterion allows a tabu move when it results in a solution with an objective value better than the current best-known solution. The algorithm stops when a certain stopping criterion is satisfied and the best found solution so far is returned as the final solution.

The exploration of the complete neighborhoods defined above can be very time-consuming. Moreover, in order to ensure some diversity in the search, full exploration is not always desirable. Therefore, we have examined the effect of introducing a truncated and/or restricted exploration strategy for the $1|r_j, prec|L_{max}$ and $1|r_j, s_{fg}|L_{max}$ problem. The results of these studies are discussed in Sections 4.2 and 4.3.

3.2 Diversification

Next to intensification, diversification plays an equally important role. When hybridizing the GA with a local search or meta-heuristic, attention should be paid to maintain a sufficient level of diversity. For that reason, the single population genetic algorithm can be extended to a dual population genetic algorithm. Extensions from a single to a dual population by taking problem specific characteristics into account can be seen as a stimulator to add diversity to the search process, which has a positive influence on the important balance between intensification and diversification. This dual population structure is borrowed from the scatter search technique. The scatter search approach (SS) has been proposed by [17] as a population-based meta-heuristic in which solutions are intelligently combined to yield better solutions. The scatter search procedure splits the population in a dual population, which contains a high quality and a diverse subset. These subsets or reference sets are generated according to a diversification generation method. After the construction of the initial reference sets, solutions are selected and combined to reproduce new solutions. In the subset generation method and solution combination method, two (or more) elements of the reference set are chosen in a systematic way to produce points both inside and outside the convex regions spanned by the reference solutions. Combining the two reference solutions from the same set stimulates intensification, while combining them from different sets stimulates diversification. After the combination method, an improvement method is applied to the obtained solutions, which are then examined to enter the population again. The reference set update method uses a dynamic update strategy that evaluates each possible reference set entrance instantly upon generation of a new solution element. A new solution may become a member of one of the reference sets either if the solution point has a better objective function value than the solution with the worst objective function value in the high quality reference set or if the solution is more diverse than the least diverse solution point in the second reference set. If the reference set update method results in no new solutions entering the reference sets, both sets are reinitialized ([19] and [39]).

The scatter search method is a very flexible approach that can be implemented in a multitude of ways and degrees of sophistication. Moreover, the genetic algorithm and the scatter search procedure have roughly the same search structure. The main difference lies in the use and construction of the population set. The generic framework of the single population genetic algorithm and the dual population scatter search methodology can be described as follows:

```
Genetic Algorithm              Scatter Search
Initial population Generation  Diversification Generation Method
While Stop Criterion not met   While Stop Criterion not met
Parent Selection                   Subset Generation Method
Crossover & Mutation Operation     Solution Combination Method
Local Search                       Improvement Method
Update Population                  Reference Set Update Method
Endwhile                       Endwhile
```

The framework indicates that every component of the GA has its counterpart in the SS. They are often only labeled differently even though they perform a similar operation. However, the divers population structure of the scatter search and the controlled way of combining pairs of reference solutions led to the study of a dual population structure for the single machine maximum lateness problem with release times [53]. In this study, the population is split into a high quality and a highly diverse population. The elements of the first population are generated randomly and seeded with a good construction heuristic as

Genetic Algorithms for Single Machine Scheduling Problems

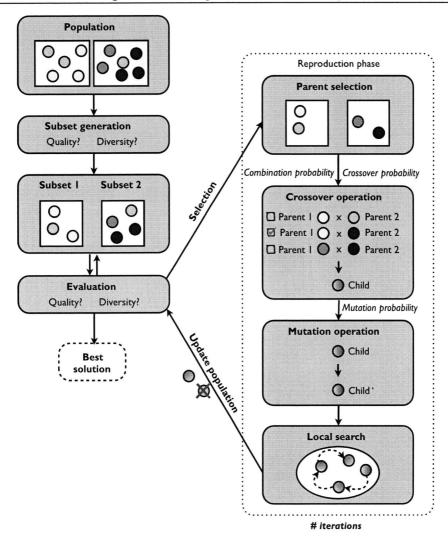

Figure 7. Outline of the dual population genetic algorithm.

described in Section 2.2. The elements of the second population are generated according to the diversification generator for permutation problems developed by [17]. From these two populations, only the best elements (in terms of solution quality) will enter the high-quality subset (subset 1). The diverse subset will contain the most diverse solutions relative to the solutions incorporated in the first subset. This means that solutions will only enter the second subset if a diversity threshold with the solutions in subset 1 is exceeded. In order to determine this diversity threshold, a distance measure is calculated between every pair of solutions. This A-distance measure of [4], is equal to the sum of all absolute differences between the positions of all items in strings p and q:

$$d_{(p,q)} = \sum_i |p_i - q_i|. \tag{1}$$

In the next step, parents can be selected from the first as well as from the second subset.

Solutions from the high quality subset are selected according to their fitness value. The better the fitness value, the higher the chance that the parent will be selected. Solutions from the second subset are selected based on their distance measure. Their chance of being selected increases with increasing distances. As a consequence, crossover can occur between solutions within the first subset or between solutions of the first and second subset. For this, we introduce a probability measure that controls the combination of high quality with highly diverse solutions. The offspring solutions replace the least fit solutions of the current population. However, in contrast to the single population GA, if the solutions are not allowed into the new population, they are checked on their diversity using equation (1). If their distance is greater than the smallest distance in the current diverse subset, the solution is accepted in the new population.

The outline of the dual population genetic algorithm is given in Figure 7. The benefits of this dual population GA and the use of a distance measure will be discussed in Section 4.1.

4 Computational Experiments

In this section, we will elaborate on the computational experiments that illustrate the accomplishment of the discussed trade-off between intensification and diversification for the genetic algorithm. Each experiment was performed on one of the single machine scheduling problems discussed above. In Section 4.1, we will show the benefit of using a dual population structure instead of the standard single population. In Section 4.2, it will be discussed how the trade-off between diversification and intensification can result in a truncation of the embedded local search algorithms. Section 4.3 will demonstrate the positive effect of cleverly restricting the neighborhood of the tabu search procedure that is hybridized with the genetic algorithm.

4.1 Single versus dual population genetic algorithm

In order to verify the benefits of the dual population structure, we compared the performance of the single population genetic algorithm described in Section 2.2, with the dual population GA of Section 3.2 for solving the single machine maximum lateness problem with release times ($1|r_j|L_{max}$) [53]. The genetic operators and parameters were fixed by means of a full factorial design experiment. These experiments showed that a relatively small *population size* (# jobs/5), a degressive *crossover probability* of 70%, which is lowered gradually to 50% as the search continues, and a progressive *mutation probability* from 10% to 20% turned out to be the best parameter settings. In addition, for the dual population GA, the *combination rate*, which determines the probability of combining a high quality with a highly diverse population element, was set to 85%. This means that there is a 85% chance of combining elements from the first subset of the population and a 15% chance of combining elements from the first with the second subset of the population. In the single population GA, the best combination of genetic operators appeared to be the *list* construction engine together with the *tournament* selection method, the *position-based* crossover and the *single swap* mutation. With the exception of the tournament selection, these GA operators were also implemented in the dual population GA. In addition, in order to show the importance of hybridizing the genetic algorithm with a local search algorithm, we also compared their

Table 3. Comparison of a single and dual population GA with and without local search (Relative performance (%))

Method	$n=100$	$n=250$	$n=500$
GA	1.1816	0.7300	0.4534
2PGA	0.7354	0.5421	0.4125
2PGA_LS	0.1159	0.2229	0.2609

performances with the dual population GA that includes the best performing local search, i.e. the *largest cost insertion* local search.

These computational experiments were performed on randomly generated data instances with up to 500 jobs, uniformly distributed processing times, release times and due dates. For more information on the data generation method, we refer the reader to [53]. The computational results are summarized in Table 3. The values in the table represent the relative performances of the three genetic algorithms over increasing problem sizes. This relative performance (or solution quality) is measured by the deviation of the objective values found by the single population (GA), dual population (2PGA) or dual population with local search (2PGA_LS) genetic algortihm with a lower bound and is equal to:

$$RP = \frac{\Sigma_{HEUR} - \Sigma_{LB}}{\Sigma_{LB}} \times 100\%, \qquad (2)$$

where Σ_{HEUR} is the sum of the objective values over the set of instances obtained by the heuristic and Σ_{LB} is the sum of the lower bounds over the same set of instances. This lower bound was obtained by combining four lower bound calculations from literature as described in [40], [5] and [9] for the $1|r_j|L_{max}$ problem. We compare the sum of objective values over a set of instances rather than individual objective values in order to avoid that RP would approach infinity when the lower bound is near zero (i.e. to avoid possible division by zero).

This table shows that, notwithstanding the similar general framework of both procedures, the dual population GA clearly outperforms the single population GA. An important reason can be found in the notion of "diversity". Although the GA uses mutation and, to a lesser extent, crossover to prevent premature convergence, it sometimes fails at preserving the required diversity level. The incorporation of a dual population structure, where a distance measure controls the quality and diversity level of the solutions, is a crucial tool that allows the GA to explore the solution space more efficiently. As such, the balance between diversification and intensification is obtained in a controlled and structured way.

The table also reveals the contribution of the hybridization with a local search algorithm. The 2PGA_LS outperforms the 2PGA, especially for the smaller problem sizes. The advantage of the LS is still present for larger problem instances, but slightly decreases over the number of jobs.

4.2 Truncation of local searches

For the problem with family setup times $(1|r_j, s_{fg}|L_{max})$, the single population GA of Section 2.2 was hybridized with the local searches discussed in Section 3.1. Again, the best combination of genetic operators and parameters was fixed by means of a full factorial design experiment. The single population GA used the tournament selection method together with the position-based crossover and the single swap mutation, with a relatively small population size (#jobs/5 → #jobs/10), a constant crossover probability of 70% and a constant mutation probability of 20% [55]. The best local search algorithm was selected after a preliminary experiment that studied the important trade-off between diversification and intensification.

Since a time limit was used as a stopping criterion, the trade-off was obtained by balancing the number of *generations* of the GA and the number of *iterations* or moves of the local search algorithms. The number of iterations is defined as the number of solutions explored by the LS algorithm during each generation of the GA. For example, for the randomized pairwise swap (RPS) local search the number of iterations is equal to the number of random swaps that are performed every generation. As already mentioned before, a full exploration of the neighborhood is not always desirable in order to ensure a certain level of diversity. Moreover, full exploration might lead to a huge number of iterations and hence a lot of CPU time for each evaluation. Consequently, the more iterations per local search run, the less generations the GA can perform within a certain time limit. On the contrary, if this number of iterations per local search run is restricted by truncating the local search algorithm at earlier stages, the GA is able to run over more generations within the same time limit.

The trade-off leads to the definition of a new parameter in the algorithm, i.e. the truncation value on the number of iterations. This truncation value defines the maximum number of neighboring solutions a local search algorithm may explore per generation of the GA. In order to determine this, we analyzed the performance of each local search algorithm embedded in the GA individually by allowing a wide range of truncation values (from no to full truncation) and examining the resulting solution quality. This is illustrated in Figure 8, where the average solution quality over increasing intensification (i.e. decreasing LS truncation) for the RPS local search is given. The solution quality was obtained by averaging out the L_{max}-value over all data instances obtained after running the hybrid GA for 1 second. These data instances were generated according to the method described in [53], extended with a rolling horizon method to generate job families and family setup times. On the left of the graph we have full truncation, resulting in no intensification and a maximum of diversification (i.e. no local search, only genetic algorithm). On the right, there are no restrictions put on the number of iterations of the LS and a maximum intensification is obtained (i.e. full neighborhood exploration). The values in between represent the trade-off between those two extremes. As can be seen in the graph, the optimal truncation value for the RPS is found near the left extreme, with a limited intensification and a fair amount of diversification.

In Figure 9, the effect of the truncation on the number of generations of the GA and the number of iterations of the RPS is illustrated. The number of generations (indicated by the bars) decreases over increasing intensification due to the increase in the number of iterations of the RPS (indicated by the line). This figures clearly demonstrates the importance of taking the trade-off between diversification and intensification into account.

This analysis was done for all the local searches described in Section 3.1. From this analysis we could observe that differences between the LS existed. Some of them were

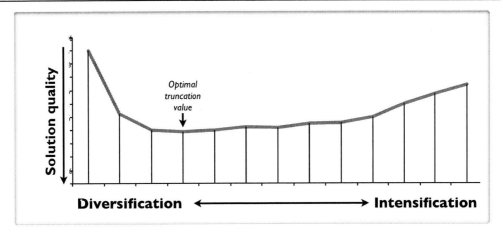

Figure 8. Evolution of average L_{max}-values over increasing intensification for the RPS local search.

truncated very early in the search process to allow more generations, while others performed best when truncation is limited to a minimum. The best local search for the problem with family setup times turned out to be a combination of a swap and a batch neighborhood local search.

4.3 Restricted search neighborhood

The inclusion of a meta-heuristic in the genetic algorithm was studied for the single machine scheduling problem with precedence relations ($1|r_j, prec|L_{max}$). Our experience have taught us that for this problem, an improvement technique that sometimes allows a deterioration in the objective function value, performs better than a pure local search algorithm [54]. This is mainly due to the characteristics of the problem. When partial orderings are imposed on the job sequences, the number of feasible solutions will drop, particularly for problem instances with a high density level. This density, or the number of precedence constraints between jobs, is determined by the order strength (OS) value, which ranges from 0% (no precedence relations) to 90% (almost all jobs linked with precedence constraints).

As already discussed above, one such improvement technique is the tabu search procedure. The tabu search implemented in our GA makes use of the general pairwise swap neighborhood, where each possible swap between jobs is examined. As a consequence, the maximum number of possible moves is equal to $\sum_{i=1}^{n-1}(n-i)$ (i.e. in the absence of precedence relations) and thus the number of neighboring solutions will increase as the number of jobs increases. Due to this large neighborhood size and the resulting computational burden, a modification has been made to restrict the neighborhood as much as possible without compromising the efficiency of the TS algorithm [58]. In order to cleverly restrict the TS neighborhood to a limited number of swaps, a computational experiment has been set up where the tabu search is used as a stand-alone search procedure to solve our problem instances. These problem instances were again generated according to the method described in [53], with the addition of the precedence relations. The stand-alone tabu search was

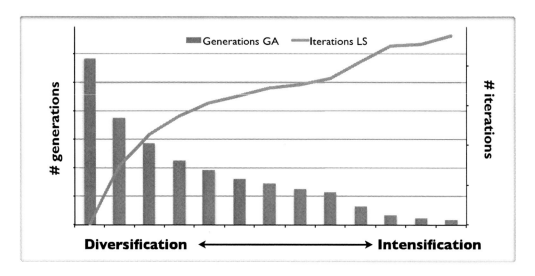

Figure 9. Evolution of average number of generations and iterations over increasing intensification for the RPS local search.

iterated a thousand times for each problem instance and at the end of the search, the frequency lists were stored. The frequency list keeps track of the move history and records information on the best swaps. As such, it represents a sort of quality measure of the swaps. By examining the frequency lists, the most promising swaps can be identified. Of course, the frequency list will vary over the different construction engines used (cf. Section 2.2). These construction engines result in different schedules and consequently also in different swaps of jobs. This is visually represented in Figure 10, where the axes range from position 1 to position n in the permutation sequence. The colored cells in the graphs show the average best moves (most executed swaps) over all the data instances. The figure clearly shows that, for all engines, the best swaps lie relatively close together (i.e. jobs situated at neighboring positions) and that this effect is the largest when using the list construction engine (i.e. most dense diagonal). For that reason, we decided to use the list engine in our TS, because this engine will be the most useful in reducing the NH to a clear value and restricting the possible swaps to those regions which contribute most to the solution quality. As such, we can efficiently reduce the search NH. In addition, the list engine is the engine requiring the smallest CPU time in constructing a feasible schedule. With this restriction, each job is swapped with a job that is located at most x positions further in the sequence. This restriction will help us to limit the computational costs of the TS without jeopardizing the quality of the examined swaps.

We have compared this stand-alone tabu search (*TS*), with the single population genetic algorithm (*GA*) and with the hybridization of both meta-heuristics (*GA/TS*). The genetic algorithm has the same parameter setting and operator combination as described in Section 4.1, for the problem without precedence relations. In order to ensure a fair comparison of the meta-heuristics, the same stopping criterion (1 sec) was used for all meta-heuristics. The results of the comparison experiment are visualized in Figure 11. In this graph, the average solution quality over all data instances is shown over the different density levels.

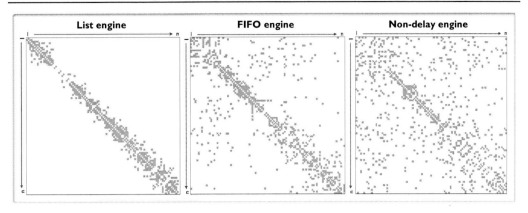

Figure 10. Frequency list of best pairwise swaps for the three construction engines.

From the figure, it can be seen that the GA/TS algorithm outperforms the GA as well as the TS for all density levels and that the difference in performance increases over increasing density. The stand-alone TS performs better than the GA, except for a density level equal to 0%. The genetic algorithm was developed for the problem without precedence constraints and thus less tailored to the problem with precedence constraints.

Moreover, in order to further validate the contribution of the genetic algorithm in the hybridization, we also compared our hybrid GA/TS with a multi-start algorithm that starts with 1,000 randomly chosen solutions and makes use of the tabu search as hill-climbing technique (*MS/TS*). This MS/TS is, despite its high CPU times (on average 1 to 2 minutes), not able to compete with any of the meta-heuristics. Its average solution quality differed remarkable from the other performance values, but this difference decreases, to some extent, over increasing density levels. This is not surprising as the number of feasible schedules drops as well. However, the average solution quality of the MS/TS for a density level of 90% is still a sixfold of the average solution quality of the worst performing GA. From this we can conclude that the inclusion of the TS alone is not a sufficient condition to ensure a good performance, but that the genetic operators have a proven contribution as well.

5 Relevance

The research studies discussed in this chapter were done to get a better understanding of the important trade-off between diversification and intensification that occurs in every meta-heuristic. The genetic algorithm was chosen because of its flexibility, robustness and ease of implementation, which allows it to works on a wide range of problems. The problems we have considered were the single machine maximum lateness problem with release times, extended with family setup times and/or precedence relations. These problems often occur as a subproblem in solving other scheduling environments such as job shops or flow shops with a makespan objective. An example is its appearance as a sub-problem in the decomposition-based approach for job shop scheduling, such as the Shifting Bottleneck Procedure of [1]. The inclusion of release times is based on the findings of [46], who stated that such single machine subproblems have a dynamic nature. Due to scheduling decisions made in pre-

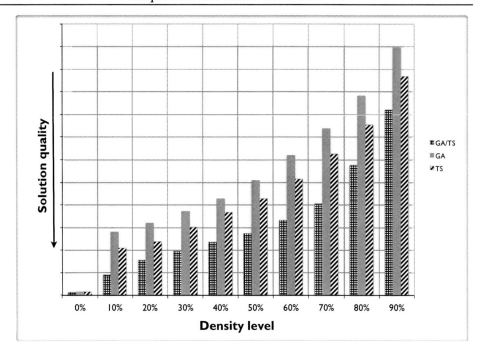

Figure 11. Meta-heuristic performance over increasing density level.

vious steps, the operations of the job arrive over time at the single machine resulting in a machine that will become available at different times. Moreover, the operations of a job have a predefined route to follow, which can be modeled by the inclusion of precedence constraints [38]. The expertise gained in solving the single machine scheduling problems helped us at constructing an effective genetic algorithm and scatter search technique for the job shop scheduling problem [51] and developing a efficient shifting bottleneck technique for a real-life job shop environment [52].

6 Conclusions

In this chapter, we have discussed several new research studies that increase the efficiency of genetic algorithms (GA). These studies were illustrated by means of the commonly known single machine scheduling problem with maximum lateness objective, extended with family setup times and/or precedence constraints. The different operators and parameters of our genetic algorithms were borrowed from the extended literature on genetic algorithms for different single machine scheduling problems and were adapted for the problems under study. The increased efficiency was obtained by the careful balance between two strategies, intensification and diversification. Intensification, on the one hand, is used to reduce the search space around promising solutions and was achieved by hybridizing our genetic algorithms with a local search (LS) algorithm or another meta-heuristic procedure. For the problems under study, we have defined a number of local search neighborhoods and exploration strategies that took the problem specific information into account. Diversification, on the other hand, is used to ensure an extensive search of the solution space. This diversity was added by means a dual population structure. The trade-off between both strategies was studied for three related single machine scheduling problems. The lessons learnt from these studies are recapitulated below and are summarized in Figure 12.

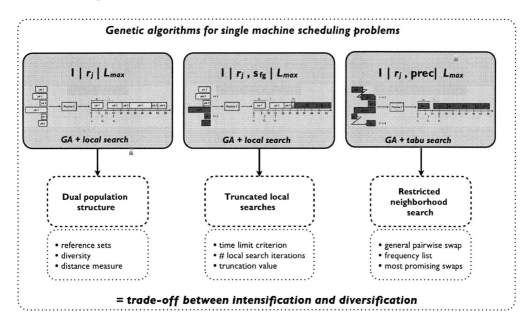

Figure 12. Genetic algorithms for SMS: lessons learnt.

The single machine maximum lateness problem with release times ($1|r_j|L_{max}$) was used to demonstrate the benefits of using a dual population structure with a corresponding distance measure to control the quality and diversity in the genetic population. This dual population structure, borrowed from the scatter search technique, splits the population into a high quality and a high diverse subset. The diversity between both sets is guaranteed by means of a distance measure. In addition, the combination of pairs of solutions within and between these subsets happens in a controlled and structured way. However, the general framework of the single and dual population GA remains the same. Notwithstanding this similar general framework, the dual population GA clearly outperformed the single population GA, which sometimes failed at preserving the required diversity level. The computational results have shown that the incorporation of a dual population structure was a crucial tool that allowed the GA to explore the solution space more efficiently, which positively influenced the balance between intensification and diversification. In addition, the advantage of embedding a local search algorithm in this dual population GA was illustrated as well.

For the single machine scheduling problem with family setup times ($1|r_j,s_{fg}|L_{max}$) it was shown that, in order to achieve a well-considered balance between diversification and intensification, a trade-off between the number of generations of the GA and the number of iterations of the LS should be made, especially when a time limit criterion is used. We have observed that a full exploration of the neighborhood is not always desirable in order to ensure a certain level of diversity. Furthermore, full exploration might lead to a huge number of iterations and hence a lot of CPU time for each evaluation. This observation led to the definition of a new parameter, the truncation value. This truncation value defined the maximum number of neighboring solutions that the local search could explore during each generation of the GA. As such, ideal truncation values were determined for every local search discussed in the chapter, increasing the efficiency of the resulting hybrid genetic algorithms.

The inclusion of a meta-heuristic in the genetic algorithm was studied for the single machine scheduling problem with precedence relations ($1|r_j, prec|L_{max}$). Due to the characteristics of the problem, the allowance of small deteriorations in the objective function value appeared to be desirable. The meta-heuristic chosen was the tabu search (TS) that made use of the general pairwise swap neighborhood. However, in order to ensure an efficient hybridization between the GA and the TS, this rather extended neighborhood was cleverly restricted to a limited number of swaps. By making use of the frequency list stored in the tabu search procedure, we could identify the most executed and thus most promising swaps. Our analyses have shown that these most promising swaps lied relatively close together. By restricting the neighborhood to these swaps, we could limit the computational costs of the TS without jeopardizing the quality of the examined swaps. Our computational results have shown that this hybrid algorithm outperformed the stand-alone genetic algorithm and tabu search procedure, which illustrates that the hybridization exploits the advantages of both individual approaches at its best.

References

[1] Adams, J., Balas, E., and Zawack, D. (1988). The shifting bottleneck procedure for job shop scheduling. *Management Science*, 34:391–401.

[2] Armentano, V. and Mazzini, R. (2000). A genetic algorithm for scheduling on a single machine with set-up times and due date. *Production Planning & Control*, 11:713–720.

[3] Baker, K. (1999). Heuristic procedures for scheduling job families with setups and due dates. *Naval Research Logistics*, 46:978–991.

[4] Campos, V., Laguna, M., and Marti, R. (2005). Context-independent scatter and tabu search for permutation problems. *Journal on Computing*, 17:111–122.

[5] Carlier, J. (1982). The one-machine sequencing problem. *European Journal of Operational Research*, 11:42–47.

[6] Chang, P.-C., Chen, S.-S., and Fan, C.-Y. (2008). Mining gene structures to inject artificial chromosomes for genetic algorithm in single machine scheduling problems. *Applied Soft Computing*, 8:767–777.

[7] Chang, P.-C., Chen, S.-S., and Mani, V. (2009). A hybrid genetic algorithm with dominance properties for single machine scheduling with dependent penalties. *Applied Mathematical Modelling*, 33:579–596.

[8] Chang, P.-C., Hsieh, J.-C., and Liu, C.-H. (2006). A case-injected genetic algorithm for single machine scheduling problems with release time. *International Journal of Production Economics*, 103:551–564.

[9] Chang, P.-C. and Su, L.-H. (2001). Scheduling n jobs on one machine to minimize the maximum lateness with a minimum number of tardy jobs. *Computers & Industrial Engineering*, 40:349–360.

[10] Chantaravarapan, S., Gupta, J., and Smith, M. (2003). A hybrid genetic algorithm for minimizing total tardiness on a single machine with family setups. *POMS meeting*.

[11] Chaudhry, I. and Drake, P. (2008). Minimizing flow-time variance in a single-machine system using genetic algorithms. *International Journal of Advanced Manufacturing Technology*, 39:355–366.

[12] Chou, F.-D., Chang, P.-C., and Wang, H.-M. (2006). A hybrid genetic algorithm to minimize makespan for the single batch machine dynamic scheduling problem. *International Journal of Advanced Manufacturing Technology*, 31:350–359.

[13] Crauwels, H., Potts, C., and Van Wassenhove, L. (1996). Local search heuristics for single-machine scheduling with batching to minimize the number of late jobs. *European Journal of Operational Research*, 90:200–213.

[14] Damodaran, P., Manjeshwar, P., and Srihari, K. (2006). Minimizing makespan on a batch-processing machine with non-identical job sizes using genetic algorithms. *International Journal of Production Economics*, 103:882–891.

[15] França, P., Mendes, A., and Moscato, P. (2001). A memetic algorithm for the total tardiness single machine scheduling problem. *European Journal of Operational Research*, 132:224–242.

[16] French, S. (1982). *Sequencing and Scheduling: An introduction to the mathematics of the Job-shop*. Ellis Horwood Limited, John Wiley & Sons.

[17] Glover, F. (1998). A template for scatter search and path relinking. *Lecture Notes in Computer Science*, 1363:13–54.

[18] Glover, F. and Laguna, M. (1997). *Tabu Search*. Kluwer Academic Publishers.

[19] Glover, F., Laguna, M., and Marti, R. (2000). Fundamentals of scatter search and path relinking. *Control and Cybernetics*, 29:653–684.

[20] Graham, R., Lawler, E., Lenstra, J., and Rinnooy Kan, A. (1979). Optimization and approximation in deterministic sequencing and scheduling: A survey. *Annals of Discrete Mathematics*, 5:287–326.

[21] Hayat, N. and Wirth, A. (1997). Genetic Algorithms and Machine Scheduling With Class Setups. *International Journal of Computer and Engineering Management*, 5(2).

[22] Hino, C., Ronconi, D., and Mendes, A. (2005). Minimizing earliness and tardiness penalties in a single-machine problem with a common due date. *European Journal of Operational Research*, 160:190–201.

[23] Holland, J. (1975). Adaptation in natural and artificial systems. *University of Michigan Press, Ann Arbor*.

[24] Hsieh, J.-C., Chang, P.-C., and Chen, S.-S. (2006). Genetic local search algorithms for single machine scheduling problems with release time. *International Federation for Information Processing*, 207:875–880.

[25] Jin, F., Song, S., and Wu, C. (2009). A simulated annealing algorithm for single machine scheduling problems with family setups. *Computers & Operations Research*, 36:2133–2138.

[26] Jolai, F., Rabbani, M., Amalnick, S., Dabaghi, A., Dehghan, M., and Parast, M. (2007). Genetic algorithm for bi-criteria single machine scheduling problem of minimizing maximum earliness and number of tardy jobs. *Applied Mathematics and Computation*, 194:552–560.

[27] Kashan, A., Karimi, B., and Jolai, F. (2006). Effective hybrid genetic algorithm for minimizing makespan on a single-batch-processing machine with non-identical job sizes. *Internatonal Journal of Production Research*, 44:2337–2360.

[28] Kellegöz, T., Toklu, B., and Wilson, J. (2008). Comparing efficiencies of genetic crossover operators for one machine total weighted tardiness problem. *Appied Mathematics and Computation*, 199:590–598.

[29] Kethley, R. and Alidaee, B. (2002). Single machine scheduling to minimize total weighted late work: A comparison of scheduling rules and search algorithms. *Computers & Industrial Engineering*, 43:509–528.

[30] Koh, S.-G., Koo, P.-H., Kim, D.-C., and Hur, W.-S. (2005). Scheduling a single batch processing machine with arbitrary job sizes and incompatible job families. *International Journal of Production Economics*, 98:81–96.

[31] Köksalan, M. and Keha, A. (2003). Using genetic algorithms for single-machine bicriteria scheduling problems. *European Journal of Operational Research*, 145:543–556.

[32] Lee, C. and Choi, J. (1995). A genetic algorithm for job sequencing problems with distinct due dates and general early-tardy penalty weights. *Computers & Operations Research*, 22:857–869.

[33] Lee, C. and Kim, S. (1995). Parallel genetic algorithms for the earliness-tardiness job scheduling problem with general penalty weights. *Computers & Industrial Engineering*, 28:231–243.

[34] Lee, S. and Asllani, A. (2004). Job scheduling with dual criteria and sequence-dependent setups: mathematical versus genetic programming. *Omega, The International Journal of Management Science*, 32:145–153.

[35] Lin, S.-W., Chou, S.-Y., and Chen, S.-C. (2007). Meta-heuristic approaches for minimizing total earliness and tardiness penalties of single-machine scheduling with a common due date. *Journal of Heuristics*, 13:151–165.

[36] Lin, S.-W. and Ying, K.-C. (2007). Solving single-machine total weighted tardiness problems with sequence-dependent setup times by meta-heuristics. *International Journal of Advanced Manufacturing Technology*, 34:1183–1190.

[37] Liu, J. and Tang, L. (1999). A modified genetic algorithm for single machine scheduling. *Computers & Industrial Engineering*, 37:43–46.

[38] Liu, Z. (2010). Single machine scheduling to minimize maximum lateness subject to release dates and precedence constraints. *Computers & Operations Research*, 37:1537–1543.

[39] Marti, R., Laguna, M., and Glover, F. (2006). Principles of Scatter Search. *European Journal of Operational Research*, 169:359–372.

[40] McMahon, G. and Florian, M. (1975). On scheduling with ready times and due dates to minimize maximum lateness. *Operations Research*, 23:475–482.

[41] M'Hallah, R. (2007). Minimizing total earliness and tardiness on a single machine using a hybrid heuristic. *Computers & Operations Research*, 34:3126–3142.

[42] Michiels, W., Aarts, E., and Korst, J. (2007). *Theoretical Aspects of Local Search Series: Monographs in Theoretical Computer Science. An EATCS Series*.

[43] Miller, B. and Goldberg, D. (1996). Genetic Algorithms, Selection Schemes, and the Varying Effects of Noise. *Evolutionary Computation*, 4, Nr.2:113–132.

[44] Miller, D., Chen, H.-C., Matson, J., and Liu, Q. (1999). A hybrid genetic algorithm for the single machine scheduling problem. *Journal of Heuristics*, 5:437–454.

[45] Monma, C. and Potts, C. (1989). On the complexity of scheduling with batch setup times. *Operations Research*, 37:798–804.

[46] Ovacik, I. and Uzsoy, R. (1994). Rolling horizon algorithms for a single-machine dynamic scheduling problem with sequence-dependent setup times. *International Journal of Production Research*, 32:1243–1263.

[47] Reeves, C. (2003). *Handbook of Metaheuristics*, chapter Genetic algorithms, pages 55–82. Kluwer Academic Publishers, Dordrecht.

[48] Rubin, P. and Ragatz, G. (1995). Scheduling in a sequence dependent setup environment with genetic search. *Computers & Operations Research*, 22:85–99.

[49] Schaller, J. (2007). Scheduling on a single machine with family setups to minimize total tardiness. *International Journal of Production Economics*, 105:329–344.

[50] Schultz, S., Hodgson, T., King, R., and Taner, M. (2004). Minimizing l_{max} for the single machine scheduling problem with family set-ups. *International Journal of Production Research*, 42:4315–4330.

[51] Sels, V., Craeymeersch, K., and Vanhoucke, M. (2011a). A hybrid single and dual population search procedure for the job shop scheduling problems. *European Journal of Operational Research*, 215:512–523.

[52] Sels, V., Steen, F., and Vanhoucke, M. (2011b). Applying a hybrid job shop procedure to a belgian manufacturing company producing industrial wheels and castors in rubber. *Computers & Industrial Engineering*, 61:697–708.

[53] Sels, V. and Vanhoucke, M. (2011a). A hybrid dual-population genetic algorithm for the single machine maximum lateness problem. *Lecture Notes in Computer Science*, 6622:14–25.

[54] Sels, V. and Vanhoucke, M. (2011b). A hybrid electromagnetism/tabu search procedure for the single machine scheduling problem with a maximum lateness objective. Technical report, Ghent University.

[55] Sels, V. and Vanhoucke, M. (2012). A hybrid genetic algorithm for the single machine maximum lateness problem with release times and family setups. *Computers & Operations Research*, to appear.

[56] Sevaux, M. and Dauzère-Pérès, S. (2003). Genetic algorithms to minimize the weighted number of late jobs on a single machine. *European Journal of Operational Research*, 151:296–306.

[57] Sevaux, M. and Sörensen, K. (2004). A genetic algorithm for robust schedules in a one-machine environment with ready times and due date. *4OR*, 2:129–147.

[58] Shin, H.-J., Kim, C.-O., and Kim, S.-S. (2002). A tabu search algorithm for single machine scheduling with release times, due dates, and sequence-dependent set-up times. *International Journal of Advanced Manufacturing Technology*, 19:859–866.

[59] Suriyaarachchi, R. and Wirth, A. (2004). Earliness/tardiness scheduling with a common due date and family setups. *Computers & Industrial Engineering*, 47:275–288.

[60] Tan, K.-C., Narasimhan, R., Rubin, P., and Ragatz, G. (2000). A comparison of four methods for minimizing total tardiness on a single processor with sequence dependent setup times. *Omega, The International Journal of Management Science*, 28:313–326.

[61] Taner, M., Hodgson, T., King, R., and Schultz, S. (2007). Satisfying due-dates in the presence of sequence dependent family setups with a special comedown structure. *Computers & Industrial Engineering*, 52:57–70.

[62] Tsai, T.-I. (2007). A genetic algorithm for solving the single machine earliness/tardiness problem wiht distinct due dates and ready times. *International Journal of Advanced Manufacturing Technology*, 31:994–1000.

[63] Valente, J. and Gonçalves, J. (2008). A genetic algorithm approach for the single machine scheduling problem with linear earliness and quadratic tardiness penalties. *FEP Working Paper*, January 28.

[64] Wang, C.-S. and Uzsoy, R. (2002). A genetic algorithm to minimize maximum lateness on a batch processing machine. *Computers & Operations Research*, 29:1621–1640.

[65] Wang, L. and Wang, M. (1997). A hybrid algorithem for earliness-tardiness scheduling problem with sequence dependent setup time-cost. *Proc of the 36th IEEE CDC San Diego*, pages 1219–1222.

[66] Webster, S., Jog, P., and Gupta, A. (1998). A genetic algorithm for scheduling job families on a single machine with arbitrary earliness/tardiness penalties and an unrestricted common due date. *International Journal of Production Research*, 36:2543–2551.

[67] Yagiura, M. and Ibaraki, T. (1996). The use of dynamic programming in genetic algorithms for permutation problems. *European Journal of Operational Research*, 92:387–401.

In: Handbook of Genetic Algorithms: New Research ISBN: 978-1-62081-158-0
Editors: A. Ramirez Muñoz and I. Garza Rodriguez © 2012 Nova Science Publishers, Inc.

Chapter 14

OPTIMAL SIZING OF ANALOG INTEGRATED CIRCUITS BY APPLYING GENETIC ALGORITHMS

S. Polanco-Martagón[1], G. Reyes-Salgado[1,2], Luis G. De la Fraga[3], E. Tlelo-Cuautle[4]*, I. Guerra-Gómez[4], G. Flores-Becerra[5] and M. Fakhfakh[6]

[1] CENIDET, Computer Department, Mexico
[2] Instituto Tecnologico De Cuautla, Mexico
[3] CINVESTAV, Computer Science Department, Mexico
[4] INAOE, Electronics Department, Mexico
[5] Instituto Tecnologico De Puebla, Mexico
[6] University of Sfax, Tunisia

Abstract

Analog signal processing applications such as filter design and oscillators require the use of different kinds of amplifiers. One kind of those amplifiers are classified to work in mixed-mode, and they can be designed by interconnection of basic analog cells. For instance, the voltage follower (VF), is quite useful in analog design, not only because it allows implementing signal conditioning circuits, but also because it can be evolved to design different kinds of mixed-mode amplifiers, namely: current conveyors, operational transresistance amplifiers, current feedback operational amplifiers, etc. Those mixed-mode integrated circuits (ICs) can be biased and sized automatically. Besides, on the one hand, IC sizing is a hard and tedious work due to the large number of parameters, constraints and performances that the designer has to handle. On the other hand, the main challenges of modern analog IC designs are oriented to solve the problem of determining the correct biases and sizes under different IC technologies. Additionally, there is a pressing need for analog circuit design automation, to meet the time to market constraints. Henceforth, this chapter shows the usefulness of the multi-objective non-dominated sorting genetic algorithm (NSGA-II) to contribute to solve the sizing problem of analog ICs. The NSGA-II is tested and linked to a circuit simulator (SPICE), to compute the optimal sizes of the analog ICs through considering several objective functions, such as: gain, bandwidth and power consumption. Additionally, a discussion on lines for future research are briefly described.

---

*E-mail address: etlelo@inaoep.mx

Keywords: Genetic Algorithm, Circuit Optimization, Analog Integrated Circuits, Voltage Follower, Metal-Oxide-Semiconductor Field-Effect-Transistor.

1 Introduction

Integrated circuits (ICs) design using metal-oxide-semiconductor field-effect-transistors (MOSFETs) imposes challenges as technology scales into nanometer regime [1, 2]. That is, the IC design does not scales with technology shrinking. This open problem is related to the biasing and sizing of the MOSFETs. Unfortunately, this is not a trivial design task due to the huge plethora of active devices [3, 4, 5, 6, 7, 8]. Each kind of active device provides different electrical characteristics, and then the same circuit under design may behave a little bit different using different active devices [9]. Besides, all kinds of active devices find applications in analog signal processing. Some illustrative examples of applications of active devices can be found in [1, 2, 9, 10, 11, 12, 13, 14].

To enhance the performances of the analog circuits, an analog IC designer should to determine the optimal width (W) and length (L) of the MOSFETs. Sometimes, the optimal performance depends on choosing the right circuit topology, an open problem yet known as circuit synthesis [1, 2, 6, 7, 8]. Further, the selected topology must accomplish an specific function which best behavior is performed by the appropriate (W/L) sizing of its circuit elements. In this manner, the sizing is the main problem encountered when trying to meet target requirements. In analog IC design this problem is more complex compared to the digital one due to the very large quantity of performance parameters which sometimes show unclear correlations/tradeoffs among them. Fortunately, evolutionary algorithms have demonstrated their usefulness to cope with this problem [15, 16, 17, 18, 19, 20]. Henceforth, this investigation describes the characteristics of a multi-objective evolutionary algorithm [21, 22], named non-dominated sorting genetic algorithm (NSGA-II) [23].

Basically, we show how does NSGA-II search for the optimal W/L sizes of the MOSFETs to accomplish target specifications. We appeal to this evolutionary algorithm, because it has been demonstrated that it gives much better results than the classically used statistic based approaches [20]. This chapter briefly discusses how to deal with the problems related to the number of parameters, the number and nature of constraints and objective functions. In the sizing process, NSGA-II links the circuit simulator SPICE (Simulation Program with Integrated Circuit Emphasis) by using standard IC technology of 0.35 μm. The cases of study are three Voltage Followers (VF), whose sizing is performed taking into account design constraints. To guarantee the appropriateness of using NSGA-II, the developed evolutionary algorithm is evaluated by applying ZDT test functions from [21].

In section 2, a general description of evolutionary algorithms for multi-objective optimization is provided. In section 3, we summarize the main characteristics of NSGA-II. In subsection 3.1, the genetic operators multiple-point crossover and single-point mutation are described. In section 4, the developed evolutionary system is tested with the functions given in [21]. In section 5, the optimization results by applying NSGA-II to three VFs is shown. Finally, the conclusions are listed in section 6.

2 Evolutionary Algorithms

Evolutionary Algorithms (EAs) are appropriate methods for approximating the Pareto-optimal front in NP-hard multi-objective optimization problems which are too complex, such as the analog circuit design in its different stages [1, 2]. This is not only because there are not numerous methods that handle the really large search space for multiple Pareto-optimal solutions, but also to their capability to search multiple points in the search space at the same time (inherent parallelism), and through recombinations, the ability to use the best characteristics of each solutions set which are able to approximate the Pareto-optimal front in a single optimization run.

A general multi-objective optimization problem consists of a number of conflicting objectives and is associated with a number of inequality and/or equality constraints. In [21] one finds some definitions which formalize concepts of this kind of problems as well as a definition of the difference between local and global Pareto-optimal sets.

In multi-objective problems, the search space is partially ordered in relation to the dominance among its individuals, in the sense that two arbitrary solutions are related to each other in two possible ways:

- one dominates the other

- neither dominates

Let us consider a multi-objective minimization problem with *m* decision variables (W/L parameters) and *n* objectives (circuit performances):

$$\begin{aligned} \text{Minimize} \quad & y = f(x) = (f_1(x), \ldots, f_n(x)) \\ \text{where} \quad & x = (x_1, \ldots, x_m) \in X \\ & y = (y_1, \ldots, y_n) \in Y \end{aligned} \quad (1)$$

where **x** is called *decision vector*, X parameter space, **y** *objective vector*, and Y *objective space*. A decision vector $\mathbf{a} \in X$ is said to *dominate* a decision vector $\mathbf{b} \in X$, $(\mathbf{a} \prec \mathbf{b})$, if and only if

$$\begin{aligned} \forall i \in \{1, \ldots, n\} : f_i(a) \leq f_i(b) \;\; \wedge \\ \exists j \in \{1, \ldots, n\} : f_j(a) < f_j(b) \end{aligned} \quad (2)$$

Once defined the dominance relationship, it is defined the non-dominated and Pareto-optimal solution

Definition 1. Let $\mathbf{a} \in X$ be an arbitrary decision vector.

1. The decision vector **a** is said to be nondominated regarding *a* set $X' \subseteq X$ if and only if there is no vector in X' which dominates **a**; formally

$$\nexists a' \in X' : a' \prec a \quad (3)$$

2. The decision vector **a** is *Pareto-optimal* if and only if **a** is nondominated regarding X.

In general, the objectives in the Pareto-optimal vector can not be improved without provoking performance degradation in another one. Above, in the terminology of multi-objective optimization, represents global optimal solution. However, at the same manner as to single-objective optimization problems, in multi-objective optimization problems may exist a local optima set which constitutes a nondominated set within a certain neighbourhood [21]. In this case, in [22] are introduced corresponding concepts for local and global Pareto-optimal.

Definition 2. Consider a set of decision vectors $X' \subseteq X$.

1. The set X' is denoted as a *local Pareto-optimal set* if and only if for every member $a' \in X'$, there isn't solution $a \in X$ which satisfies $||a - a'|| \leq \epsilon$ and $||f(a) - f(a')|| < \delta$, where ϵ denotes a small positive number (in principle, a' is obtained by perturbing a in a small neighbourhood), which dominates any member in the set X', then the solutions belonging to the set X' constitute a *local Pareto-optimal set*, and formally is[1]

$$\forall a' \in X' : \nexists a \in X : a \prec a' \wedge ||a - a'|| < \epsilon \wedge ||f(a) - f(a')|| < \delta \quad (4)$$

where $|| \cdot ||$ is a corresponding distance metric and $\epsilon > 0, \delta > 0$.

2. The set X' is called a *global Pareto-optimal set* if and only if there isn't solution $a \in X$ which dominates $a' \in X'$, then the solutions belonging to the set X' constitute a *global Pareto-optimal set*; formally is

$$\forall a' \in X : \nexists a \in X : a \prec a' \quad (5)$$

It can be seen that a global pareto-optimal set does not necesarily contains all Pareto-optimal solutions. When making reference to the whole Pareto-optimal solutions only, the set is called "Pareto-optimal set"; the corresponding set of objective vectors is denoted as "Pareto-optimal front".

3 Non-dominated Sorting Genetic Algorithm - NSGA-II

The non-dominated sorting genetic algorithm (NSGA) was one of the first evolutionary algorithms (EAs) devoted to find multiple-optimal solutions, preserving diversity, with emphasis for moving towards the true Pareto-optimal region in one single simulation run. This algorithm was a popular method for multi-objective optimization, nevertheless it was mainly criticized for the following reasons

1. *Computational complexity:* The non-dominated sorting procedure used in each generation had a high complexity for large populations.

2. *Lack of elitism:* Elitism within EAs can accelerate the performance and to prevent the loss of good solutions once found [21].

---
[1]The definition of *local Pareto-optimal set* was slightly modified in [21].

3. *Optimal sharing parameter* σ_{share}: One of the main problems with the NSGA was the choosing and specification of the optimal parameter value for the sharing parameter σ_{share}.

An improved version of the NSGA, called NSGA-II, was developed in [23] with a better non-dominated sorting mechanism, incorporating elitism and without using a sharing parameter *a priori*.

The objective of the NSGA-II is the improving of the adaptive fit of a population of candidate solutions to a Pareto front constrained by a set of objective functions. This improvement is performed by sorting and ranking all the population into a hierarchical sub-populations, which are selected according to their fitness in order to create offsprings, i.e. by ordering all the population, based on the order of Pareto dominance, in different Pareto sub-fronts it is possible to select those which have better performance. In this algorithm it has been contemplated the form to select, between two candidates, a solution without causing the loss of diversity. In this manner it is possible to obtain the best part of a population while maintaining diversity among the solutions.

The NSGA-II algorithm is based in two methods: *fast non-dominated sort* and *crowding distance assignment*. These methods, coupled with the fact that constraints can be added easily, guarantee the convergence to the Pareto optimal set.

Algorithm 1 provides a pseudocode listing for the NSGA-II Algorithm. At the beginning the population is initialized as usual, i.e. the population is randomly initialized within the feasible region and then individuals are evaluated against objective functions. Once the population is initialized, through the Fast Non-dominated sort method the solutions are ranked and classified into sub-fronts. The first sub front being completely a non-dominant set in the current population and the second front being dominated only by the individuals in the first sub-front, the third is only dominated by the second and the subsequent sub-fronts continues at the same way. In this manner, each individual in each sub front is assigned a rank (fitness) value based on the sub-front in which it belongs to. Individuals in the first sub front are assigned the value of 1, and individuals in the second sub-front are assigned the value of 2, and so on.

In addition to the fitness value, a new parameter, called *crowding distance*, is calculated for each individual. The crowding distance is a measure of how close an individual is to its neighbours. A large value on the distance will result in a best diversity in the population. Thus, solutions that are far from the rest are selected; this selection is performed by the crowding-comparison operator ($\prec_n$) which assumes that each solution has two attributes, the nondomination rank and the crowding distance, then the crowded-comparison operator is defined as:

$$i \prec_n j \; if \, (i_{rank} < j_{rank}) \; or \; \left[(i_{rannk} = j_{rank}) \; and \; (i_{distance} > i_{distance}) \right] \quad (6)$$

Parents are selected from the population by using binary tournament selection based on the rank and/or crowding distance. In the selection process, if two solutions belong to the same sub front, a solution is preferred against the other when its rank is lesser than or if its crowding distance is greater than the other solution. The selected population generates offsprings from typical crossover and mutation operators. The current population and its

Algorithm 1 NSGA-II

Require: Pop_{size} : set, $P_{crossover}$, $P_{mutatuion}$: $real$
Ensure: $Children$: set
1: $Pop \leftarrow InitializeRandomly(Pop_{size})$
2: $Evaluate(Pop)$
3: $FastNondominatedSort(Pop)$
4: $Selected \leftarrow SelectByRank(Pop, Pop_{size})$
5: $Children \leftarrow CrossoverAndMutation(Selected, P_{crossover}, P_{mutation})$
6: **while** $\neg StopCriteria()$ **do**
7:    $Evaluate(Children)$
8:    $Union \leftarrow Pop \cup Children$
9:    $Fronts \leftarrow FastNondominatedSort(Union)$
10:   $Parents \leftarrow \emptyset$
11:   $Front_L \leftarrow \emptyset$
12:   **for each** $Front_i \in Fronts$ **do**
13:      $CrowdingDistanceAssignment(Front_i)$
14:      **if** $|Parents| + |Front_i| > Pop_{size}$ **then**
15:         $Front_L \leftarrow i$
16:         $Break()$
17:      **else**
18:         $Parents \leftarrow Parents \cup Front_i$
19:      **end if**
20:   **end for**
21:   **if** $|Parents| < Pop_{size}$ **then**
22:      $Front_L \leftarrow SortByRankAndDistance(Front_L)$
23:      $Parents \leftarrow Parents \cup Front_L[1 : Pop_{size} - |Front_L|]$
24:   **end if**
25:   $Selected \leftarrow SelectByRankAndDistance(Parents, Pop_{size})$
26:   $Pop \leftarrow Children$
27:   $Children \leftarrow CrossoverAndMutation(Selected, P_{crossover}, P_{mutation})$
28: **end while**
29: **return** $Children$

offsprings are merged into an only one, this new population is sorted again based on non-domination and only N individuals (from 2N) with the best performance are selected for the next generation, where N is the size population. Thereby, the last subfront could have more than the necessary individuals which will be selected based on the rank and the crowding distance. Thus, the evolution process is performed following the above steps.

3.0.1 Fast Nondominated Sort

This is the method that is in charge to classify and to rank the population into a set of sub fronts, the classification of solutions is based on the Pareto non-domination. Algorithm 2 lists the steps of this method.

The method begins with a population (P) of solutions in the search space, and two

Algorithm 2 FastNondominateSort
Require: P
1: **for each** $p \in P$ **do**
2:     $S_p \leftarrow \emptyset$
3:     $n_p \leftarrow 0$
4:     **for each** $q \in P$ **do**
5:        **if** $p \prec q$ **then**
6:           $S_p \leftarrow \cup \{q\}$
7:        **else if** $q \prec p$ **then**
8:           $n_p \leftarrow n_p + 1$
9:        **end if**
10:     **if** $n_p \leftarrow 0$ **then**
11:        $P_{rank} \leftarrow 1$
12:        $F_1 \leftarrow F_1 \cup \{p\}$
13:     **end if**
14:     **end for**
15: **end for**
16: $i \leftarrow 0$
17: **while** $F_i \neq \emptyset$ **do**
18:     $Q \leftarrow \emptyset$
19:     **for each** $p \in F_i$ **do**
20:        **for each** $q \in S_p$ **do**
21:           $n_q \leftarrow n_q - 1$
22:           **if** $n_q \leftarrow 0$ **then**
23:              $q_{rank} \leftarrow i + 1$
24:              $Q \leftarrow Q \cup \{q\}$
25:           **end if**
26:        **end for**
27:     **end for**
28:     $i \leftarrow i + 1$
29:     $F_i \leftarrow Q$
30: **end while**

entities for each solution in P: 1) the number of the solution which dominates the solution p-th (n_p), and 2) a set of solutions that the solution p-th dominates (S_p).

Each solution in P is compared against each other to find the non-dominated solutions. The group of solutions, that have their domination count as zero, will be labelled as the non-dominated solutions and they will be separated temporally from P and placed into the first sub front (F_1). Now, for each p-th solution in F_1 will be visited each solution (q) of the set S_p and its domination count is reduced by one. Then, solutions which their domination count becomes zero will be separated and placed into the second sub front (F_2). Next, the aforementioned steps are followed for each member in F_2 to find the third sub front, and so on until all sub fronts are identified.

3.0.2 Crowding Distance Assignment

After that the non-dominated sort is completed, the crowding distance is computed. Since solutions are selected based on rank and crowding distance, this value will be assigned to all individuals from each sub front. This value takes effect when there exists the necessity of choosing the last individuals of the new population which will be selected to perform offspring. In this way, the crowding distance helps to choose solutions that belong to the same sub front. In Algorithm 3 is shown the psudocode of the computation and assignation of the crowding distance.

The basic idea about the crowding distance is finding the euclidian distance between each individual in a front based on their m objectives in the m dimensional hyper space. The crowding distance is calculated as follows: First, the sets of solutions have to be sorted in an ascending order for each objective function, then those solutions that are in the boundary (smallest and largest function value) are assigned an infinite distance value. Solutions located within the boundary are assigned to a distance value equal to the normalized difference in the function values of two adjacent solutions, this is done for each objective function. The total crowding-distance value is the sum of individual distance values corresponding to each objective function.

Algorithm 3 CrowdingDistanceAssignment

Require: T: set
1: $l \leftarrow |T|$
2: **for each** i **do** {algo}
3:     set $T[i]_{distance} \leftarrow 0$
4:     **for each** $objective\ m$ **do**
5:         $T \leftarrow sort(T, m)$
6:         $T[1]_{distance} \leftarrow T[l]_{distance} \leftarrow \infty$
7:         **for** $i \leftarrow 1 \rightarrow (l-1)$ **do**
8:             $T[i]_{distance} \leftarrow T[i]_{distance} + (T[i+1].m - T[i-1].m)/(f_m^{max} - f_m^{min})$
9:         **end for**
10:     **end for**
11: **end for**

3.1 Genetic Operators

The genetic operators mission is to recombine the existing genetic information along all the population of solutions and then generate new individuals that could allow the exploration of new regions in the search space. Traditional Genetic Algorithms use 1-point crossover that is the simplest way to combine information among individuals as well as the use of bit-to-bit mutation with the probability of changing several points along the genomic string.

Our version of NSGA-II has been developed following the list code from Algorithms 1, 2 and 3, and by using a binary-coded string to codify the variables. At the same time, the NSGA-II has been developed using a multiple-point crossover and a single-bit mutation along the binary string.

3.1.1 Multiple-point Crossover

Since the 1-point crossover is the simplest way to form new individuals, it has some inconveniences, like the fact that it does not permit any combination of the genetic material between parents. For this reason, it has been used the multi-point crossover. In this alternative genetic operator, two points are chosen for cutting the genetic string of the parents, then the obtained parts are interchanged between them defined by the cutting points as is shown in Figure 1.

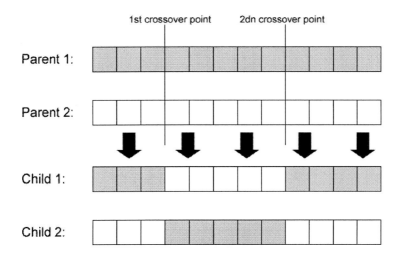

Figure 1. Two-point crossover.

3.1.2 Single-bit Mutation

The simplest way of mutation which can be applied to an individual is to change the value in a single position along the genomic string, i.e. if the position value is one, then it becomes zero, else if the position value is zero, then it becomes one, like in Figure 2.

Unlike the traditional bit-to-bit mutation, where each position in the genomic string could change depending on a mutation rate, the single-bit mutation only takes into account the change of only one position along the genomic string. This is because the mutation may break the information relationships among genes that have formed with the evolution.

4 Behavior of NSGA-II on Test Functions ZDT

In this section, a set of well known test functions are employed to evaluate the performance of the NSGA-II algorithm. These test functions have been used in [21] as a comparison of various evolutionary approaches to multi-objective optimization, each one contemplates available possibilities in [22], and at the same time each one contemplates a particular

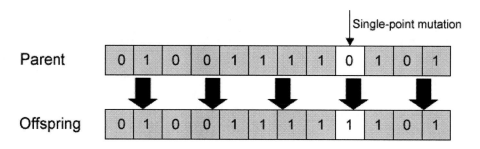

Figure 2. Single-bit mutation.

feature that causes difficulty in the evolutionary optimization process, mainly to converging in the Pareto-optimal front.

The set of test functions, as it can be seen, are restricted to two objectives to minimize. This is because in [21] it is considered what two objectives are the simplest and complete manner of reflect the essential aspects of the multi-objective optimization. Each test function is structured in the same manner over three functions f_1, g and h:

$$\begin{aligned} \text{Minimize} \quad & F(X) = (f_1(x_1), f_2(X)) \\ \text{subject to} \quad & f_2(X) = g(x_2, \ldots, x_n) \cdot h(f_1(x_1), g(x_2, \ldots, x_n)) \\ \text{where} \quad & X = (x_1, \ldots, x_m) \end{aligned} \quad (7)$$

The function f_1 is only dependent on the first decision variable, g is a function of the remaining $n-1$ variables, and the parameters of h are the function values of f_1 and g. The test functions differ in these functions as well as in the number of variables n and in the values the variables may take. Next, the five test functions used in this work are described,

- The test function ZDT1 has 15 variables in the range [0,1]. Its Pareto-front is convex, continuous and a uniform distribution along the front. It is the simplest problem in the set of test functions.

$$\begin{aligned} f_1(x_1) &= x_1 \\ g(x_2, \ldots, x_n) &= 1 + 9 \cdot \sum_{i=2}^{n} \frac{x_i}{(n-1)} \\ h(f_1, g) &= 1 - \sqrt{\frac{f_1}{g}} \end{aligned} \quad (8)$$

where $n = 15$, and $x_i \in [0, 1]$. The Pareto-optimal front is formed with $g(X) = 1$

- The function ZDT2 has 15 variables in the range [0,1], and its Pareto-front is the nonconvex counterpart to ZDT1. The distribution of solutions along the front is uniform.

$$f_1(x_1) = x_1$$
$$g(x_2,\ldots,x_n) = 1 + 9 \cdot \sum_{i=2}^{n} \frac{x_i}{(n-1)} \qquad (9)$$
$$h(f,g) = 1 - \left(\frac{f_1}{g}\right)^2$$

where $n = 15$ and $x_i \in [0, 1]$. Its Pareto-optimal front is formed with $g(X) = 1$.

- The test function ZDT3 has 15 variables in the range [0,1] and it represents the discreteness feature; its Pareto-optimal front consists of several non-contiguous convex parts. The introduction of the sine function in h causes discontinuity in the Pareto-optimal front. However, there is no discontinuity in the parameter space.

$$f_1(x_1) = x_1$$
$$g(x_2,\ldots,x_n) = 1 + 9 \cdot \sum_{i=2}^{n} \frac{x_i}{(n-1)} \qquad (10)$$
$$h(f_1,g) = 1 - \sqrt{\frac{f_1}{g}} - \left(\frac{f_1}{g}\right)\sin(10\pi f_1)$$

where $n = 15$ and $x_i \in [0, 1]$. The Pareto-optimal front is formed with $g(X) = 1$.

- The function ZDT4 has 10 variables in the range [0,1]. The Pareto-front is convex. The problem complexity is that it has 21^9 Pareto-optimal local fronts and therefore, it test to the gentic algorithm in relation with their capacity to deal with multimodality.

$$f_1(x_1) = x_1$$
$$g(x_2,\ldots,x_n) = 1 + 10 \cdot (n-1) + \sum_{i=2}^{n}(x_i^2 - 10\cos(4\pi x_i)) \qquad (11)$$
$$h(f_1,g) = 1 - \sqrt{\frac{f_1}{g}}$$

where $n = 10$, $x_1 \in [0, 1]$ and $x_2,\ldots,x_n \in [-5, 5]$. The global Pareto-front is formed with $g(X) = 1$.

- The function ZDT6 has 10 variables in the range [0,1]. Its Pareto-front is nonconvex. The problem complexity is given by the combination of the nonconvex shape of the Pareto-front and the non-uniform distribution along the front.

$$f_1(x_1) = 1 - e^{(-4x_1)}\sin^6(6\pi x_1)$$
$$g(x_2,\ldots,x_n) = 1 + 9 \cdot \left(\frac{(\sum_{i=2}^{n} x_i)}{(n-1)}\right)^{\frac{1}{4}} \qquad (12)$$
$$h(f_1,g) = 1 - \left(\frac{f_1}{g}\right)^2$$

where $n = 10$ and $x_i \in [0, 1]$. The Pareto-front is formed with $g(X) = 1$.

The test functions were evaluated (in a PC with an Intel(R) Core$^{(TM)}$ i5 3.2 GHz x4, and a 6GHz RAM) over ten runs with a population size of 100 along 150 generations, by using multi-point crossover and single-bit mutation as genetic operators. A similar work has been presented in [20]. Figures 3-7 depict the real Pareto-optimal front of each test function together with the behavior of the NSGA-II against the test function. Table 1 shows statistical results using the NSGA-II against the test functions, in this table are shown the following values: the maximum value (MAX), minimum value (MIN), average value (AVG), and the standard deviation (STD).

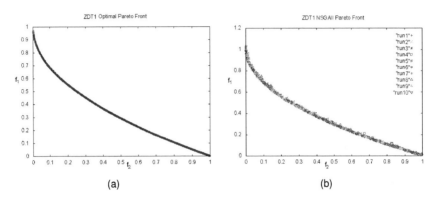

Figure 3. ZDT1 real Pareto-optimal front (a) and Pareto-optimal front by NSGA-II (b).

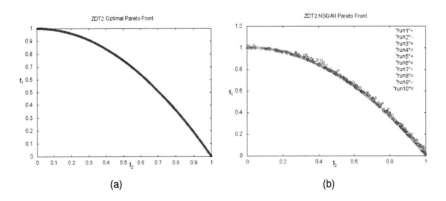

Figure 4. ZDT2 real Pareto-optimal front (a) and Pareto-optimal front by NSGA-II (b).

5 Sizing Optimization Results

The circuits sized by applying NSGA-II are the voltage followers depicted in Figure 8 [6]. The encoding of the three VFs that were sized by using standard CMOS technology of 0.35 μ m, and Vss = -Vdd = 1.65. It is worthy to mention that also these VFs have been synthesized by genetic algorithms [6, 8] .

Optimal Sizing of Analog Integrated Circuits by Appying Genetic Algorithms   307

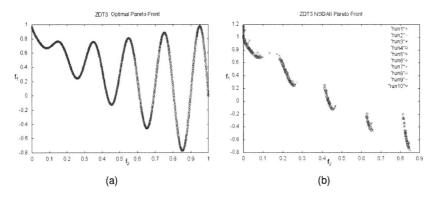

Figure 5. ZDT3 real Pareto-optimal front (a) and Pareto-optimal front by NSGA-II (b).

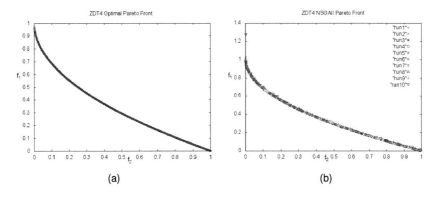

Figure 6. ZDT4 real Pareto-optimal front (a) and Pareto-optimal front by NSGA-II (b).

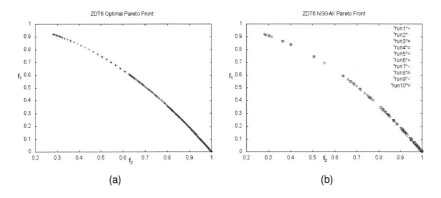

Figure 7. ZDT6 real Pareto-optimal front (a) and Pareto-optimal front by NSGA-II (b).

The optimization system was programmed in C and the circuit simulations are made with TopSPICE^{TM}, by modifying each transistor width (W_i), lenth (L) and current bias (I), , and recollecting results from the output listing. In generating the feasible solutions, it was verified that the P-MOSFETs W/L ratios were greater than or equal to the N-MOSFETs. It means: $W_{M_1,M_2} \geq W_{M_3,M_4}$ for Figure 8(a), $W_{M_1,M_4} \geq W_{M_2,M_3}$ for Figure 8(b), and $W_{M_3,M_4} \geq W_{M_1,M_2}$ for Fig. 8(c). These circuits were optimized along 100 generations

Table 1. Results on test functions

Function	f_1 MAX	f_1 MIN	f_1 AVG	f_1 STD	f_2 MAX	f_2 MIN	f_2 AVG	f_2 STD
ZDT1	9.97E-01	0.00E+00	4.448E-01	2.98E-01	1.03E+00	4.76E-03	3.925E-01	2.59E-01
ZDT2	1.00E+00	0.00E+00	5.90E-01	2.67E-01	1.17E+00	6.19E-03	5.91E-01	2.97E-01
ZDT3	8.77E-01	0.00E+00	3.16E-01	2.72E-01	1.18E+00	-7.63E-01	2.79E-01	5.13E-01
ZDT4	1.00E+00	0.00E+00	4.32E-01	3.01E-01	1.28E+00	2.02E-03	4.02E-01	2.68E-01
ZDT6	1.00E+00	2.81E-01	9.01E-01	1.73E-01	9.21E-01	0.00E+00	1.58E-01	2.48E-01

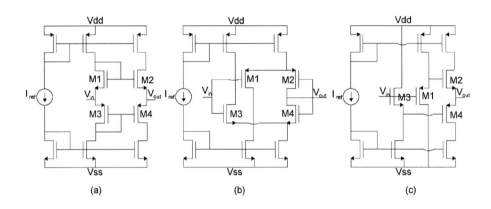

Figure 8. Voltage Followers.

over 10 runs.

5.1 Voltage Follower of Figure 8(a)

The VF depicted in Figure 8(a) is encoded with six design variables: current bias (I), transistor length (L) and transistor width (W_i), where i represents a specific transistor (or transistors which share the same width) of the circuit, as Table 2 shows.

Table 2. Fig. 8(a) VF encoding

gene	design variable	Encoding Transistor
x_1	I	$M_1, \ldots, M_{10}$
x_2	L	$M_1, \ldots, M_{10}$
x_3	W_1	M_3, M_4
x_4	W_2	M_1, M_2
x_5	W_3	M_5, M_7, M_9
x_6	W_4	M_6, M_8, M_{10}

The optimization problem is expressed as:

$$\begin{aligned}minimize\quad &\mathbf{f}(x) = [f_1(x), f_2(x)]^T\\ subject\ to\quad &h_k(x) \geq 0, \quad k = 1, \ldots, 10\\ where\quad &x \in X\end{aligned} \qquad (13)$$

where $X : \mathbb{R}^6 \mid x_1 \in \{10, 20, \ldots, 100\}\mu A$, $x_2 \in \{0.4, 0.7, 1, 1.05, 1.4\}\mu m$ and $x_i \in \{0.35, 0.7, \ldots, 700\}\mu m$, $i = 3, \ldots, 6$ is the decision space for the variables $x = [x_1, \ldots, x_6]$. $\mathbf{f}(x)$ is the vector formed by two objectives:

- $f_1(x)$ = Voltage gain
- $f_2(x)$ = Voltage bandwidth

Finally, h_k, k = 1, . . . ,10 are performance constraints.

Once encoding the variables and targeting the objectives, it has been proceeded to perform the sizing optimization of the VF from Figure 8(a). Thus, the results are shown in Figure 9, and the statistical results are listed in Table 3 and Table 4.

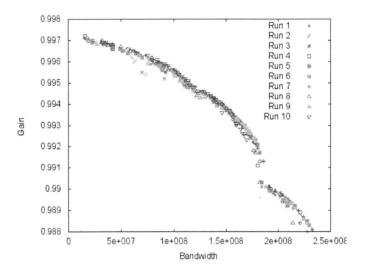

Figure 9. NSGA-II optimization for the VF in Figure 8(a).

5.2 Voltage Follower of Figure 8(b)

The VF depicted in Figure 8(b) is encoded with eight design variables: current bias (I), transistor length (L) and transistor width (W_i), where i represents a specific transistor, as Table 5 shows.

The optimization problem is expressed as:

$$\begin{aligned}minimize\quad &\mathbf{f}(x) = [f_1(x), f_2(x)]^T\\ subject\ to\quad &h_k(x) \geq 0, \quad k = 1, \ldots, 10\\ where\quad &x \in X\end{aligned} \qquad (14)$$

Table 3. NSGA-II optimization measurements for the VF in Figure 8(a)

Measure	Gain	Bandwidth
MAX	9.97E-01	2.32E+08
MIN	9.88E-01	1.51E+07
AVG	9.93E-01	1.43E+08
STD	2.11E-03	4.60E+07

Table 4. NSGA-II best objective results for the VF in Figure 8(a)

	Best for:	
	Gain	Bandwidth
Gain	**9.97E-01**	9.88E-01
Bandwidht	1.59E+07	**2.32E+08**
Variable Values		
I	10	100
L	1.4	0.7
W_1	242.90	101.15
W_2	600.60	101.15
W_3	55.30	33.25
W_4	5.95	34.30

Table 5. 8(b) VF encoding

gene	design variable	Encoding Transistor
x_1	I	$M_1, \ldots, M_{10}$
x_2	L	$M_1, \ldots, M_{10}$
x_3	W_1	M_1, M_2
x_4	W_2	M_3, M_4
x_5	W_3	M_5, M_9
x_6	W_4	M_6, M_{10}
x_7	W_5	M_7
x_8	W_6	M_8

where $X : \mathbb{R}^8 \mid x_1 \in \{10, 20, \ldots, 100\} \mu A$, $x_2 \in \{0.4, 0.7, 1, 1.05, 1.4\} \mu m$, $x_i \in \{0.35, 0.7, \ldots, 700\} \mu m$ for $i = 3, \ldots, 6$, $x_7 = (x_5/2)$ and $x_8 = (x_6/2)$ is the decision space for the variables $x = [x_1, \ldots, x_8]$. $\mathbf{f}(x)$ is the vector formed by two objectives:

- $f_1(x)$ = Voltage gain

- $f_2(x)$ = Voltage bandwidth

Finally, h_k, k = 1, . . . ,10 are performance constraints. Figure 10 depicts the Pareto-front obtained with the NSGA-II algorithm for the VF in Figure 8(b), and Tables 6 and 7 show the statistical results.

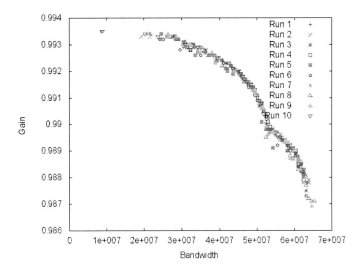

Figure 10. NSGA-II optimization for the VF in Figure 8(b).

Table 6. NSGA-II optimization measurements for the VF in Figure 8(b)

Measure	Gain	Bandwidth
MAX	9.94E-01	6.53E+07
MIN	9.87E-01	8.61E+06
AVG	9.91E-01	4.85E+07
STD	1.68E-03	1.12E+07

5.3 Voltage Follower of Figure 8(c)

The VF depicted in Figure 8(c) is encoded with six design variables: current bias (I), transistor length (L) and transistor width (W_i), where i represents a specific transistor (or transistors which share the same width) of the circuit, as Table 2 shows.

The optimization problem is expressed as:

$$\begin{aligned} minimize \quad & \mathbf{f}(x) = [f_1(x), f_2(x)]^T \\ subject\ to \quad & h_k(x) \geq 0, \quad k = 1, \dots, 10 \\ where \quad & x \in X \end{aligned} \quad (15)$$

where $X : \mathbb{R}^6 \mid x_1 \in \{10, 20, \dots, 100\} \mu A$, $x_2 \in \{0.4, 0.7, 1, 1.05, 1.4\} \mu m$ and $x_i \in \{0.35, 0.7, \dots, 700\} \mu m$, $i = 3, \dots, 6$ is the decision space for the variables $x = [x_1, \dots, x_6]$. $\mathbf{f}(x)$ is the vector formed by two objectives:

Table 7. NSGA-II best objective results for the VF in Figure 8(b)

	Best for:	
	Gain	Bandwidth
Gain	**9.94E-01**	9.87E-01
Bandwidht	8.91E+06	**6.53E+07**
Variable Values		
I	20	100
L	1.4	1.0
W_1	584.15	79.80
W_2	116.20	79.80
W_3	155.05	73.85
W_4	54.95	40.60
W_5	77.52	36.92
W_6	27.47	20.30

Table 8. 8(c) VF encoding

gene	design variable	Encoding Transistor
x_1	I	$M_1, \ldots, M_{10}$
x_2	L	$M_1, \ldots, M_{10}$
x_3	W_1	M_1, M_4
x_4	W_2	M_2, M_3
x_5	W_3	M_5, M_7, M_9
x_6	W_4	M_6, M_8, M_{10}

- $f_1(x)$ = Voltage gain
- $f_2(x)$ = Voltage bandwidth

Finally, h_k, $k = 1, \ldots, 10$ are performance constraints. Table 9 and Table 10 lists the results obtained from the sizing optimization of the VF of Figure 8(c) which are reflected in Figure 11.

Table 9. NSGA-II optimization measurements for the VF in Figure 8(c)

Measure	Gain	Bandwidth
MAX	9.96E-01	1.62E+08
MIN	9.81E-01	8.81E+06
AVG	9.90E-01	1.11E+08
STD	3.91E-03	2.78E+07

Optimal Sizing of Analog Integrated Circuits by Appying Genetic Algorithms 313

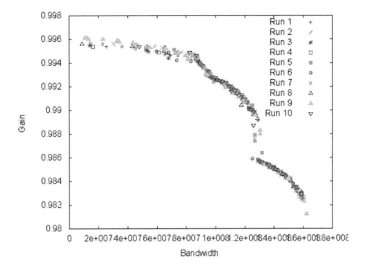

Figure 11. NSGA-II optimization for the VF in Figure 8(c).

Table 10. NSGA-II best objective results for the VF in Figure 8(c)

	Best for:	
	Gain	Bandwidth
Gain	**9.96E-01**	9.81E-01
Bandwidht	1.26E+07	**1.62E+08**
Variable Values		
I	10	100
L	1.4	0.7
W_1	504.00	70.35
W_2	67.90	67.90
W_3	96.25	96.25
W_4	3.85	96.25

5.4 Discussion of Results

Before discussing results, it is important to highlight and to remember that each variable values, within the three Voltage Followers, correspond to the physical dimensions of each codified transistor. And, in the integrated circuit design, small dimensions for the L and W values of the transistors are required to reduce the silicon area.

It is easy to see what actual shapes of Figure 9, Figure 10 and Figure 11 correspond to a multi-objective problem, because while the gain is closer to unity, the bandwidth is diminished; This is clearly ascertainable because the correlation of each VF correspond to 0.94 for the VF in Figure 9, 0.94 for the VF in Figure 10 and 0.89 for the VF in Figure 11.

Figures 9, 10 and 11 are the corresponding Pareto-Fronts obtained from the NSGA-II algorithm. Among the three cases, the VF of Figure 8(a) provides the best results of Gain and Bandwidth as shown in Table 3. However, the results obtained with the VF of Figure 8(b) are more homogenise with respect to the gain, this shows a better convergence to the objective "closer to unity". The virtue that can be attributed to the VF of Figure 8(c) is that a section of its Pareto-front is almost horizontal, allowing to get a good gain without losing lot of bandwidth.

From the obtained results, one can apply fuzzy sets [24] to select the desired sizes which best accomplishes the required target specifications.

6 Conclusion

This investigation showed the evolutionary algorithm NSGA-II having the capability to deal with constrained multi-objective problems with several variables. Likewise, it has been displayed two alternatives of genetic operators, such as the single-point mutation and the multiple-point crossover. These two genetic operators were used to realize five synthetic multiple-objective problems which are the ZDT functions, showing good results without the need of performing lot of generations in the evolution process.

For the circuit sizing optimization, we sized three voltage followers whose variables were the transistor width and length, and the current bias. The constraints were the saturation condition for each transistor in the circuit. In general, the programmed NSGA-II has found optimized results with a good distribution among solutions. It was programmed in C and linked with TopSPICE as circuit evaluator.

Compared to a multi-objective evolutionary algorithm based on decomposition (MOEA/D) [20], NSGA-II has the ability to find the best average results. The coded algorithm has the capability to work with a great number of transistors to explore the best design. Also it is possible to define the bounds of the search space to ensure that the "optimal solutions are feasible".

Acknowledgements

This work has been partially supported by CONACyT-MEXICO under the projects 81604-R and 131839-Y.

References

[1] Tlelo-Cuautle, E. (2011). Analog Circuits: Applications, Design and Performance: Nova Science Publishers Inc

[2] Tlelo-Cuautle, E. (2011). Advances in Analog Circuits: InTech. Open Access http://www.intechweb.org/books/show/title/advances-in-analog-circuits

[3] Sanchez-López, C., Fernández, F., V. , Tlelo-Cuautle, E., & Tan, S., X. D. (2011). Pathological Element-Based Active Device Models and Their Application to Symbolic Analysis. *IEEE Trans Circuits Syst I Regul Pap*, 58(1), 1382-1395. doi: 10.1109/tcsi.2010.2097696

[4] Sanchez-López, C., Martínez-Romero, E., & Tlelo-Cuautle, E. (2011). Symbolic Analysis of OTRAs-Based Circuits. *Journal of Applied Research and Technology*, 9(1), 69-80.

[5] Khateb, F., & Biolek, D. (2011). Bulk-Driven Current Differencing Transconductance Amplifier. *Circuits, Systems, and Signal Processing*, 30(5), 1071-1089. doi: 10.1007/s00034-010-9254-9

[6] Tlelo-Cuautle, E., Duarte-Villaseñor, M. A., & Guerra Gómez, I. (2008). Automatic Synthesis of VFs and VMs by Applying Genetic Algorithms. *Circuits, Systems, and Signal Processing*, 27(3), 391-403. doi: 10.1007/s00034-008-9030-2

[7] Tlelo-Cuautle, E., Moro-Frías, D. Sánchez-López, C., & Duarte-Villaseñor, M. A. (2008). Synthesis of CCII-s by superimposing VFs and CFs through genetic operations. *IEICE Electronics Express*, 5(11), 411-417.

[8] Duarte-Villaseñor, M., Tlelo-Cuautle, E., & de la Fraga, L. G. (2011). Binary Genetic Encoding for the Synthesis of Mixed-Mode Circuit Topologies. *Circuits, Systems, and Signal Processing*, 1-15. doi: 10.1007/s00034-011-9353-2

[9] Muñoz Pacheco, J. M., Campos López, W., Tlelo-Cuautle, E., & Sanchez López, C. (2012). OpAmp-, CFOA- and OTA-Based Configurations to Design Multi-Scroll Chaotic Oscillators. *Trends in Applied Sciences Research*, 7(2), 1-7. doi: 10.3923/tasr.2012

[10] Pathak, J. K., Singh, A. K., & Senani, R. (2011). Systematic realisation of quadrature oscillators using current differencing buffered amplifiers. *IET Circuits, Devices & Systems*, 5(3), 203-211.

[11] Swamy, M. (2011). Mutators, Generalized Impedance Converters and Inverters, and Their Realization Using Generalized Current Conveyors. *Circuits, Systems, and Signal Processing*, 30(1), 209-232. doi: 10.1007/s00034-010-9208-2

[12] Sanchez-López, C., Castro-Hernández, A., & Pérez-Trejo, A. (2008). Experimental verification of the Chua's circuit designed with UGCs. *IEICE Electronics Express*, 5(17), 657-661. doi: 10.1587/elex.5.657

[13] Trejo-Guerra, R., Tlelo-Cuautle, E., Cruz-Hernández, C., & Sanchez-López, C. (2009). Chaotic communication system using Chua's ocillators realized with CCII+s. *International Journal of Bifurcation and Chaos (IJBC)*, 19(12), 4217-4226. doi: 10.1142/S0218127409025304

[14] Trejo-Guerra, R., Tlelo-Cuautle, E., Jiménez-Fuentes, J. M., & Sanchez-López, C. (2011). Multiscroll Floating Gate Based Integrated Chaotic Oscillator. *International Journal of Circuit Theory and Applications*. DOI: 10.1002/cta.821

[15] Bo, L., Yan, W., Zhiping, Y., Leibo, L., Miao, L., Zheng, W., Jing, L. Fernández F.V. (2009). Analog circuit optimization system based on hybrid evolutionary algorithms. *Integr. VLSI J.*, 42(2), 137-148. doi: 10.1016/j.vlsi.2008.04.003

[16] Liu, B., Fernández, F., V., & Gielen, G. G. E. (2011). Efficient and Accurate Statistical Analog Yield Optimization and Variation-Aware Circuit Sizing Based on Computational Intelligence Techniques. *IEEE Trans Comput Aided Des of Integrated Circuits and Systems*, 30(1), 793-805. doi: 10.1109/tcad.2011.2106850

[17] Cristian, F., & Alex, D. (2011). Measuring the uniqueness and variety of analog circuit design features. *Integr. VLSI J.*, 44(1), 39-50. doi: 10.1016/j.vlsi.2010.06.003

[18] Fakhfakh, M., Cooren, Y., Sallem, A., Loulou, M., & Siarry, P. (2010). Analog circuit design optimization through the particle swarm optimization technique. *Analog Integrated Circuits and Signal Processing*, 63(1), 71-82. doi: 10.1007/s10470-009-9361-3

[19] Chatterjee, A., Fakhfakh, M., & Siarry, P. (2010). Design of second-generation current conveyors employing bacterial foraging optimization. *Microelectron. J.*, 41(10), 616-626. doi: 10.1016/j.mejo.2010.06.013

[20] Tlelo-Cuautle, E., Guerra-Gómez, I., de la Fraga, L. G., Flores-Becerra, G., Polanco-Martagón, S., Fakhfakh, M., Reyes-Garcia, C.A. Rodrguez-Gmez, G. Reyes-Salgado, G. (2011). Evolutionary Algorithms in the Optimal Sizing of Analog Circuits in *Intelligent Computational Optimization in Engineering*. In M. Kppen, G. Schaefer & A. Abraham (Eds.), (Vol. 366, pp. 109-138): Springer Berlin / Heidelberg. DOI: 10.1007/978-3-642-2S1705-0_5

[21] Zitzler, E., Deb, K., & Thiele, L. (2000). Comparison of Multiobjective Evolutionary Algorithms: Empirical Results. *Evol. Comput.*, 8(2), 173-195. doi: 10.1162/106365600568202

[22] Deb, K. (1999). Multi-Objective Genetic Algorithms: Problem Difficulties and Construction of Test Problems. *Evolutionary Computation*, 7(3), 205-230. doi: 10.1162/evco.1999.7.3.205

[23] Kalyanmoy, D., Pratap, A., & Agarwal, S. (2002). A Fast and Elitist Multi-Objective Genetic Algorithm: NSGA-II. *IEEE Transactions on Evolutionary Computation*, 6(2), 182-197.

[24] Polanco-Martagón, S., Reyes-Salgado, G., Flores-Becerra, G., Guerra-Gómez, I., Tlelo-Cuautle, E., de la Fraga, L. G., & Duarte-Villaseñor, M. A. (2012). Selection of MOSFET Sizes by Fuzzy Sets Intersection in the Feasible Solutions Space. *Journal of Applied Research and Technology*, 10(1), 1-10.

In: Handbook of Genetic Algorithms: New Research ISBN: 978-1-62081-158-0
Editors: A. Ramirez Muñoz and I. Garza Rodriguez © 2012 Nova Science Publishers, Inc.

Chapter 15

GENETIC ALGORITHMS TO MAXIMIZE THE LYAPUNOV EXPONENT IN CHAOTIC OSCILLATORS

Luis Gerardo de la Fraga[1] *and Esteban Tlelo-Cuautle*[2]
[1] Cinvestav, Computer Science Department
Av. IPN 2508, 07260 Mexico City, Mexico
[2] INAOE, Department of Electronics
Luis Enrique Erro No. 1, Tonantzintla, Pue., 72840 Mexico

Abstract

In this article two different genetic algorithms, one traditional population based with binary representation, and other steady state with real representation, are applied to solve the problem of maximize the Lyapunov exponent in a chaotic oscillator. The studied oscillator is one based on saturated nonlinear function series. We compute the positive Lyapunov exponent oscillators with 2 to 6 scrolls. We show that both genetic algorithms are suitable to maximize the positive Lyapunov exponent. As a result, the phase diagrams show that for a low value of the positive Lyapunov exponent the attractors are well defined, while for its maximum value the attractors are not well appreciated, but the higher value increases the unpredictability grade of the chaotic system. Both algorithms report almost the same results but using the steady state genetic algorithm, a reduction of eight times in execution time is obtained.

1. Introduction

A chaotic oscillator [1] can be generated from the solution of the set of differential equations:

$$\begin{aligned}\dot{x} &= y, \\ \dot{y} &= z, \\ \dot{z} &= -ax - by - cz - d_1 f(x),\end{aligned} \quad (1)$$

function $f(x)$ controls the oscillator and will be defined later.

The unpredictability of the oscillations increases when the Lyapunov exponent of the system in (1) also increases, or in other words, the system is more chaotic as the Lyapunov exponent increases. Therefore, in this work we are interested in the maximization of the

Lyapunov exponent of the chaotic oscillator represented by (1), this is, it is necessary to find the value of the four variables a, b, c and d_1 which maximizes the Lyapunov exponent value. Those four variables can have values within the range $[0.0, 1.0]$. We want also to keep the value for those variables within four decimal places, from 0.0000 to 1.0000 in order to obtain an easy implementation of the oscillator in hardware [2]. The search space for a single variable is $2 \times 10 \times 10 \times 10 \times 10$ (first digit can take two values 0 or 1, and each one of the other four decimal places can take ten values, from 0 to 9), equal to 2×10^4. The total search space is (for four variables) $(2 \times 10^4)^4$, or equal to 1.6×10^9. And the execution time for one run of the evaluation of the Lyapunov exponent is in the order of minutes (due the integration step to solve (1) and to estimate the Lyapunov exponent). Because these two reasons, huge search space and high execution time, the exhaustive search becomes impractical and we decided to use two versions of a genetic algorithm [3], one with the traditional binary representation to encode a solution of the problem with a string of binary variables, and other with real numbers representation [4].

2. Genetic Algorithm

A GA, and in general any population based evolutionary algorithm, starts with a population of individuals (also called chromosomes). The individuals are strings which represent possible solutions of the optimization problem to be solved. One can compute *fitness* of each individual which is a measure of the "goodness" of the individual as a solution to the problem. These individuals are combined to produce a next generation of offspring. The offspring's are reproduced the operations of mutation and cross-over. This process is repeated to get subsequent generations. The general structure of a GA is described in the following pseudocode:

BEGIN-General-GA

1. Randomly create an initial population size P.

2. Do:

    (a) Compute individuals' fitness.

    (b) Select two individuals (called parents) to be reproduced.

    (c) With probability p_c, apply crossover operator between parents.

    (d) Apply the mutation operator to the children with a probability p_m.

    (e) Children will form the next generation population.

    (f) Apply elitism mechanism.

3. While iterations is less than a given number of *generations*.

4. Report the individual with the best fitness.

END-General-GA.

The coding of the representation is one of the most important issues in the GA design. One individual must code one solution of the problem, and for the general GA the code must be a binary string.

Table 1. Caption

Decimal number	Binary number	Grey code
0	0000	0000
1	0001	0001
2	0010	0011
3	0011	0010
4	0100	0110
5	0101	0111
6	0110	0101
7	0111	0100
8	1000	1100
9	1001	1101
10	1010	1111
11	1011	1110
12	1100	1010
13	1101	1011
14	1110	1001
15	1111	1000

We are going to analyze how to code the input variables of our problem. A variable can get values from 0.0000 to 1.0000 (we need four variables to represent the input in our problem), this same variable can be coded with five genes ($g_4 g_3 g_2 g_1 g_0$); the first gene g_4 code the significant digit of a variable, as because this gene can have values of 0 or 1, it is also a binary variable. The other four genes g_3, g_2, g_1 and g_0 must code each one value of the four decimal digits of the variable, therefore each one can have values from 0 to 9 (merging the four genes one can represent values from 0000 to 9999). Now, each of these genes can be represented with four binary variables in the form ($b_3 b_2 b_1 b_0$). All the values that these last genes can have are in the second column of Table 1. As an example, if $g_3 = 9$ have the binary variables values 0101, corresponding each one the values of ($b_3 b_2 b_1 b_0$).

There are two problems in our previous codification: (a) there are values that can have genes g_3-g_0 that are nor useful for us (these values are 10 to 15), and (b) the bits that change from one to other values of the gene: their values change more that in one place for consecutive numbers, for example 7 in binary is 0111, and 8 is 1000, the four binary values change totally for the consecutive numbers 7 and 8!

Now we are going to solve these two problems.

2.1. Chromosome reparation

A genetic algorithm can be used over a codification if there exist a way to repair the chromosomes of its population.

In our problem, the reparation is relatively simple: we need to change the values of invalid binary numbers 10-14 to some valid values within the range 0-9.

Thinking that the invalid numbers 10-14 must fall into an array of numbers 0-9, in our implementation we decided to change the values of invalid numbers as is described

Table 2. Each invalid number in a gene can be replaced by the number in the corresponding second column

Invalid number	Valid number
10	0
11	2
12	4
13	6
14	8
15	9

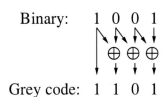

Figure 1. Example to generate the Grey code of binary number 1001.

in Table 2. Of course, many other choices of values in the second column in Table 2 are possible and valid.

2.2. Grey codes

The use of Grey codes solves the problem of bit changing in two consecutive binary numbers, in fact the change is only one single bit between any two consecutive Grey code numbers.

The algorithm to convert any binary number to its Grey code number is as follows: (1) the most significant bit (the most to the left) pass without change, (2) the next bits are calculated with the xor operation between bits b_i and b_{i-1}. Figure 1 shows an example of converting the number 1001 (9 in decimal) to the Grey code 1101.

The genes within the chromosome store their values as Grey code numbers. To repair them, it is not necessary to calculate their respective decimal values, a lookuptable with values as Grey codes (as it is shown in Tab. 3) can be used for this purpose. Actually, Tab. 3 is the same that Tab. 2 but with Grey code values.

As a summary about coding: chromosomes are initialized randomly; then chromosome is repaired and with the repaired chromosome the fitness function is evaluated. After their chromosomes evaluations, selection, crossover and mutation operations are applied to them. Now these operations will be briefly explained.

2.3. Selection, crossover, mutation and elitism operators

After all individuals are evaluated, a selection process takes place to choose the individuals (presumably, the fittest) that will become the parents of the following generation. A variety of selection schemes exist [5], including roulette wheel selection [6], stochastic remainder

Table 3. Values as Grey code number to repair the genes

Grey code	Decimal value	Value to repair the gene
1111	15	0
1110	14	3
1010	10	6
1011	11	5
1001	9	12
1000	8	13

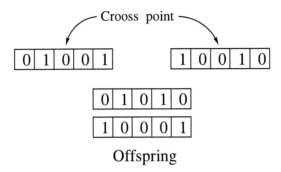

Figure 2. Use of a single-point crossover between two chromosomes. Notice that each pair of chromosomes produces two descendants for the next generation.

selection [7, 8], stochastic universal selection [9, 10], ranking selection [11] and tournament selection.

After being selected, crossover takes place. During this stage, the genetic material of a pair of individuals is exchanged in order to create the population of the next generation. This operator is applied with a certain probability p_c to pairs of individuals selected to be parents (p_c is normally set between 60% and 100%). When using binary encoding, there are three main ways of performing crossover:

1. Single-point crossover: A position of the chromosome is randomly selected as the crossover point as indicated in Fig. 2.

2. Two-point crossover: Two positions of the chromosome are randomly selected as to exchange chromosomic material, as indicated in Fig. 3.

3. Uniform crossover: This operator was proposed by Syswerda [12] and can be seen as a generalization of the two previous crossover techniques. In this case, for each bit in the first offspring it decides (with some probability p_c) which parent will contribute its value in that position. The second offspring would receive the bit from the other parent. This operator is illustrated in Fig. 4. Although for some problems uniform crossover presents several advantages over other crossover techniques [12], in general, one-point crossover seems to be a bad choice, but there is no clear winner between two-point and uniform crossover [13, 14].

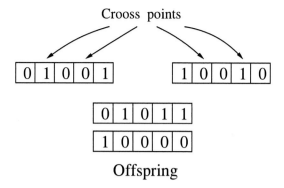

Figure 3. Use of two-point crossover between two chromosomes. In this case the genes at the extremes are kept, and those in the middle part are exchanged.

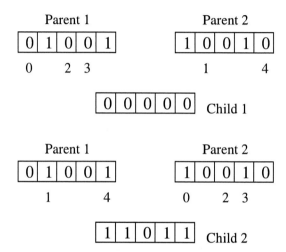

Figure 4. Use of 0.5-uniform crossover (i.e., adopting a 50% probability of crossover) between two chromosomes. The number below the parent's bits indicates to where the value at that position is moved into the child.

In our implementation, tournament selection and two-point crossover were used.

The offspring generated by the crossover operator are subject to mutation, which is a genetic operator that randomly changes a gene of a chromosome. As we use a binary representation, a mutation changes a 0 to 1 and viceversa. This operator is applied with a probability p_m to each binary variable in a chromosome (p_m normally adopts a low probability that goes from 1% up to 10% as a maximum). The use of this operator allows the introduction of new chromosomic material to the population and, from the theoretical perspective, it assures that–given any population–the entire search space is connected [14].

Finally, the individual with the highest fitness in the population is retained, and it passes intact to the following generation (i.e., it is not subject to either crossover or mutation). This operator is called elitism and its use is required to guarantee convergence of a simple GA, under certain assumptions (see [93] for details).

The binary representation for the problem could appears the optimum codifications to solve the problem, but also we want to compare the bGA with the result obtained with a real coded genetic algorithm, such that the one called G3 in [4]; in this GA the variables values will be rounded to the nearest number within four decimal places.

3. G3 Genetic Algorithm

Deb introduced the *generalized generation gap* (G3) model for a genetic algorithm with real representation. This genetic algorithm is called an incremental/steady state genetic algorithm [15, 16], which is different to the generational model described in the previous section, in that there is typically one single new member inserted into the new population at every generation. All the population keep itself all the knowledge of the solution to the problem. A replacement/deletion strategy defines which member of the population will be replaced by the new offspring. The pseudocode of G3 is as follows:

1. Randomly create an initial population of size P.

2. Evaluate each individual.

3. From the population P, select the best parent and other $\mu - 1$ parents randomly.

4. Generate λ offspring from the chosen μ parents using the parent centric recombination (PCX) scheme.

5. Evaluate the two created offspring.

6. Chose two parents at random from the population P.

7. From a combined subpopulation of the two parents chosen in step 3 and the λ offspring created in step 2, choose the best two solutions and replace the chosen two parents with these solutions.

For our results, we choose the suggested values in [4] of $\mu = 3$ and $\lambda = 2$. This means that two parents are selected randomly from the population in step 3, and three parents participate in the recombination.

Here it is important to note the differences between the binary coded GA (bGA) an the real coded GA (G3GA): (1) In a generation of G3GA only two evaluations of the fitness function are performed. In bGA, P evaluations, equal to the population size, need be performed. (2) In G3GA the offspring is generated by a recombination operation with the real values of parents; in bGA, offspring is generated by selection and mutation operands over binary string.

The parent centric recombination (PCX) scheme that uses G3GA is performed as follows:

1. First the mean vector $\vec{g}$, from the three vector parents $\vec{x}_i$ for $i = 1, 2, 3$, is computed.

2. For each offspring

    (a) Choose one parent, $\vec{x}_p$, with equal probability.

(b) Calculate the direction vector $\vec{d}$:

$$\vec{d} = \vec{g} - \vec{x}_p.$$

(c) Compute the perpendicular distances, D_i, of the other two parents to the vector $\vec{d}$

$$D_i = \|\vec{d}\| \sqrt{1 - \left(\frac{\vec{d}_i \cdot \vec{d}}{\|\vec{d}_i\| \|\vec{d}\|}\right)^2}.$$

(d) If some $d_j < \epsilon$, stop the execution; d_j is any element in vectors $\vec{d}$ or $\vec{d}_i$.
(e) Calculate the average of distances $\bar{D} = (D_1 + D_2)/2$.
(f) Calculate offspring, $\vec{y}$, as

$$\vec{y} = \vec{x} + \omega_1 \vec{d} + \vec{v} - \frac{\vec{v} \cdot \vec{d}}{\|\vec{d}\|} \vec{d}, \qquad (2)$$

where each element in vector $\vec{v}$ is calculated as $v_i = \omega_2 D_i$.

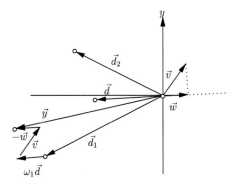

Figure 5. Diagram showing an example of PCX. Vector $\vec{w}$ is the projection of $\vec{v}$ on vector $\vec{d}$, $\vec{w} = \vec{d}(\vec{v} \cdot \vec{d})/\|\vec{d}\|$.

In expressions (2) and to calculate $\vec{v}$, ω_1 and ω_2 are zero mean normally distributed variables with variance σ_1^2 and σ_2^2, respectively. The complexity of PCX operator is linear with respect to the involved number of parents. This procedure creates offspring around the parents with the idea that these solutions are also potential good candidates or are nearer to the local optimum: Figs. 5 is shown how the vectors in PCX are calculated and how the final offspring is around to the bottom-right father in the same Fig. 5; in Fig. 6 500 offsprings are generated using each of the three parents, clearly we can see that offspring are generated around parents.

The fourth term in the right part of Eq. 2 is programmed in the G3GA source code available at http://www.iitk.ac.in/kangal/codes.shtml but it is not described in the article [4]. For the PCX example in Fig. 6 in two dimensions and in our results, the variances were set to $\sigma_1^2 = \sigma_2^2 = 0.01$. The used value for ϵ was 10^{-10}.

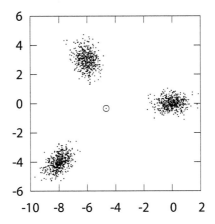

Figure 6. PCX real example: 500 children were generated per parent at positions $(0,0)$, $(-6,3)$ and $(-8,4)$.

4. Chaotic Oscillator

The oscillator represented by the differential equation in Eq. 1 is controlled by a piecewise-linear (PWL) approximation called series of a saturated function f described in Eq. (3) [17], where $k \geq 2$ is the slope of the saturated function and a multiplier factor to saturated plateaus, plateau= $\pm nk$, being n an odd integer to even-scrolls or n an even integer to obtain odd-scrolls. h is the saturated delay of the center of the slopes, and must agree with $h_i = \pm mk$, where $i = 1, \ldots, [(scrolls - 2)/2]$ and $m = 2, 4, \ldots, (scrolls - 2)$ to even-scrolls; and $i = 1, \ldots, [(scrolls - 1)/2]$ and $m = 1, 3, \ldots, (scrolls - 2)$ to odd-scrolls; p and q are positive integers. For examples, to design a four scrolls oscillator

$$f(x;k,h,p,q) = \begin{cases} (2q+1)k & x > qh+1 \\ k(x-ih) + 2ik & |x - ih| \leq 1 \\ & -p \leq i \leq q \\ (2i+1)k & ih+1 < x < (i+1)h-1 \\ & -p \leq i \leq q-1 \\ -(2p+1)k & x < -ph-1 \end{cases} \qquad (3)$$

As an example about how is the form of this PWL function in Eq. (3), in Fig. 7 is shown the PWL function for a 4-scroll oscillator.

4.1. Measuring the Lyapunov coefficient

The Lyapunov exponents give the most characteristic description of the presence of a deterministic nonperiodic flow. Therefore, Lyapunov exponents are asymptotic measures characterizing the average rate of growth (or shrinking) of small perturbations to the solutions of a dynamical system. Lyapunov exponents provide quantitative measures of response sensitivity of a dynamical system to small changes in initial conditions [18]. The number of Lyapunov exponents equals the number of state variables, and if at least one is positive, this is an indication of chaos [18, 2, 19]. That way, an algorithm capable of computing the

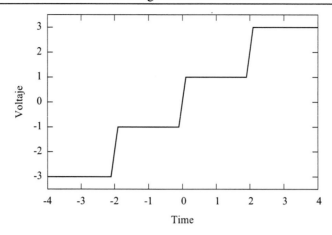

Figure 7. Example of PWL function in Eq. (3) for a 4-scroll oscillator.

Lyapunov exponents in a simple fashion is very much in need to guarantee chaotic regime. In order to measure the Lyapunov exponents, the initial state is set to

$$\vec{y}_0 \in \mathbb{R}^{12}$$
$$\vec{y}_0 = [\vec{x}_0^T, \vec{e}_1^T, \vec{e}_2^T, \vec{e}_3^T]^T$$

where $[\vec{e}_1, \vec{e}_2, \vec{e}_3] = I$, and I is the identity matrix of size 3×3. Thus, $\vec{e}_i$, for $i = 1, 2, 3$, are each unitary column vector of the identity matrix I.

The original system in (1) is observed by attaching other three systems to it. If $\vec{x} = [\dot{x}, \dot{y}, \dot{z}]^T$ represents one state of the system in (1) at any $t > 0$, the state of the new observational system will be $\vec{y} = [\vec{x}, \vec{x}_1, \vec{x}_2, \vec{x}_3]^T$. Note in this last expression for $\vec{y}$, the addition of three new systems $\vec{x}_i$ for $i = 1, 2, 3$ to the original system $\vec{x}$.

The observational system is integrated by several steps until an orthonormalization period T_O is reached. After this, the state of the variational system is orthonormalized by using the standard Gram-Schmidt method [20]. The next integration is carried out by using the new orthonormalized vectors as initial conditions.

The Lyapunov exponents measure the long time sensitivity of the flow in $\vec{x}$ with respect to the initial data $\vec{x}_0$ at the directions of every orthogonalized vector. This measure is taken when the variational system is orthonormalized, if $\vec{y} = [\vec{x}, \vec{p}_1, \vec{p}_2, \vec{p}_3]^T$ is the state after the matrix $[\vec{x}_1, \vec{x}_2, \vec{x}_3]$ is orthonormalized, the Lyapunov exponent λ_i, for $i = 1, 2, 3$ is given by

$$\lambda_i \approx \frac{1}{T} \sum_{j=T_O}^{T} \ln \|\vec{p}_i\| \qquad (4)$$

For instance, in [2] the time step selection was made by using the minimum absolute value of all the eigenvalues of the system λ_{min} [21], and ψ was chosen well above the sample theorem as 50,

$$t_{\text{step}} = \frac{1}{\lambda_{min}\psi}$$

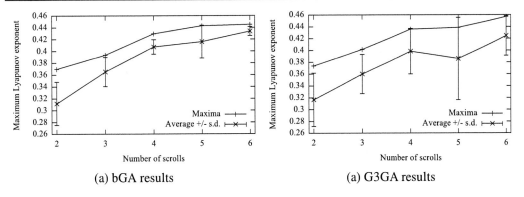

Figure 8. The mean ± standard deviation of 30 runs of both algorithm.

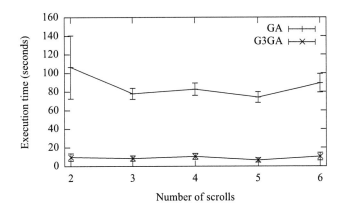

Figure 9. The average execution time (in seconds) of the 30 runs of bGA and G3GA algorithms.

5. Results

The calculation of the Lyapunov exponents for the saturated nonlinear function series based chaotic oscillator described by Eqs. (1) and (3), can be performed by simply setting: $a = b = c = d_1 = 0.7, k = 10, h = 2$ [22, 23]; p and q are adjusted to generate several scrolls [2].

In most of the work reported using saturated nonlinear function series based chaotic oscillator [22, 23, 2] the variables of the system are fixed to 0.7, but the positive Lyapunov exponent is relatively small. Furthermore, in this article we compare this value for the fixed and equal constants with the values obtained with bGA and G3GA to maximize the positive Lyapunov exponent, as shown in Table 4.

bGA was executed with a population size of 40 individuals and during 100 generation. Also the probability mutation was set to 0.15 and probability crossover to 0.8. For G3GA algorithm, we used all the recommended values in [4]: 100 individuals and variances for PCX equal to 0.01. The comparison of statistics for 30 executions of both algorithms are in Fig. 8.

The execution times for both algorithms are shown in Fig. 9 [1]. GAG3 algorithm takes

---
[1]Programs were compiled with gcc and -O2 flags, over a MacBook Pro laptop

around eight times less execution time than the bGA algorithm, and furthermore GAG3 produce slightly better results, as the reader can verify in the results shown in Tab. 4.

Table 4. Maxima Lyapunov exponents and their associated coefficient values for 2-6 scrolls obtained with bGA and G3GA algorithms

	bGA results	
Scrolls	Coefficients	Lyapunov exp.
2	0.2514 0.6850 0.1102 0.7972	0.369845
3	0.7946 0.9263 0.1942 0.8697	0.393587
4	0.7049 0.9882 0.1046 0.6859	0.429539
5	0.5226 0.5471 0.1050 0.5300	0.442864
6	0.8210 0.7521 0.1041 0.8040	0.444982
	G3GA results	
Scrolls	Coefficients	Lyapunov exp.
2	0.4319 0.9825 0.1286 0.8932	0.373822
3	0.7437 0.9378 0.1694 0.6978	0.401654
4	0.4167 0.9566 0.1264 0.4341	0.435834
5	0.8238 0.7386 0.1166 0.8478	0.438589
6	0.2862 0.7284 0.0702 0.2846	0.456974

Figure 10 shows the phase diagram for the cases listed in Table 4. It can be appreciated that the dynamic behavior of the chaotic system is more complex as the positive Lyapunov exponent increases, because it achieves greater unpredictability of the chaotic behavior.

Our bGA implementation was coded in C programming language. We use the G3GA (it is also in C) source code available at http://www.iitk.ac.in/kangal/codes.shtml. The Runge-Kutta method of four order was used to solve (1) and also to calculate the Lyapunov exponents. The integration step was fixed to 0.01. For all simulations the initial condition was set to $\vec{x}_0 = [0.1, 0.0, 0.0]^T$.

6. Conclusions

The positive Lyapunov exponent of PWL-function-based chaotic attractors has been maximized by applying a GA with binary representation (bGA) and a GA with real representation (called GAG3). GAG3 is a steady state GA: two the generated offsepring replace two parents in the population, instead the bGA that is a population based algorithm. In this application, G3GA gives a slightly better results, consuming eight times lesser execution time than the bGA algorithm.

From the obtained results it was observed that the variable c is the most sensitive in the system. Then, by selecting small values for c and keeping a, b, d with large values one obtains higher positive Lyapunov exponents.

Genetic Algorithms to Maximize the Lyapunov Exponent in Chaotic Oscillators

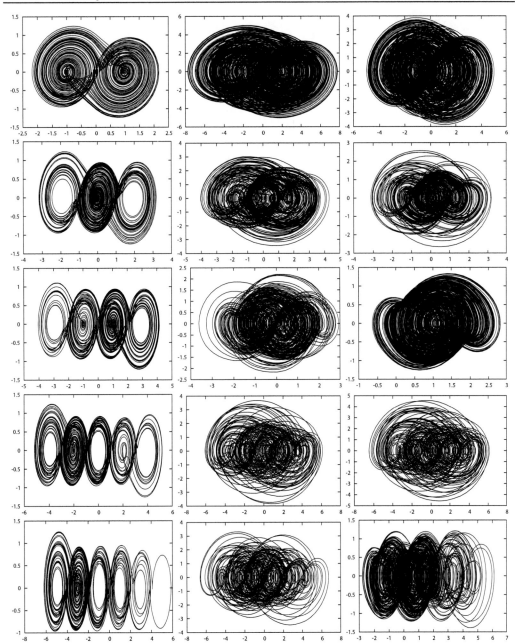

Figure 10. Attractors for the optimized oscillators: First column with fixed constants, second column shows GA results, and third column shows G3GA results; row 1 to 5 show results for 2, 3, 4, 5 and 6 scrolls, respectively.

References

[1] J. Lu, G. Chen, and X. Yu. Generating chaotic attractors with multiple merged basins of attactions: a switching piecewise-linear control approach. *IEEE Trans. Circuits Syst.-I*, 50:198–207, 2003.

[2] E. Trejo-Guerra, E. Tlelo-Cuautle, J.M. Muñoz-Pacheco, C. Sánchez-López, and C. Cruz-Hernández. On the relation between the number of scrolls and the Lyapunov exponents in PWL-functions-based n-scroll chaotic oscillators. *International Journal of Nonlinear Sciences & Numerical Simulations*, 11(11):903–910, Nov 2010.

[3] D.B. Fogel. An introduction to simulated evolutionary optimization. *IEEE Transactions on Neural Networks*, 5(1):3–14, 1994.

[4] K. Deb, A. Anand, and D. Joshi. A computationally efficient evolutionary algorithm for real-parameter optimization. *Evolutionary Computation*, 10(4):371–395, 2002.

[5] D.E. Golberg and K. Deb. *Foundations of Genetic Algorithms*, chapter A comparison of selection schemes used in genetic algorithms, pages 69–93. Morgan Kaufmann, San Mateo, California, 1991.

[6] A.K. De Jong. *An Analysis of the Behavior of a Class of Genetic Adaptive Systems*. PhD thesis, University of Michigan, Ann Arbor, Michigan, USA, 1975.

[7] L.B. Booker. *Intelligent Behavior as an Adaptation to the Task Environment*. PhD thesis, University of Michigan, AnArbor, Michigan, USA, 1982. Logic of Computers Group.

[8] A. Brindle. *Genetic Algorithms for Function Optimization*. PhD thesis, University of Alberta, Edmonton, Alberta, Canada, 1981. Department of Computer Science.

[9] J.E. Baker. Reducing bias and inefficiency in the selection algorithm. In J.J. Grefenstette, editor, *Proceedings of the Second International Conference on Genetic Algorithms*, pages 14–22, July 1987.

[10] J.J. Grefenstette and J.E. Baker. How genetic algorithms work: A critical look at implicit parallelism. In J.D. Schaffer, editor, *Proceedings of the Third International Conference on Genetic Algorithms*, pages 20–27, June 1989.

[11] J.E. Baker. Adaptive selection methods for genetic algorithms. In J.J. Grefenstette, editor, *Proceedings of the First International Conference on Genetic Algorithms*, pages 101–111, 1985.

[12] G. Syswerda. Uniform crossover in genetic algorithms. In J.D. Schaffer, editor, *Proceedings of the Third International Conference on Genetic Algorithms*, pages 2–9, 1989.

[13] M. Mitchell. *An Introduction to Genetic Algorithms*. The MIT Press, Cambridge, Massachusetts, 1996.

[14] Z. Michalewicz. *Genetic Algorithms + Data Structures = Evolution Programs*. Springer-Verlag, New York, USA, 3rd edition, 1996.

[15] A. Rogers and A. Prügel-Bennet. *Foundations of Genetic Algorithms 5*, chapter Modeling the dynamics of a steady state genetic algorithm, pages 57–68. Springer, 1999.

[16] F. Vavak and T.C. Fogarty. Comparison of steady state and generational genetic algorithms for use in nonstationary environments. In *Proceedings of IEEE International Conference on Evolutionary Computation 1996*, pages 192–195, May 1996.

[17] J. Lu and G. Chen. Generating multiscroll chaotic attractors: Theories, methods and applications. *International Journal of Bifurcation and Chaos*, 16:775–858, 2006.

[18] L. Dieci. Jacobian free computation of Lyapunov exponents. *Journal of Dynamics and Differential Equations*, 14(3):697–717, 2002.

[19] T.S. Parker and L.O. Chua. *Practical Numerical Algorithms for Chaotic Systems*. Springer-Verlag, NY, 1989.

[20] G.H. Golub and C.V. Loan. *Matrix Computations*. The Johns Hopkins University Press, 3rd ed, 1996.

[21] E. Tlelo-Cuautle, J.M. Muñoz-Pacheco, and J. Martínez-Carballido. Frequency-scaling simulation of Chua's circuit by automatic determination and control of step-size. *Applied Mathematics and Computation*, 194:486–491, 2007.

[22] J. Lu, S. Yu, H. Leung, and G. Chen. Experimental verification of multidirectional multiscroll chaotic attractors. *IEEE Transactions on Circuits and Systems I*, 56:149–165, 2006.

[23] J.M. Muñoz-Pacheco and E. Tlelo-Cuautle. *Electronic Design Automation of Multiscroll Chaos Generators*. Bentham Science Publishers Ltd, USA, 2010.

In: Handbook of Genetic Algorithms: New Research
Editors: A. Ramirez Muñoz and I. Garza Rodriguez © 2012 Nova Science Publishers, Inc.

ISBN: 978-1-62081-158-0

Chapter 16

APPLICATION OF PARTICLE SWARM OPTIMIZATION TO PACKING PROBLEM

Eisuke Kita[*] *and Young-Bin Shin*
Graduate School of Information Science,
Nagoya University, Japan

Abstract

Packing problem is a class of optimization problems which involve attempting to pack the items together inside a container, as densely as possible. This research focuses on the application of particle swarm optimization (PSO) for solving two-dimensional packing problems at the arbitrary polygon-shaped packing region. Total number of items and the position vector of the item center are taken as the design variables. Then, total number of the items is maximized when all objects are included inside a two-dimensional domain without their overlapping. The problem is solved by two algorithms; standard and improved PSOs. In the standard PSO, the particle position vector is updated by the best particle position in all particles (global best particle position) and the best position in previous positions of each particle (personal best position). The improved PSO utilizes, in addition to them, the second best particle position in all particles (second global best particle position) in the stochastic way. In the numerical example, the algorithms are applied for three problems. The results show that the improved PSO can pack more items than the standard PSO and success rate is also improved.

Keywords: Packing Problem, Particle Swarm Optimization, Global Best Position, Second Global Best Position, Personal Best Position.

1. Introduction

Packing problems are a class of optimization problems in mathematics which involve attempting to pack objects together (often inside a container), as densely as possible. There are many variations of this problem, such as two-dimensional packing, linear packing, packing by weight, packing by cost, and so on. They have many applications, such as filling up

---
[*]E-mail address: kita@is.nagoya-u.ac.jp

containers, loading trucks with weight capacity, creating file backup in removable media and technology mapping in Field-programmable gate array semiconductor chip design.

We focus on the two-dimensional packing problems. Popular problems in two-dimensional packing are to packing circles or squares in a circle or a square. The problems are studied analytically and the maximum numbers of items are determined[1, 2, 3]. In this study, we consider that the packing regions are the arbitrary polygon-shaped packing region and then, same items are packed in the region without their overlap. The typical example in the steel industry is to stamp same polygonal figures from a rectangular board. The aim of this job is to minimize the remainder region on board. Since the packing problem is one of typical NP-hard problems, it is quite difficult to find optimal solution in the polynomial time.

For solving NP-hard problems, some researchers have applied evolutionary computations such as Genetic Algorithm (GA)[4], Simulated Annealing (SA)[5], Particle Swarm Optimization (PSO)[6] and so on. In this study, we will apply PSO for solving two-dimensional packing problems. PSO, which has been presented in 1995 by Kennedy and Eberhart[6], is based on a metaphor of social interaction such as bird flocking and fish schooling. PSO is a population-based optimization algorithm, which could be implemented and applied easily to solve various function optimizations problem, or the problems that can be transformed to the function minimization or maximization problem.

The application of PSO for solving packing problem has been presented by some researchers[7, 8, 9, 10]. Liu et al.[7] presented evolutionary PSO for solving bin packing problem. Zhao et al.[8, 9] applied the discrete PSO for solving rectangular packing problem. Thapatsuwan et al.[10] compared GA and PSO for solving multiple container packing problems. They focus on the packing problem of container in the storage or the ship cabin. Since the storage and the ship cabin are designed so that their sizes are equal to the integral multiple of the container sizes, it is assumed that the items are placed every certain intervals. On the other hand, we will consider that the packing region is arbitrarily polygon-shaped. Since, in this case, the packing region sizes do not depend on the item sizes, the problems to be solved are much more difficult than the previous studies.

In this study, PSO is applied for solving the packing problems which have arbitrarily polygon-shaped regions. The design objective is to maximize the total number of the items packed in the region without the item overlap. The total number of items and the position vectors of the item centers are taken as the design variables. The problem is solved by the standard and the improved PSOs. In the PSO, the potential solutions of the optimization problem to be solved are defined as the particle position vectors. Then, the particle positions are updated by PSO update rules. In the standard PSO, the particle position vector is updated by the best position of all particles (global best position) and the local best position in previous positions of each particle (personal best position). The improved PSO utilizes, in addition to them, the second best position of all particles (second global best position)[11].

The remaining part of this paper is organized as follows. The PSO algorithms and the optimization problem are explained in section 2. and 3., respectively. In section 4., the packing problem in two-dimensional regions is solved. Finally, the conclusions are summarized again in section 5..

2. PSO Formulation

2.1. Standard PSO

2.1.1. Update Rule

Particle swarm optimization (PSO) is one of the latest evolutionary computations presented by Eberhart and Kennedy[6]. PSO mimics social interaction of animals

In the PSO algorithm, the particles represent potential solutions of the optimization problem. Each particle in the swarm has a position vector $\boldsymbol{x}_i(t)$ ($i = 1, 2, \ldots, N$) and the velocity vector $\boldsymbol{v}_i(t)$ in the search space at time t. The particle position vector is defined by the design variable set of the optimization problem. Each particle also has memory and hence, can remember the best position in search space it ever visited. The satisfaction of the particle i for the design objective is estimated by the objective function or the fitness function $f(\boldsymbol{x}_i(t))$.

The position at which each particle takes the best fitness function is known as the personal best position $\boldsymbol{x}_i^{pbest}(t)$ and the overall best out of all particles in the swarm is as global best position $\boldsymbol{x}^{gbest}(t)$. The velocity and the position are updated according to the following formulas

$$\boldsymbol{x}_i(t+1) = \boldsymbol{x}_i(t) + \boldsymbol{v}_i(t+1) \tag{1}$$

$$\boldsymbol{v}_i(t+1) = w \cdot \boldsymbol{v}_i(t) + c_1 \cdot rand(1) \times (\boldsymbol{x}_i^{pbest} - \boldsymbol{x}_i(t)) \\ + c_2 \cdot rand(2) \times (\boldsymbol{x}^{gbest} - \boldsymbol{x}_i(t)) \tag{2}$$

where w is the inertia weight, c_1 and c_2 are acceleration coefficient, and t is the iteration time and, $rand(1)$ and $rand(2)$ are random numbers in the interval $[0, 1]$.

The inertia weight w governs how much of the velocity should be retained from the previous time step. Generally the inertia weight is not fixed but varied as the algorithm progresses. The inertia weight w, in this study, is generally updated by self-adapting formula as

$$w = w_{\max} - (w_{\max} - w_{\min}) \times \frac{t}{t_{\max}} \tag{3}$$

where the parameter $w_{\max}$ and $w_{\min}$ denote the maximum and minimum inertia weight, respectively. The parameter t and $t_{\max}$ are the iteration step and the maximum iteration steps in the simulation, respectively.

The parameters c_1 and c_2 determine the relative pull of $\boldsymbol{x}_i^{pbest}$ and $\boldsymbol{x}^{gbest}$. According to the recent work done by Clerc[12], the parameters are given as

$$c_1 = c_2 = 1.5. \tag{4}$$

2.1.2. Algorithm

Total number of the particles in the swarm is N. The PSO process is as follows.

Step 1. Randomly initialize the position and velocity vectors of particles.

Step 2. Set $t = 1$.

Step 3. For $i = 1, \ldots, N$:

    Step a. Evaluate fitness function $f(\boldsymbol{x}_i(t))$ for the particle i.

    Step b. If $f(\boldsymbol{x}_i(t)) > f(\boldsymbol{x}_i^{pbest})$, set $\boldsymbol{x}_i^{pbest} = \boldsymbol{x}_i(t)$.

Step 4. Find the best particle $\boldsymbol{x}^1$ among $\boldsymbol{x}_i(t)$ $(i = 1, 2, \ldots, N)$ according to their fitness.

Step 5. If $\boldsymbol{x}^1 > \boldsymbol{x}^{gbest}$, set $\boldsymbol{x}^{gbest} = \boldsymbol{x}^1$.

Step 6. For $i = 1, \ldots, N$, update the velocity and position vectors of the particles according to equations (2) and (1), respectively.

Step 7. Set $t = t + 1$ and go to step 3 if $t \leq t_{\max}$.

2.2. Improved PSO

2.2.1. Update Rule With Second Global Best Position

The standard PSO have no handling mechanism for the local optimization. The standard PSO reduce the chance of local optimization to make use of $\boldsymbol{x}_i^{pbest}$ in one way. In this section, we introduce improved PSO that each particle remembers the position of the second global best particle $\boldsymbol{x}^{sgbest}$. In using $\boldsymbol{x}^{sgbest}$, PSO attempts to search diversity and change movement of particles. The velocity and position of the improved PSO are updated according to the following formulas.

$$\boldsymbol{x}_i(t+1) = \boldsymbol{x}_i(t) + \boldsymbol{v}_i(t+1) \tag{5}$$
$$\begin{aligned}\boldsymbol{v}_i(t+1) = & w \cdot \boldsymbol{v}_i(t) + c_1 \cdot rand(1) \times (\boldsymbol{x}_i^{pbest} - \boldsymbol{x}_i(t)) \\ & + c_2 \cdot rand(2) \times (\boldsymbol{x}^{gbest} - \boldsymbol{x}_i(t)) \\ & + c_3 \cdot rand(3) \times (\boldsymbol{x}^{sgbest} - \boldsymbol{x}_i(t)) \end{aligned} \tag{6}$$

where w is the inertia weight, c_1, c_2 and c_3 are acceleration coefficient, and t is the iteration time, and $rand(1), rand(2)$ and $rand(3)$ are random numbers distributed in the interval $[0, 1]$. The parameter c_1 and c_2 are taken as the same values in the standard PSO; $c_1 = c_2 = 1.5$. The parameter c_3 is determined from some numerical experiments, which is given as $c_3 = 1.9$.

2.2.2. Algorithm

The improved PSO share the information of $\boldsymbol{x}_i^{pbest}$, $\boldsymbol{x}^{gbest}$ and $\boldsymbol{x}^{gsbet}$. Obviously, $\boldsymbol{x}^{sgbest}$ is worse than $\boldsymbol{x}^{gbest}$. If only equation (6) is used for updating particle positions, the obtained result is similar to worse than that of standard PSO. Therefore, both update rules of the standard and improved PSOs are used in the present algorithm. Here, we present the following update rule in which the PSO of equation (2) and the PSO of equation (6) are randomly used in probability P_s. The process is as follows:

Step 1. Randomly initialize the position and velocity vectors of particles.

Step 2. $t = 1$.

Step 3. For $i = 1, 2, \ldots, N$:

   Step a. Evaluate fitness function $f(x_i(t))$ for the particle i.
   Step b. If $f(x_i(t)) > f(x_i^{pbest})$, $x_i^{pbest} = x_i(t)$.

Step 4. Find the first best particle x^1 and the second best particle x^2 from $x_i(t)$ ($i = 1, 2, \ldots, N$).

Step 5. Update x^{gbest} and x^{sgbest}:

   Step a. If $x^1 > x^{gbest}$, set $x^{gbest} = x^1$.
   Step b. If $x^1 < x^{gbest}$ and $x^1 > x^{sgbest}$, set $x^{sgbest} = x^1$.
   Step c. If $x^2 > x^{sgbest}$, set $x^{sgbest} = x^2$.

Step 6. Generate a random number r in the interval $[0, 1]$.

Step 7. If $r > P_s$, update the velocity and position vectors of all particles according to equations (2) and (1), respectively.

Step 8. If $r \leq P_s$, update the velocity and position vectors of all particles according to equations (6) and (5), respectively.

Step 9. Set $t = t + 1$ and go to step 3 if $t \leq t_{\max}$.

3. Packing Problem

3.1. Optimization Problem

The packing problem can be formulated to maximize the number of items N included into a two-dimensional polygonal region P as follows:

$$\begin{aligned}
\max \quad & z & (7)\\
\text{subject to} \quad & g_1(i, P) = 0 \\
& g_2(i, j) = 0 \\
& 0.5w \leq p_x^i \leq W - 0.5w \\
& 0.5h \leq p_y^i \leq H - 0.5h \\
& i = 1, 2, \ldots, z; j = 1, 2, \ldots, z
\end{aligned}$$

where the vector $\{p_x^i, p_y^i\}$ denotes the center position vector of the item i, w and h are item sizes, and W and H are feasible space sizes. The function $g_1(i, P)$ evaluates the inclusion of the item i in the region P, which is defined as follows:

$$g_1(i, P) = \begin{cases} 0 & \text{The item } i \text{ is perfectly included in the region } P. \\ 1 & \text{The item } i \text{ is not included in the region } P \text{ at all.} \end{cases} \quad (8)$$

The function $g_2(i, j)$ evaluates the overlap between the item i and the item j, which is defined as follows:

$$g_2(i, j) = \begin{cases} 0 & \text{The item } i \text{ and } j \text{ are not overlapped.} \\ 1 & \text{The item } i \text{ and } j \text{ are overlapped.} \end{cases} \quad (9)$$

3.2. PSO Implementation

When the number of the items is given, PSO is applied for solving the item packing problem within the packing region without violating the constraint conditions. The optimization problem is defined as follows.

- Fitness function

$$f(\boldsymbol{x}_i) = \frac{1}{1 + \sum_{i=1}^{z} \left\{ g_1(i, P) + \sum_{j=1}^{z, i \neq j} g_2(i, j) \right\}} \quad (10)$$

- Design variable vector

$$\boldsymbol{x}_i = \{p_x^1, p_y^1, \ldots, p_x^i, p_y^i, \ldots, p_x^z, p_y^z\}^T \quad (11)$$

- Side constraint for design variable

$$0.5w \leq p_x^i \leq W - 0.5w \quad (i = 1, 2, \ldots, z)$$
$$0.5h \leq p_y^i \leq H - 0.5h \quad (i = 1, 2, \ldots, z) \quad (12)$$

3.3. Optimization Process

The process of the present algorithm is shown in Fig.1 and summarized as follows.

Step 1. Set $z = 0$.

Step 2. Set the values of the threshold P_s and the maximum step size $t_{\max}$.

Step 3. Update the item number by $z = z + 1$.

Step 4. Perform PSO algorithm:

  (a) Set $t = 0$

  (b) Randomly initialize the random position and velocity vectors of particle distributed over the design space.

  (c) Evaluate the fitness function for each particle $f(\boldsymbol{x}_i)$.

  (d) Determine the particles $\boldsymbol{x}_i^{pbest}$, $\boldsymbol{x}^{gbest}$ and $\boldsymbol{x}^{sgbest}$.

  (e) If $f(\boldsymbol{x}^{gbest}) = 0$, go to Step 3.

  (f) Generate a random number r in the interval $[0, 1]$.

  (g) If $r \leq P_s$, update particles by equation (6), otherwise updated by equation (2).

  (h) Set $t = t + 1$.

  (i) If $t \leq t_{max}$, return to Step 4c.

Step 5. If $g_1(i, P) = g_2(i, j) = 0$, return to Step 3.

Step 6. Stop by $z = z - 1$.

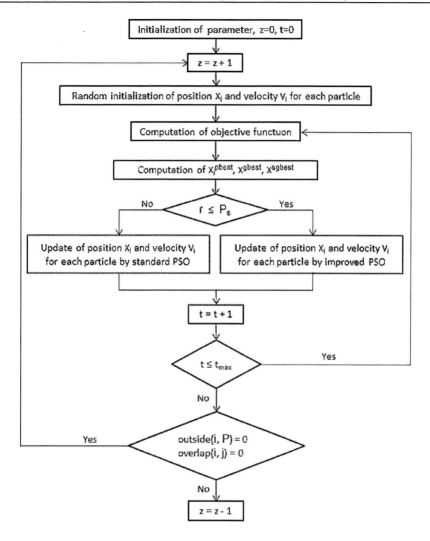

Figure 1. Flowchart of the improved PSO algorithm for solving 2D packing problems.

Table 1. PSO parameters

Number of particles	$N = 200$
Maximum iteration step	$t_{max} = 2000$
Parameters	$w_{\max} = 0.9, w_{\min} = 0.4, c_1 = 1.5, c_2 = 1.5, c_3 = 1.9$

4. Numerical Examples

4.1. Case A

The packing problem in two-dimensional polygonal regions is considered as a numerical example. The packing region of case A is shown in Figure 2. PSO parameters are shown

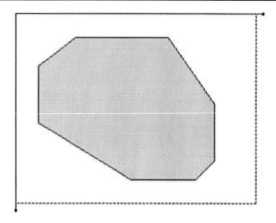

Figure 2. Packing region in test case A.

Table 2. Comparison of final results obtained with standard and improved PSOs in case A.

	Standard PSO	Improved PSO
Average value of packed item number	11.144	12.984
Average CPU time (seconds)	35.007	59.753
Success rate in $z_{\max} \geq 14$	18.4%	73.2%

in Table 1. Number of particles and maximum iteration steps are specified as $N = 200$ and $t_{\max} = 2000$, respectively. The other parameters are taken as $w = 0.9$, $c_1 = 1.5$, $c_2 = 1.5$, $c_3 = 1.9$, and $P_s = 0.1$.

Five hundred simulations are performed from different initial conditions. Maximum item numbers for case A are shown in Figure 3. The figures are plotted with the run number as the horizontal axis and the item number z as the vertical axis, respectively.

The results by the standard and the improved PSOs are compared in Table 2. The average item number and the average CPU time denote the average values of the maximum item numbers and CPU time in five hundred runs, respectively. The success rate means the percentage of the runs in which the maximum item number $z_{\max}$ is greater than 13. The average item number is 11.144 in case of the standard PSO and 12.984 in improved PSO. The average CPU time is 35.007 and 59.753, respectively. The success rate is 18.4% and 73.2%, respectively. The use of the improved PSO can increase the item number and

Table 3. Effect of selection probability P_s in case A

P_s	0.1	0.2	0.3	0.4	0.5	0.6
Average item number z	12.97	12.81	12.69	12.06	12.72	11.46
Average CPU time (seconds)	58.82	67.11	64.99	72.47	86.94	144.53

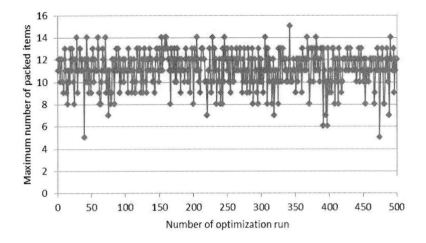

(a) Standard PSO

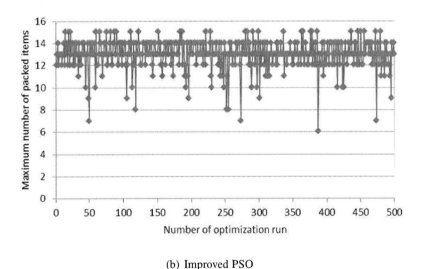

(b) Improved PSO

Figure 3. Maximum item numbers in case A.

improve the success rate although the CPU time is increased.

Figure 4 shows the fitness function $f(x^{gbest})$ at $z = 8, 10, 12, 14$ and 15. In case of item number $z = 8, 10, 12$ and 14, fitness functions almost converge to 1 at 400, 700, 800 and 1000 iterations, respectively. In case of item number $z = 15$, fitness cannot converge to 1. Therefore, in this case, maximum item number is concluded to be $z = 14$. Figure 5 shows the item placement in case of the improved PSO. We can observe from Figure 5 that the items overlap in case of $z = 15$.

Next, the effect of the parameter P_s is discussed. Table 3 shows the maximum number of items and the CPU times for the different parameter P_s. The results show that the item number is maximized at $P_s = 0.1$ and CPU time is also shortest.

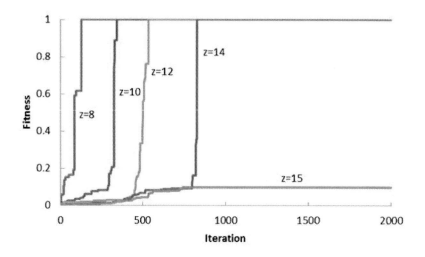

Figure 4. Fitness convergence of improved PSO in case A.

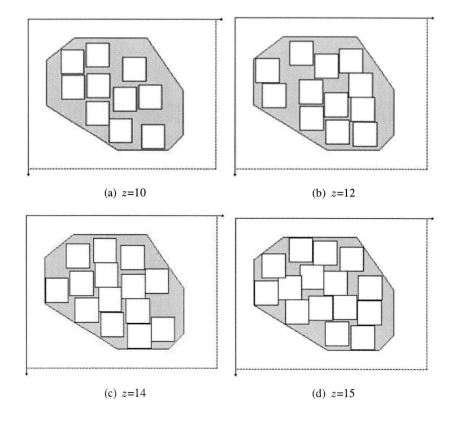

Figure 5. Item placement by improved PSO in case A.

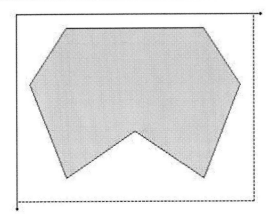

Figure 6. Packing region in Case B.

Table 4. Comparison of standard and improved PSO in case B

	standard PSO	Improved PSO
Average value of packed item number z	12.07	13.99
Average CPU time (seconds)	35.198	65.124
Success rate in $z \geq 14$	19.8%	75.2%

4.2. Case B

The packing region of case B is shown in Figure 6. PSO parameters are identical to the case A (Table 1). Number of particles and maximum iteration steps are specified as $N = 200$ and $t_{max} = 2000$, respectively. The other parameters are taken as $w = 0.9$, $c_1 = 1.5$, $c_2 = 1.5$, $c_3 = 1.9$, and $P_s = 0.1$.

The results are shown in Figure 7 and Table 4. The average item number is 12.07 in case of the standard PSO and 13.99 in improved PSO. The average CPU time is 35.198 and 65.124, respectively. The success rate is 19.8% and 75.2%, respectively. The use of the improved PSO can increase the item number improve the success rate although the CPU time is increased.

Next, the effect of the parameter P_s to the convergence property is discussed. The maximum number of items and the CPU times for the different parameter P_s are listed in

Table 5. Effect of selection probability P_s in case B

P_s	0.1	0.2	0.3	0.4	0.5	0.6
Average item number	14.1	13.83	13.97	13.76	13.37	13.39
Average CPU time (seconds)	66.57	77.00	81.80	85.56	95.18	108.65

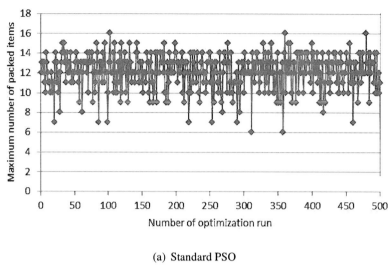

(a) Standard PSO

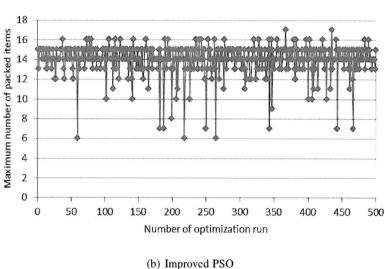

(b) Improved PSO

Figure 7. Maximum item numbers in Case B.

Table 5. The results show that, at $P_s = 0.1$, the item number is largest and CPU time is shortest.

5. Conclusions

PSO solution of the two-dimensional packing problem was presented in this study. Since the storage and the ship cabin are designed so that their sizes are equal to the integral multiple of the container sizes, it is assumed that the items are placed every certain intervals. We considered in this study that the packing region is arbitrarily polygon-shaped. The problem was solved by the standard and improved PSOs. In the standard PSO, the particle

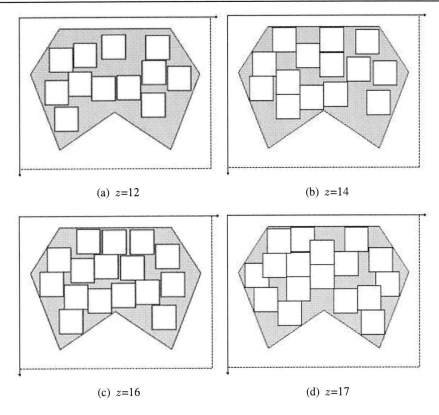

Figure 8. Item placement by improved PSO in case A.

position vectors are updated by the global and the personal best positions. The improved PSO utilizes, in addition to them, the second global best position of all particles. The use of the second global best position is determined in the probabilistic way.

The algorithms were compared in two numerical examples. The design objective is to maximize the number of items contained in the packing region without the item overlap. The results showed that the improved PSO algorithm could find better solutions than the standard PSO. The maximum item number in the improved PSO is bigger by one or two items than that in the standard PSO. In case of the improved PSO, the CPU time and the maximum item number depends on the probability to switch the standard PSO and the PSO with second global best position of particles. Therefore, in the next study, we would like to discuss the adequate parameter design.

References

[1] Hallard T Croft, Falconer Kenneth J., and Guy Richard K. *Unsolved Problems in Geometry*. Springer-Verlag, 1991.

[2] J. Melissen. Packing 16, 17 or 18 circles in an equilateral triangle. *Discrete Mathematics*, 145:333–342, 1995.

[3] Erich Friedman. Packing unit squares in squares: a survey and new results. *The Electronic Journal of Combinatorics*, DS7, 2005.

[4] J.H. Holland. *Adaptation in Natural and Artificial Systems*. University of Michigan Press, 1st. edition, 1975.

[5] S. Kirkpatrick, C.D. Gelatt Jr., and M.P. Vecchi. Optimization by simulated annealing. *Science*, 220(4598):671–680, 1983.

[6] J. Kennedy and R.C. Eberhart. Particle swarm optimization. In *Proceedings of IEEE the International Conference on Neural Networks*, volume 6, pages 1942–1948, 1995.

[7] D.S. Liu, K.C. Tan, S.Y. Huang, C.K. Goh, and W.K. Ho. On solving multiobjective bin packing problems using evolutionary particle swarm optimization. *European Journal of Operational Research*, 190(2):357 – 382, 2008.

[8] Chen Zhao, Liu Lin, Cheng Hao, and Liu Xinbao. Solving the rectangular packing problem of the discrete particle swarm algorithm. In *Business and Information Management, 2008. ISBIM '08. International Seminar on*, volume 2, pages 26 –29, 2008.

[9] Chuan He, Yuan-Biao Zhang, Jian-Wen Wu, and Cheng Chang. Research of three-dimensional container-packing problems based on discrete particle swarm optimization algorithm. In *Test and Measurement, 2009. ICTM '09. International Conference on*, volume 2, pages 425 –428, dec. 2009.

[10] P. Thapatsuwan, J. Sepsirisuk, W. Chainate, and P. Pongcharoen. Modifying particle swarm optimisation and genetic algorithm for solving multiple container packing problems. In *Computer and Automation Engineering, 2009. ICCAE '09. International Conference on*, pages 137 –141, march 2009.

[11] Ryan Forbes and Mohammad Nayeem Teli. Particle swarm optimization on multi-funnel functions.

[12] M. Clerc. The swarm and the queen: towards a deterministic and adaptive particle swarm optimization. In *Proceedings of 1999 Congress on Evolutionary Computation*, volume 3, pages 1951–1957, 1999.

In: Handbook of Genetic Algorithms: New Research ISBN: 978-1-62081-158-0
Editors: A. Ramirez Muñoz and I. Garza Rodriguez © 2012 Nova Science Publishers, Inc.

Chapter 17

APPLICATION OF ADVANCED GRAMMATICAL EVOLUTION TO FUNCTION IDENTIFICATION PROBLEM

Eisuke Kita[*] *and Hideyuki Sugiura*
Graduate School of Information Science, Nagoya University, JAPAN

Abstract

The aim of the function identification problem is to find the unknown function representation for the given data set. Grammatical Evolution is one of the evolutionary computations which can find the function representation by using the one-dimensional chromosome and the translation rule described in Backus Naur Form (BNF). This paper describes the application of an advanced Grammatical Evolution (GE) to function identification problem. The advanced Grammatical Evolution uses two-dimensional chromosome, instead of the one-dimensional chromosome employed in the original GE. The continuous and discontinuous functions are taken as the numerical examples. The results show that Grammatical Evolution with one-dimensional chromosome can find the continuous function faster than the Genetic Programming, and that the advanced Grammatical Evolution with two-dimensional chromosome is more effective than that with one-dimensional chromosome for finding the discontinuous function.

Keywords: Grammatical Evolution (GE), Function Identification, Two-Dimensional Chromosome, Polynomial Function, Step Function, Genetic Programing (GP).

1. Introduction

Evolutionary computations are techniques implementing mechanisms inspired by biological evolution such as reproduction, mutation, recombination, natural selection and survival of the fittest; Genetic Algorithms (GA) [1, 2, 3], Simulated Annealing (SA) [4], Evolutionary Programming [5], Genetic Programming (GP) [6, 7], Particle Swarm Optimization [8, 9] and so on. In this study, we will focus on Grammatical Evolution (GE) [10, 11, 12].

---
[*]E-mail address: kita@is.nagoya-u.ac.jp

Genetic Algorithms (GA) [1, 2, 3] is very popular algorithm in the evolutionary computation field. In GA [1, 2, 3], a population of chromosomes of candidate solutions to an optimization problem evolves toward better solutions by applying the genetic operators such as selection, crossover, mutation and so on. On the other, Genetic Programming and Grammatical Evolution are designed for the different object. Their object is to find function representations for the unknown data sets and computer programs that perform a user-defined task.

Genetic Programing [6, 7] represents computer programs in memory as tree structures and then, evolves the tree structures toward the programs or functions to be desired by applying the genetic operators. The genetic operators often generates the tree structures which are invalid from the view-point of the syntax of functions or programs. For overcoming this difficulty, Grammatical Evolution uses the translation rules defined in the Backus Naur Form (BNF). The original GE [10, 11, 12] starts from the syntax definition in Backus Naur Form (BNF) which translates chromosomes to functions or programs. Chromosome in binary number is translated to that in decimal number every few bits. The rule to be used are selected from the BNF syntax list according to the remainder of the decimal numbers with respect to the total number of candidate rules. The original Grammatical Evolution still has one difficult when it is applied for finding the discontinuous function and the program with the conditional branching such as "if-else" statement. Since the conditional branching cannot defined explicitly in the BNF syntax, the individuals described in an invalid grammar are often generated and therefore, the search performance becomes worse. For overcoming this difficulty, we adopt the two-dimensional chromosome. In the two-dimensional chromosome, the "if-else" statement, the condition statements, and the executing statements are defined in the different rows of the chromosome. Therefore, the conditional branching structure is not destroyed through the search process. The validity is discussed in the function identification problem of the function with the conditional branching.

The remaining part of this paper is as follows, The original GE and the advanced GE are explained in section 2. and 3., respectively. In section 4., numerical results are shown. Finally, the discussion is summarized in section 5..

2. Grammatical Evolution

2.1. Search Process

The algorithm of the original Grammatical Evolution is as follows.

1. Define a syntax which translates genotype (chromosome) to phenotype (function) in Backus Naur Form (BNF).

2. Randomly generate initial individuals.

3. Translate chromosomes to functions according to the syntax.

4. Estimate fitness functions of chromosomes.

5. Apply selection, crossover and mutation operators to update the population.

Table 1. BNF syntax of simple example

(A)	`<expr> ::= <expr><op><expr>`	(A0)
	`| <num>`	(A1)
	`| <var>`	(A2)
(B)	`<op> ::= +`	(B0)
	`| -`	(B1)
	`| *`	(B2)
	`| /`	(B3)
(C)	`<x> ::= x`	(C0)
	`| y`	(C1)
(D)	`<num> ::= 1`	(D)

6. Terminate the process if the criterion is satisfied.

7. Go to step 3.

Except for BNF syntax definition and the translation from chromosomes to functions, the Grammatical Evolution algorithm is same as the Genetic Algorithm.

The translation from genotype (chromosome) to phenotype (function) is performed as follows.

1. Translate a chromosome from binary number to decimal number every n-bits.

2. Define a leftmost decimal as β.

3. Define a leftmost no-terminal symbol as α and the number of candidate rules related to α as n_α.

4. Calculate the remainder γ of the number β with respect to n_α.

5. Replace α with γ-th rule in the rules for α.

6. Go to step 2 if terminal symbols still exits.

2.2. Translation From Chromosome to Function

We would like to explain the translation from chromosome to function by a simple example.

The BNF syntax is shown in Table 1. We notice that the symbol `<expr>` has three candidate rules; `<expr><op><expr>`, `<num>` and `<var>`. This means that the symbol `<expr>` can be replaced with `<expr><op><expr>`, `<num>` or `<var>`. The symbol `<op>`, `<x>` and `<num>` have four, two and one candidate rules, respectively.

We would like to explain how the chromosome (bit string) "1101001001" is translated to the function "1+y".

First, the chromosome is translated into the decimal number every 2-bits;

$$11\ 01\ 00\ 10\ 01 \rightarrow 3\ 1\ 0\ 2\ 1$$

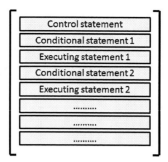

Figure 1. Two-dimensional chromosome

The first 2-bit is $\beta = 3$. The start symbol $\alpha = $ <expr> has three candidate rules; $n_\alpha = 3$. The remainder of $\beta = 3$ with respect to $n_\alpha = 3$ is $\gamma = 0$. Then, the symbol $\alpha = $ <expr> is replaced with 0-th candidate rule <expr><op><expr> (A0).

Next, we will consider the leftmost symbol $\alpha = $<expr> of the symbol <expr><op><expr>. The second 2-bit is $\beta = 1$. The symbol $\alpha = $ <expr> has three candidate rules; $n_\alpha = 3$. The remainder of $\beta = 1$ with respect to $n_\alpha = 3$ is $\gamma = 1$. The symbol $\alpha = $ <expr> is replaced with <num> (A1) to generate the symbol <num><op><expr>.

According to the algorithm, the bit string is translated finally as follows.

$$11\ 01\ 00\ 10\ 01 \rightarrow 3\ 1\ 0\ 2\ 1 \rightarrow 1\text{+y}$$

2.3. Difficulty of Original GE

One of the familiar conditional statement is described as follows.

```
if( Conditional part ){
    Executing part 1
else{
    Executing part 2
}
```

This statement is composed of "if( )", "else", conditional part, executing part 1, and executing part 2 and the validity of the statement depends on their order. If "if" and "else" or the conditional and the executing parts are exchanged, the statement is invalid in the meaning of the programing language.

In one-dimensional chromosome, it is difficult to define the order of the statement and statement parts explicitly. Therefore, invalid programs are often generated.

3. Advanced GE

3.1. Two-dimensional Chromosome

For overcoming the difficulty of the original GE, the advanced GE adopts the two-dimensional chromosome representation.

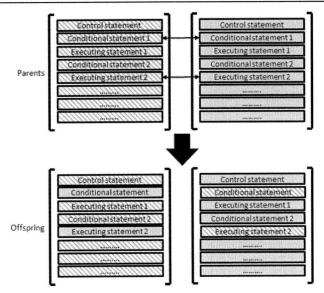

Figure 2. Crossover operation

The two-dimensional chromosome is defined as the collection of one-dimensional chromosomes (Fig.1). Each row is similar to the one-dimensional chromosome in the original GE.

The "Control part" determines whether the two-dimensional chromosome is "if statement" or not and the total number of the pairs of the conditional and the executing statements.

The "Conditional part" is the condition statement for the following "Executing statement." If "Conditional part 1" is true, the "Executing part 1" is executed. If not so, the process goes to the following statements.

In the two-dimensional chromosome, the statement part order in the conditional statement is preserved exactly. Therefore, the conditional statement structure is not destroyed.

3.2. Genetic Operations

3.2.1. Selection

Tournament selection operation is employed. Parents are randomly selected from the population and the best individual among them is preserved to the new population.

3.2.2. Crossover

After parents are selected from the population, the crossover operation replaces the rows of the parents chromosomes at the crossover rate η_c by means of the uniform crossover.

At this time, the rows with same functions are replaced each other. For example, in the two-dimensional chromosome shown in Fig.1, the "Conditional sentence" of one parent is replaced with the "Conditional sentence" of another parent (Fig.2).

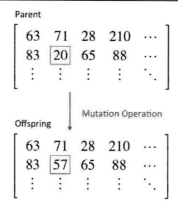

Figure 3. Mutation operation

Table 2. BNF syntax for continuous function identification problem

(A)	`<expr> ::= <expr><expr><op>`	(A0)
	`      | <x>`	(A1)
(B)	`<op>  ::= +`	(B0)
	`      | -`	(B1)
	`      | *`	(B2)
	`      | /`	(B3)
(C)	`<x>   ::= x`	(C0)

3.2.3. Mutation

In Fig.3, the integer numbers, e.g., 63, 71, 28, and so on, denote the values specified at the genes of the two-dimensional chromosome. Mutation operation changes, at the mutation rate η_m, the gene value within the value range defined in advance. In this case, the value 20 is changed to 57.

4. Numerical Example

4.1. Continuous Function

As a first example, the following test function is considered.

$$f(x) = x^4 + x^3 + x^2 + x \tag{1}$$

Pair of data is generated by estimating the function $y_i = f(x_i)$ at the point $\{x_i\} = \{-1.0, -0.9, -0.8, \cdots, 0.9, 1.0\}$. Therefore, the number of the pair of data is $N = 21$. The advanced GE is compared with the GP and the original GE.

The BNF syntax of this problem is shown in Table 2. The start symbol is `<expr>`.

Table 3. GE parameters for continuous4 function identification problem

Generation	1000
Population size	100
Chromosome length	10000
Selection	Tournament
Tournament size	5
Elitist size	1
Crossover	Uniform crossover
Crossover rate	0.9
Mutation rate	0.1

Table 4. GP Parameters for continuous function identification problem

Generation	1000
Population size	100
Selection	Roulette
Elitist size	1
Crossover	One-point crossover
Crossover rate	0.9

The fitness function of individuals is defined as follows

$$E = \sqrt{\frac{1}{N}\sum_{i=1}^{N}(f_{(x_i)} - \bar{f}_{(x_i)})^2}. \qquad (2)$$

where $f(x_i)$ and $\bar{f}(x_i)$ are the values of an exact and a predicted functions at the point $\{x_i\} = \{-1.0, -0.9, -0.8, \cdots, 0.9, 1.0\}$, respectively. Besides, the value N denotes the number of data set, which is taken as $N = 21$.

The parameters of GP and GE are shown in Table 3 and 4, respectively. Tournament selection, one-elitist strategy and uniform crossover are employed. Simulation is performed 50 times at different initial population. The crossover rate and the mutation rate strongly affect the performance of GP and GE. Numerical experiments were performed by GP and GE at several values of the crossover and mutation rates. The parameters for the best search performance are shown in Table 3 and 4,

The fitness convergence history of the best individual is shown in Fig.4. The abscissa and the ordinate denote the generation and the average fitness.

At early generation, the convergence speed of the GE is slower than that of the GP. According to the development of the generation, GE overtakes GP. The slow convergence of GE could be due to BNF shown in Table 2 because the development of GE depends only on the BNF.

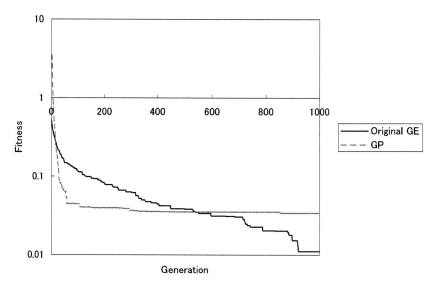

Figure 4. Fitness convergence history of best individual in continuous function identification problem

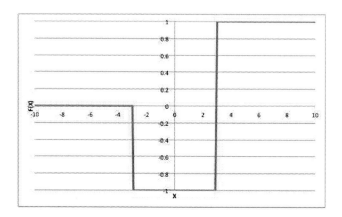

Figure 5. Discontinuous test function

4.2. Discontinuous Function

As a second example, the following test function is considered.

$$f(x) = \begin{cases} 0 & (x < -3) \\ -1 & (-3 \leq x \leq 3) \\ 1 & (3 < x) \end{cases} \quad (3)$$

The function is illustrated in Fig.5. Pair of data is generated by estimating the function $y_i = f(x_i)$ at the point $\{x_i\} = \{-10.0, -9.9, -9.8, \cdots, 9.9, 10.0\}$. Therefore, the number of the pair of data is $N = 201$. The advanced GE is compared with the original GE.

The BNF syntax of this problem is shown in Table 5. The start symbol is <elm>.

Table 5. BNF syntax for discontinuous function identification problem

(A)	`<elm> ::= if(<expr><cond><expr>){<expr>}`	(A0)
	`         else{<elm>}| <expr>|`	(A1)
(B)	`<cond> ::= <  |`	(B0)
	`	>`	(B1)
(C)	`<expr> ::= <expr><expr><op>`	(C0)
	`	<var>`	(A1)
(D)	`<var> ::= <X>`	(D0)
	`	<num>`	(D1)
(E)	`<op> ::= +`	(E0)
	`	-`	(E1)
	`	*`	(E2)
	`	/`	(E3)
(F)	`<X> ::= x`	(F0)
(G)	`<num> ::= 0`	(G0)
	`	1`	(G1)
	`	2`	(G2)
	`	3`	(G3)
	`	4`	(G4)
	`	5`	(G5)
	`	6`	(G6)
	`	7`	(G7)
	`	8`	(G8)
	`	9`	(G9)

Table 6. GE parameters

Generation	1000
Population	100
Selection	Tournament
Tournament size	5
Elitist size	5
Crossover	Uniform
Crossover rate	0.9
Mutation rate	0.1

The fitness function of individuals is defined as follows

$$E = \sqrt{\frac{1}{N}\sum_{i=1}^{N}(f_{(x_i)} - \bar{f}_{(x_i)})^2} \quad (4)$$

where $f(x_i)$ and $\bar{f}(x_i)$ are the values of an exact and a predicted functions at the point $\{x_i\} = \{-10.0, -9.9, -9.8, \cdots, 9.9, 10.0\}$. Besides, the value N denotes the number of data set, which is taken as $N = 201$.

Table 7. Chromosome definition

(a) Original GE

Chromosome length	10000

(b) Advanced GE

Chromosome length	200
Number of condition sentences	5

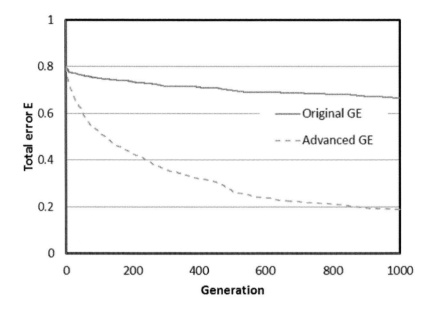

Figure 6. Fitness convergence history of best individuals in discontinuous function identification problem

The GE parameters are shown in Table 7. Tournament selection, one-elitist strategy and one-point crossover are employed. Simulation is performed 50 times at different initial population. The crossover rate and the mutation rate strongly affect the performance of GP and GE. Numerical experiments were performed by GP and GE at several values of the crossover and mutation rates.

The convergence history of the fitness of the best individual is shown in Fig.6. The figure is plotted with the generation as the horizontal axis and the average fitness as the vertical axis. We notice the convergence speed of the advanced GE is much faster than that of the original GE. The test function, step function, cannot be defined without the conditional statement. This problem is so hard for the original GE because the original GE tends to destroy the conditional statement. Therefore, the advanced GE shows the faster convergence than the original GE.

5. Conclusion

Grammatical Evolution (GE) is one of the evolutionary algorithms to find functions and programs, which can deal with tree structure by one-dimensional chromosome of Genetic algorithm, according to the Backus Naur Form (BNF) syntax. When the original Grammatical Evolution is applied for finding the discontinuous function and the program with the conditional branching such as "if-else" statement, the conditional branching cannot be defined explicitly in the BNF syntax. For overcoming this difficulty, the two-dimensional chromosome was presented in this study. In the two-dimensional chromosome, the "if-else" statement, the condition statements, and the executing statements are defined in the different rows of the chromosome. Therefore, the order of the parts of the conditional branching structure is not destroyed through the search process.

The Grammatical Evolution was applied to the function identification problems. The continuous function was taken as the first example. The original Grammatical Evolution was compared with the Genetic Programming. The results showed that the original Grammatical Evolution could find the solution much faster than the Genetic Programming. Next, the discontinuous function was taken as the second example. The result showed that the advanced GE could find the discontinuous function much faster than the original Grammatical Evolution. Since the discontinuous function had to be represented with the conditional branching, the two-dimensional chromosome of the advanced GE was very effective for that problem.

References

[1] J. H. Holland. *Adaptation in Natural and Artificial Systems*. The University of Michigan Press, 1 edition, 1975.

[2] D. E. Goldberg. *Genetic Algorithms in Search, Optimization and Machine Learning*. Addison Wesley, 1 edition, 1989.

[3] L. Davis. *Handbook of Genetic Algorithms*. Van Nostrand Reinhold, 1 edition, 1991.

[4] S. Kirkpatrick, C.D. Gelatt Jr., and M.P. Vecchi. Optimization by simulated annealing. *Science*, 220:671–680, **1983**.

[5] D. B. Fogel and J. W. Atmar. *Proc. 1.st annual Conference on Evolutionary Programming*. Evolutionary Programming Society, 1992.

[6] J. R. Koza, editor. *Genetic Programming II*. The MIT Press, 1994.

[7] J. R. Koza, F. H. Bennett III, D. Andre, and M. A. Keane, editors. *Genewtic Programming III*. Morgan Kaufmann Pub., 1999.

[8] J. Kennedy and R.C. Eberhart. Particle swarm optimization. In *Proceedings of IEEE the International Conference on Neural Networks*, volume 6, pages 1942–1948, **1995**.

[9] Ryan Forbes and Mohammad Nayeem Teli. Particle swarm optimization on multi-funnel functions.

[10] C.Ryan, J.J.Collins, and M.O'Neill. Grammatical evolution: Evolving programs for an arbitrary language. In *Proceedings of 1st European Workshop on Genetic Programming*, pages 83–95. Springer-Verlag, 1998.

[11] C.Ryan and M.O'Neill. Crossover in grammatical evolution: A smooth operator? In *Proceedings of the European Conference on Genetic Programming*, pages 149–162. Springer-Verlag, 2000.

[12] C.Ryan and M.O'Neill. *Grammatical Evolution: Evolutionary Automatic Programming in an Arbitrary Language*. Springer-Verlag, 2003.

INDEX

A

abatement, x, 159, 162, 173, 186, 187
Abraham, 316
adaptation, 44
additives, 14, 15
adenosine, 56
adjustment, vii, viii, 1, 2, 3, 7, 8, 9, 10, 11, 12, 35, 36, 41, 49, 162, 272
advancement, 62, 66
adverse effects, 215
Africa, 240
age, 15, 94
Air Force, 66
air pollutants, 112
air quality, viii, 93, 94, 98, 109, 112
allele, 232
alters, 5, 199
amino, 59
amplitude, viii, 41, 43, 46, 48, 50, 58
annealing, 62, 64, 186, 187, 244, 269, 290, 346, 357
APC, 32, 33, 34
aptitude, 254, 255
arithmetic, 45, 48, 58, 205, 231, 234, 248
aspiration, 277
assessment, 3, 11, 31, 162, 163, 185, 248, 251
asymmetry, 225
automation, xii, 249, 295
automobiles, 89, 111, 112

B

bandwidth, viii, xii, 61, 70, 76, 83, 87, 88, 295, 309, 310, 312, 313, 314
banking, 95
barriers, 187
base, vii, 25, 97, 115, 116, 215, 232, 234, 251
beams, viii, 61, 62, 63, 67, 78, 254

Belgium, 157, 265
bending, 257
benefits, 248, 280, 287
benzene, 110, 111
bias, 36, 49, 55, 56, 307, 308, 309, 311, 314, 330
biodiesel, ix, 93, 96, 111
bioinformatics, 230
biological processes, 64
biomass, 235
bleaching, 53, 54
bonds, 45
bounds, 7, 8, 17, 19, 24, 30, 31, 36, 281, 314
branching, 348, 357
Brazil, 251
breakdown, 66
breeding, viii, 41, 47, 50, 51, 52
Broadband, 180
building blocks, x, 189, 191, 192, 193, 194, 198, 199, 205
building code, xi, 253, 254, 257, 259, 261, 262, 263
butadiene, 110
by-products, 14

C

calculus, 63
calibration, 97
Canary Islands, 156
candidates, 44, 246, 247, 299, 324
carbon, viii, 93, 94, 109, 110, 111
carbon dioxide, viii, 93, 94
carbon monoxide, viii, 93, 94, 109, 110
case studies, 130
case study, 131
CDC, 293
CEC, 156
certification, 263
challenges, xii, 53, 295, 296

chaos, 325
chaotic behavior, 328
chemical, ix, x, 14, 45, 113, 114, 116, 189, 211, 218
chemical bonds, 45
chemical industry, 114
Chicago, 263
children, 25, 94, 111, 112, 256, 318, 325
China, 110
chromatography, 44, 59
chromosome, xiii, 46, 57, 65, 137, 138, 141, 143, 146, 147, 149, 151, 152, 154, 174, 177, 190, 191, 192, 193, 198, 232, 267, 270, 271, 320, 321, 322, 347, 348, 349, 350, 351, 352, 357
circulation, 95, 111
City, 110, 111, 317
classes, 232, 247, 272, 273
classification, 95, 111, 112, 237, 245, 266, 300
CO2, 94, 95, 96, 99, 100, 101, 102, 103, 104, 105, 106, 107, 108, 109
coding, 45, 46, 71, 90, 270, 318, 320
combustion, 215, 218
commercial, 255, 263
communication, 62, 315
comparative analysis, 130
compensation, 186
competition, 4, 118, 119, 120
complement, 192
complexity, ix, 113, 234, 262, 291, 298, 305, 324
composition, xi, 15, 21, 22, 35, 122, 123, 127, 213, 221
compressibility, 216
compression, viii, 41, 46, 259
computation, x, xi, 48, 189, 191, 241, 243, 244, 247, 248, 271, 302, 331
computer, 14, 24, 28, 31, 35, 66, 221, 348
computing, 9, 24, 251, 325
conceptual model, 142, 146
conditioning, xii, 94, 95, 295
conduction, 217
conference, 156, 251
configuration, viii, ix, 5, 23, 24, 25, 26, 27, 29, 30, 31, 32, 61, 94, 114, 116, 125, 127, 129, 215, 222
construction, 143, 152, 154, 155, 271, 272, 278, 280, 284, 285
consumers, 62
consumption, xii, 114, 115, 117, 124, 125, 127, 128, 129, 131, 295
containers, 334
contaminant, 94, 95, 96, 97, 98, 99, 107, 108, 109, 111
contour, 218

contradiction, 29, 30, 95
convergence, x, 12, 19, 48, 65, 72, 73, 88, 189, 191, 194, 197, 198, 205, 210, 215, 245, 254, 281, 299, 314, 322, 342, 343, 353, 354, 356
cooling, 122, 215, 217, 225
correlation, 7, 45, 59, 107, 231, 236, 238, 313
correlation coefficient, 231, 236, 238
correlations, 296
corrosion, 22
cost, vii, ix, xii, 1, 2, 3, 14, 16, 17, 19, 20, 21, 22, 25, 26, 28, 32, 34, 35, 36, 62, 63, 64, 65, 67, 71, 82, 113, 115, 123, 124, 125, 126, 127, 128, 129, 130, 131, 197, 253, 256, 263, 276, 281, 293, 333
cost saving, ix, 113
covering, 82, 232
CPU, 282, 284, 285, 288, 340, 341, 343, 344, 345
critical value, 49
cross-validation, 245
crown, 231, 234
crystalline, 31, 64
crystallization, ix, 113, 114, 115, 116, 117, 118, 119, 120, 121, 122, 131
crystals, 115
cycling, 277
Cyprus, 1

D

damping, 43, 161, 164
Darwin, Charles, 136
data generation, 281
data processing, 82
data set, viii, xiii, 41, 43, 44, 45, 46, 47, 48, 49, 50, 52, 53, 54, 55, 56, 57, 234, 347, 348, 353, 355
data structure, 45, 57
database, ix, 97, 107, 135, 136
decay, viii, 12, 13, 41, 42, 46, 54, 55, 56, 58
decoding, 143, 147, 151, 174
decomposition, 44, 285, 314
deconvolution, vii, viii, 41, 42, 43, 44, 46, 51, 52, 53, 54, 55, 56, 57, 58, 59
degradation, 63, 298
Denmark, 90
Department of Defense, 230
Department of Education, 239
Department of Energy, 132
Department of Transportation (DOT), 109
dependent variable, 233
depth, 160, 238, 242
designers, 262
destruction, 276

detection, 42, 43, 45, 53, 59
deviation, 9, 10, 11, 14, 20, 26, 32, 46, 47, 48, 49, 50, 145, 148, 149, 153, 154, 197, 199, 204, 251, 281
diesel fuel, 111, 112
dimensionality, 25
directional antennas, 67
discontinuity, 305
discs, 38
dispersion, 254, 255
displacement, 27, 222
distillation, ix, 113, 114, 115, 116, 117, 118, 119, 120, 121, 122, 123, 124, 125, 126, 127, 128, 129, 130, 131, 132
distributed load, 254
distribution, vii, 1, 2, 3, 4, 5, 6, 7, 8, 10, 23, 24, 25, 26, 27, 28, 29, 32, 34, 36, 42, 50, 56, 62, 173, 181, 182, 184, 185, 186, 215, 223, 225, 304, 305, 314
diversification, xii, 138, 265, 266, 275, 277, 278, 279, 280, 281, 282, 285, 286, 287
diversity, xii, 11, 25, 124, 131, 191, 232, 236, 265, 266, 267, 274, 277, 278, 279, 280, 281, 282, 287, 298, 299, 336
DOI, 250, 251, 315, 316
dominance, 116, 120, 191, 270, 297, 299
dosage, 17, 21, 22
dynamic scaling, 48

E

economics, 230
editors, 357
education, 95
election, 11, 230
electric field, 81
electromagnetic, 66, 79, 90
electromagnetic fields, 90
electromagnetism, 292
e-mail, 113
emission, 95, 110, 215
employees, 109
encoding, 46, 174, 190, 191, 193, 270, 273, 274, 275, 306, 308, 309, 310, 312, 321
energy, ix, 45, 63, 114, 115, 116, 117, 124, 125, 127, 128, 129, 131, 162, 186, 214, 215, 216
energy consumption, 114, 115, 117, 124, 125, 127, 128, 129, 131
engineering, x, 23, 35, 116, 156, 187, 189, 211, 216, 230, 238, 249
England, 37, 227
environment, viii, 35, 61, 62, 95, 110, 266, 286, 292

Environmental Protection Agency (EPA), 96, 112, 161
environments, 229, 243
equality, 17, 19, 297
equilibrium, 11, 12, 13, 35, 115, 122, 218
equipment, x, 114, 117, 159, 160, 161, 163, 171, 173, 187
Euclidean space, 17, 25
European market, 259
evidence, 31
evolution, vii, 2, 3, 10, 51, 52, 62, 64, 226, 230, 267, 300, 303, 314, 347, 358
evolutionary computation, xiii, 335
exchange rate, 107
excitation, 45
execution, xiii, 78, 151, 248, 317, 318, 324, 327, 328
experimental condition, 58
expertise, 286
exploitation, 275
exposure, 94, 95, 109, 110, 111, 112
extraction, 114, 192, 193

F

fabrication, 63
families, 266, 269, 282, 288, 290, 293
FBI, 154
ferrite, 63
fiber, 160, 161, 164, 166
filters, 97, 226
financial, 95, 131, 186, 205, 239
financial support, 131, 186, 205, 239
finite element method, 89
Finland, 90
fish, 334
fitness, vii, viii, 1, 2, 3, 4, 5, 6, 7, 8, 9, 10, 11, 12, 13, 14, 18, 25, 35, 41, 47, 48, 49, 50, 52, 65, 66, 72, 73, 82, 137, 138, 174, 221, 230, 232, 247, 248, 267, 270, 272, 273, 280, 299, 318, 320, 322, 323, 335, 336, 337, 338, 341, 348, 353, 355, 356
flatness, 242, 247
flexibility, 114, 285
fluctuations, 10, 54
fluid, 215, 218, 225
fluidized bed, 215
fluorescence, 42, 53, 55, 56, 57, 58
fluorescence decay, 42, 53, 55, 56, 58
foils, 59
force, 48, 142, 146, 150, 216, 244
forecasting, 95, 110, 112
formation, 39, 45, 276

formula, 5, 6, 17, 166
fractal dimension, vii, 1, 24, 29, 31, 34
fractal properties, vii, 1, 24, 29, 30, 31, 34, 35, 39
freedom, 63, 116
frequency distribution, 27, 28, 29, 34, 36
function values, 191, 199, 302, 304
funding, 109
fuzzy sets, 314

G

genes, 45, 46, 57, 58, 138, 143, 147, 151, 174, 191, 192, 233, 247, 248, 270, 303, 319, 320, 321, 322, 352
genetic diversity, 232, 236
genetic information, 302
genetic programming, 263, 291
genetics, ix, 64, 135, 136, 191, 267
genome, 230
genotype, xi, 3, 6, 8, 11, 12, 213, 221, 230, 348, 349
geometrical parameters, xi, 213, 221, 226
geometry, xi, 66, 67, 79, 213, 215, 217, 218, 219, 220, 221, 223, 224, 225, 226, 244
Germany, 155, 226, 227
GPS, vi, vii, xi, 96, 229, 230, 231, 234, 235, 238, 239, 240, 250, 251, 276
graph, 282, 284
greenhouse, 116
greenhouse gas, 116
greenhouse gases, 116
grids, 24
growth, vii, 1, 23, 24, 38, 39, 62, 94, 325
growth mechanism, vii, 1
Guangzhou, 110
guidelines, 115

H

haploid, 46
harvesting, 239
health, 94, 95, 112
health care, 95
hearing loss, 186
heat transfer, 215, 225
height, 81, 95, 163, 170, 231, 234
high strength, 22
history, 284, 353, 354, 356
Hong Kong, 110, 112
House, 125, 130, 239
human, ix, xi, 90, 112, 114, 135, 136, 214, 226
human activity, 112

humidity, ix, 93, 95, 96, 171
Hungary, 41
hybrid, vii, ix, x, xii, 59, 90, 113, 114, 115, 116, 117, 118, 119, 121, 122, 123, 127, 129, 130, 131, 155, 156, 157, 162, 189, 194, 200, 210, 265, 268, 269, 270, 282, 285, 288, 289, 290, 291, 292, 293, 315
hybridization, xii, 265, 275, 281, 284, 285, 288
hypothesis, 217

I

IAM, 59
ideal, 216, 222, 243, 288
identification, xiii, 96, 347, 348, 352, 353, 355, 357
identification problem, xiii, 347, 348, 352, 353, 355, 357
identity, 326
image, 42, 43, 44, 46, 47, 49, 52, 54, 55, 56, 57, 59, 160, 161, 221, 222
images, 45, 46, 222
improvements, xii, 253, 259, 274
impurities, 115
incidence, 167, 168, 170
independence, 217
independent variable, ix, 93, 94, 95, 109
India, 132
individuals, xi, 4, 5, 6, 8, 18, 19, 35, 45, 46, 47, 48, 49, 50, 51, 52, 120, 121, 137, 213, 220, 221, 222, 226, 230, 232, 233, 236, 246, 247, 248, 254, 256, 258, 259, 261, 262, 263, 297, 299, 300, 302, 303, 318, 320, 321, 327, 348, 353, 355, 356
industrial experience, 210
industries, 114
industry, 62, 114
inefficiency, 36, 330
inequality, 17, 70, 297
inertia, 335, 336
information exchange, ix, 135, 136
ingredients, 155
inheritance, 230
initial state, 326
INS, 276, 277
insertion, 63, 160, 171, 192, 193, 243, 274, 275, 276, 281
integrated circuits, vii, xii, 295
integration, 88, 318, 326, 328
intelligence, 156, 272
interface, 51, 220
interference, viii, 61, 62, 63, 66, 73, 88, 243
inversion, viii, 41, 274

investment, ix, 114, 115, 116
isomers, 115, 116, 121, 122, 123, 127, 131
Israel, v, 113
issues, 318
Italy, 241, 251
iteration, viii, x, 8, 41, 48, 49, 52, 53, 54, 55, 56, 57, 147, 160, 189, 190, 192, 195, 196, 198, 230, 237, 248, 267, 335, 339, 340, 343

J

Japan, 333

K

kinetics, 39
knots, 233
Korea, 111

L

lakes, 95
landscape, 14, 48, 50
laptop, 327
laws, 31
lead, 20, 94, 96, 107, 263, 282, 287
leakage, 95
learning, vii, 2, 24, 90, 229
light, 27, 45
linear model, 95, 230, 233
linear programming, 244
liquid chromatography, 59
locus, 70
logging, 234

M

machine learning, vii, 2, 90
machinery, 239
magnetic resonance, 59
magnitude, 198, 218, 225, 230
majority, 24
management, 109
MANETs, 88
manipulation, 131
manufacturing, 95, 114, 230, 241, 242, 243, 245, 248, 251, 292
mapping, 3, 174, 196
marketing, 95
mass, 114, 215, 234
materials, 15, 23, 35, 187

mathematical programming, 115
mathematics, 230, 251, 289, 333
matrix, 26, 27, 66, 164, 166, 168, 169, 326
matter, iv, viii, 93, 94, 109, 110, 111, 199
measurements, vii, xi, 42, 44, 45, 53, 110, 229, 230, 238, 239, 310, 311, 312
media, 38, 334
melt, ix, 113, 114, 115, 116, 117, 118, 119, 120, 121, 122, 131
melting, 64
melting temperature, 64
memory, 277, 335, 348
metals, 23
metaphor, 334
meter, 173, 181
methodology, x, 98, 131, 213, 239, 264, 278
Mexico, 110, 111, 131, 295, 317
microenvironments, 109
mission, 302
mixing, x, 6, 7, 189, 190, 194, 200, 201, 202, 204, 205
modelling, 80, 238
models, 2, 14, 15, 16, 22, 24, 26, 27, 29, 30, 31, 34, 35, 37, 44, 63, 95, 98, 107, 109, 115, 116, 122, 163, 230, 233, 236, 238, 239
modifications, 49, 248, 254, 255
modulus, 160, 255
molecules, 45
momentum, 215
Monte Carlo method, 24, 156
morphometric, xi, 213, 222
motivation, 194, 210
multimedia, 90
multiple regression, 95
multiples, 63
multiplier, 325
mutation, vii, viii, xi, 2, 3, 5, 6, 8, 9, 11, 12, 18, 19, 20, 25, 35, 41, 46, 48, 57, 65, 72, 137, 138, 143, 146, 148, 152, 160, 174, 178, 190, 191, 213, 221, 222, 226, 230, 232, 236, 247, 248, 258, 259, 270, 274, 275, 280, 281, 282, 299, 300, 302, 303, 304, 306, 318, 320, 322, 323, 327, 347, 348, 352, 353, 356
mutation rate, 3, 12, 19, 35, 65, 233, 274, 303, 353, 356

N

nanometer, 296
natural evolution, 64, 230
natural resources, 14
natural selection, ix, 65, 135, 136, 174, 191, 267, 347

Index

navigation system, 230
Netherlands, 112, 133, 155
neural network, 38, 242
neural networks, 38, 242
next generation, 25, 49, 65, 143, 148, 152, 232, 248, 267, 272, 275, 300, 318, 321
nitric oxide, viii, 93, 96, 215
nitrogen, viii, 93, 96
nitrogen dioxide, viii, 93, 96
NOAA, 97
nodes, 219, 254
normal distribution, 50
NPC, 14, 15, 16, 17, 20, 22
NPL, 248
nuclear magnetic resonance, 59
numerical tool, 116

O

open spaces, 230
operations, ix, x, 57, 113, 114, 115, 117, 139, 140, 141, 142, 143, 144, 145, 189, 191, 192, 193, 199, 201, 205, 239, 251, 286, 318, 320
opportunities, 114
optical properties, 45
optimization method, ix, x, 63, 88, 116, 135, 136, 189, 191, 215, 226
organ, 90
organic compounds, 110, 111, 112
organize, 25
oscillation, 43, 73
oscillators, xii
overlap, 334, 337, 341, 345
ozone, 95, 110, 111, 112

P

Pacific, 110
parallel, ix, 90, 135, 136, 139, 246, 270
parallelism, 37, 297, 330
parameter estimation, 56
parents, 6, 7, 25, 48, 49, 174, 232, 247, 248, 256, 267, 272, 273, 279, 303, 318, 320, 321, 323, 324, 328, 351
Pareto, 116, 117, 121, 123, 124, 125, 127, 129, 131, 132, 215, 297, 298, 299, 300, 304, 305, 306, 307, 311, 313, 314
Pareto optimal, 299
parity, 205
penalties, 269, 270, 288, 290, 291, 293
permeability, 38
permit, 63, 215, 303

Perth, 37
pesticide, 95, 109
petroleum, 114
pharmacokinetics, 44
phase diagram, xii, 317, 328
phenotype, 3, 221, 254, 256, 348, 349
phenotypes, 3, 46, 230, 256
physics, 219, 230
plants, 114
playing, 94
point mutation, 58, 296, 314
poison, 161, 164
polar, 70
polarization, 68, 79, 81
pollutants, 112
pollution, 94, 95, 112
polydispersity, 23
population growth, 94
population size, 82, 143, 147, 152, 177, 194, 232, 247, 258, 259, 261, 274, 282, 306, 318, 323, 327
population structure, 278, 280, 281, 287
porosity, vii, 1, 23, 29, 34, 36, 160, 161
porous media, 38
porous space, 29, 34
Portugal, 135, 156
positive correlation, 107
precipitation, ix, 93, 97
preparation, iv
principles, ix, 113, 136, 191, 267
probability, ix, 5, 7, 9, 18, 25, 42, 50, 65, 82, 135, 136, 137, 138, 143, 148, 152, 174, 190, 191, 232, 240, 248, 254, 255, 256, 258, 259, 271, 273, 279, 280, 282, 302, 318, 321, 322, 323, 327, 336, 340, 343, 345
probability distribution, 42, 50
problem-solving, 232
process control, 241
programming, 14, 115, 232, 244, 264, 270, 293, 328
project, x, 50, 51, 78, 87, 97, 131, 136, 146, 147, 148, 149, 150, 151, 152, 154, 157, 249
propagation, viii, 61, 62, 160
protection, 112
prototype, 87
pruning, 233
public domain, 66
pulp, 200, 202, 210, 211
purification, ix, 113, 114, 115
purity, 121, 123, 127

Q

quality improvement, 241
quantization, 24

R

radiation, viii, 61, 62, 63, 66, 67, 73, 77, 78, 79, 81, 82, 83
Radiation, 74, 77, 89
radius, 23, 24, 25, 26, 27, 28, 32, 34, 68, 79, 81, 234, 248
rainfall, 95, 111
random numbers, 46, 50, 52, 143, 147, 151, 335, 336
reactant, 53, 54, 55, 56
reaction mechanism, 48
reaction time, 45
real numbers, 318
real time, viii, 61, 62, 67, 96
recall, 45
recombination, 65, 138, 232, 323, 347
recommendations, iv
recovery, 59
regression, viii, 59, 93, 94, 95, 96, 98, 107, 108, 109, 110, 112, 229, 230, 231, 233, 238, 239, 242
regression analysis, 94, 96, 109, 231, 233
regression equation, 96
rejection, 73, 242, 256, 258
reliability, x, 159, 173, 178, 199
rent, 280
repair, 319, 320, 321
reparation, 319
repetitions, 236
reproduction, 25, 46, 65, 137, 174, 232, 247, 267, 347
requirements, 82, 114, 117, 241, 247, 296
researchers, 97, 269, 334
reserves, 114
residual error, 49, 55, 56, 230, 234
resistance, 22, 71, 161, 217, 259
resolution, 18, 44, 52, 55, 59
resource availability, 155
resources, ix, 14, 15, 28, 135, 146, 150, 152, 153, 155, 266
response, 42, 45, 46, 48, 57, 66, 73, 78, 83, 230, 233, 263, 325
restrictions, ix, 66, 113, 116, 127, 128, 266, 282
risk, 15, 114, 272
root, 201
routes, 110, 112
rubber, 292
rules, 66, 115, 121, 143, 246, 259, 263, 264, 269, 290, 334, 336, 348, 349, 350
Russia, 157

S

safety, 254, 257, 259
SAP, 23, 24, 29, 30, 31
savings, ix, 14, 114, 116
scaling, 4, 24, 28, 29, 30, 48, 273, 331
scatter, 278, 286, 287, 288, 289
schema, 137, 244
school, 94, 110, 111, 112, 249
schooling, 334
science, 23, 35, 45, 230
scripts, 51
search space, vii, ix, xii, 1, 2, 3, 7, 10, 11, 12, 14, 17, 19, 25, 35, 36, 64, 71, 116, 135, 136, 148, 191, 221, 226, 232, 246, 247, 265, 274, 275, 286, 297, 300, 302, 314, 318, 322, 335
seed, 199
semiconductor, 296, 334
sensitivity, 325, 326
sensors, 97
sequencing, 268, 288, 289, 290
sexual reproduction, 65
shape, x, xi, 4, 14, 19, 44, 47, 56, 57, 67, 69, 159, 173, 187, 213, 215, 218, 221, 224, 225, 226, 305
shear, 257
showing, 53, 74, 75, 76, 127, 314, 324
signals, vii, viii, 41, 42, 43, 44, 45, 46, 53, 55, 56, 57, 58, 61, 62, 63, 94, 230
signal-to-noise ratio, xi, 229
silica, 14, 15
silicon, 313
simulation, xi, 13, 24, 26, 28, 31, 33, 34, 36, 38, 55, 73, 83, 213, 215, 216, 218, 221, 222, 226, 256, 298, 331, 335
simulations, 9, 10, 27, 28, 30, 35, 121, 307, 328, 340
Singapore, 90
slag, 14, 15
smoking, 107, 108
smoothing, 48, 54
SMS, 266, 268, 269, 270, 287
software, ix, xi, 19, 83, 93, 96, 98, 111, 213, 218, 242, 248, 251
SOI, viii, 61, 62
solubility, 122
solution space, 2, 35, 46, 48, 267, 272, 274, 275, 281, 287

sound speed, 160
South Africa, 240
space-time, 90
Spain, 156, 157, 229, 240, 253, 263
species, 53, 54, 56
specifications, 241, 242, 266, 296, 314
spectroscopy, 44
speed of light, 45
spelling, 239, 263
Spring, 104, 105
stability, 73, 217
standard deviation, 9, 10, 11, 14, 20, 26, 32, 46, 47, 48, 49, 50, 199, 200, 204, 306, 327
standard error, 248
state, xii, 36, 56, 59, 70, 143, 146, 150, 203, 215, 216, 317, 323, 325, 326, 328, 331, 351
states, xi, 69, 253
statistical inference, viii, 41
statistics, 2, 49, 52, 58, 327
steel, xi, 253, 254, 255, 257, 263, 264, 334
steel industry, 334
stock, 210
storage, 334, 344
strategy use, 29, 30
stress, 216, 255, 257
structure, xi, 45, 57, 64, 66, 69, 116, 253, 254, 255, 257, 259, 261, 262, 278, 280, 287, 292, 318, 348, 351, 357
substitution, 59
success rate, xiii, 333, 340, 341, 343
sulfur, viii, 93, 94, 96
sulfur dioxide, viii, 93, 94
suppression, 57
surplus, 262
survival, ix, 135, 136, 137, 267, 347
suspensions, 200
Sustainable Development, 251
Switzerland, 250, 251
symmetry, 225
synthesis, ix, 63, 89, 113, 115, 122, 126, 127, 296
system analysis, 88

T

Taiwan, 112, 159
target, 19, 82, 176, 178, 245, 296, 314
taxis, 95
techniques, vii, viii, xii, 2, 36, 59, 61, 62, 63, 67, 87, 90, 95, 96, 107, 109, 110, 115, 116, 230, 244, 245, 251, 264, 265, 266, 275, 321, 347
technologies, 62
technology, 14, 62, 114, 132, 239, 296, 306, 334
teeth, 251

telecommunications, 95
telephones, 90
temperature, ix, xi, 64, 93, 94, 96, 108, 122, 123, 213, 215, 216, 217, 218, 221, 222, 223, 224, 225, 226
temporal variation, 111
testing, 16, 56
thermal properties, 217
thermal resistance, 217
three-dimensional space, 24
time increment, 139
time series, 42, 52, 95, 112
time variables, 97, 112
topology, 296
torsion, 257
total energy, 114, 117
trade, xii, 25, 35, 65, 265, 275, 280, 282, 285, 287
trade-off, xii, 25, 35, 65
training, 234
traits, 199
transformation, 10, 43, 174, 244, 257
transformations, 53
transistor, 307, 308, 309, 311, 313, 314
translation, xiii, 347, 348, 349
transmission, 187
transport, ix, 93, 94, 109, 111, 112, 216
transportation, 95, 110, 112
trial, 20, 26, 47
turbulence, 214, 215, 216
turbulent flows, 216

U

uniform, xi, xii, 6, 7, 48, 50, 62, 67, 71, 73, 82, 88, 89, 174, 213, 215, 222, 223, 226, 265, 273, 304, 305, 321, 322, 351, 353
United, 96, 109, 112, 158, 187, 230, 265
United Kingdom (UK), 37, 61, 109, 110, 132, 156, 158, 251, 264, 265
United States (USA), 1, 37, 96, 109, 112, 133, 157, 187, 210, 226, 227, 230, 234, 251, 330, 331
updating, 271, 336
urban, 95, 110, 111

V

Valencia, 157
validation, 234, 245
valuation, 271
vapor, 115, 122

variable factor, 15, 16
variables, ix, xi, xiii, 2, 10, 14, 17, 18, 19, 59, 64, 65, 93, 94, 95, 96, 97, 98, 99, 107, 109, 111, 112, 116, 117, 118, 119, 120, 131, 229, 231, 233, 234, 235, 236, 237, 238, 239, 254, 255, 256, 263, 297, 302, 304, 305, 308, 309, 310, 311, 314, 318, 319, 323, 324, 325, 327, 333, 334
variations, 65, 70, 96, 137, 243, 259, 333
vector, xiii, 64, 68, 116, 142, 143, 146, 147, 150, 151, 297, 298, 309, 310, 311, 323, 324, 326, 333, 334, 335, 337, 338
vehicles, 79, 94, 95, 110, 111, 112
velocity, xi, 161, 164, 166, 167, 169, 213, 215, 216, 218, 221, 223, 224, 225, 226, 335, 336, 337, 338
ventilation, 94, 95, 96, 107, 108, 187
vibration, 187
viscosity, 161, 214, 216
vision, x, 213, 242
volatile organic compounds, 110, 111, 112

W

Washington, 112
waste, 242
water, 122, 215
wave number, 160, 163, 166
wear, 243
wind speed, ix, 93, 94, 95, 97, 108, 215
wind speeds, 108
wind turbines, 215
windows, 94
wireless technology, 62
wires, 66
Wisconsin, 1
wood, xi, 229, 235
workers, 186
workplace, 93
worldwide, 114

Y

yield, x, 115, 154, 189, 190, 191, 193, 199, 255, 278